Probability on Discrete Sample Spaces
with applications

Probability on Discrete Sample Spaces

with applications

ANNE E. SCHEERER

Associate Professor of Mathematics
Dean of the Summer Session
Creighton University

INTERNATIONAL TEXTBOOK COMPANY

Scranton, Pennsylvania

Library of Congress Catalog Card Number: 69-15357

Standard Book Number 7002 2236 7

To
my father and my mother

Preface

Applications of the theory of probability today appear in even the most elementary expositions of topics in the life, physical and social sciences. Hence there is need for texts which introduce students with limited mathematical backgrounds to the fundamental concepts of the theory. This book attempts to do just that; in its preparation I set out to do three things not commonly found in elementary texts and which, I think, have particular value in the development of probability.

First of all, I wanted to introduce the concept of a probability space and to emphasize and reemphasize that every problem in probability must be solved in the context of such a space. I hoped that by doing this the meaning of the probability of an event would lose its mystery, and also that students whose background included some calculus could easily move into the study of continuous sample spaces.

Secondly, I have considered examples in which the sample space is not finite. Although doing this introduces serious difficulties, I feel that the presentation of the countable case often makes for more interesting use of the basic concepts and provides greater insight into the underlying principles. The major difficulty comes from not assuming a knowledge of limiting processes. Instead I illustrate the "reasonableness" of the limits I need and then define the limit under discussion. (An appendix at the end of the text fills in the details for those with the requisite mathematical background.)

Finally, I wanted a text in which the applications were not statistical—not because I consider statistics unimportant (quite the contrary) but rather because this has already been done so many times and frequently very well. However, there are other applications which are modern and interesting but which, to my knowledge, have not appeared in an elementary textbook. Having said this, I must back off and acknowledge that my first application, Markov chains, has been discussed in several beginning texts. The one uncommon thing I have done here is to get vector and matrix multiplication from probabilistic considerations. My other applications are built around basic ideas in queueing theory, branching processes, and information theory. I cannot explore any of them in depth because of the constraints imposed by the assumed mathematical background, but I hope the discussions will encourage the more able and better prepared students to dig further.

I have made several references to the assumed mathematical background. What is it? It is that body of classical algebra and computational techniques usually taught in the secondary schools. Basic topics which have special application in probability theory are included in the text. In one section of Chapter 10 (Sec. 10-2d) I violate my own rules and use the elementary derivative in order to include a particularly interesting result; however no subsequent material is dependent upon it and the violation is acknowledged when it is committed.

In organizing the text, I have divided it into three major parts: (1) preliminaries, (2) theory, and (3) applications. Part I, Some Basic Tools, includes an introduction to set theory, mathematical induction, combinatorics, and the use of the sigma notation. It was originally planned that this material would be contained in an appendix, but after discussing it with colleagues and having tried to cover it hurriedly with one class of students, I decided the material was too necessary to be relegated to the back of the book. In using the text with a group which has recently covered this material, it can easily be passed over. Part II, Probability Theory, treats the basic concepts of probability in an essentially standard way. The notion of independence is not introduced until after the fundamentals including random variables and distribution functions have been completely studied. Part III, Applications, uses Part II in its development. In Chapter 8, Markov Chains, I have used electronic computers available to the students of Boston University to carry out the computations involved in some of the illustrative examples. This was particularly useful from my point of view, because in problems that led to a limiting matrix or vector the repetitive multiplication involved could be carried out often enough so that the result guaranteed by the theory was actually found. Neither the text nor the problems depend upon the reader's familiarity with computers; the sole purpose is to illustrate certain results. Students and classes knowledgeable in computer usage may enjoy programming other problems to study the behavior of chains.

A word concerning the organization of the material would perhaps be helpful. Each chapter is divided into several major paragraphs; for example, there are four in Chapter 5: 5-1, 5-2, 5-3 and 5-4. Each of these major subdivisions is in turn further broken down into sections designated by lower case letters. We have in paragraph 5-3 sections 5-3a, 5-3b and 5-3c. Within each section significant definitions, theorems, properties and figures are numbered sequentially. In 5-3a we have $5\text{-}3a_1$, $5\text{-}3a_2$, Figure and $5\text{-}3a_3$, Figure. When definitions, theorems or properties are used in the establishment of subsequent results, location of the initial statement is simplified since its identification, say, $5\text{-}3a_1$, places it in a chapter, a paragraph and a section. Only infrequently does a section have more than five or six numbered items.

Every author is grateful for the Preface, for in it one can express one's appreciation to those who have assisted in the realization of the manuscript. My first thanks go to the President and Directors of Georgetown University who granted me the sabbatical year in which I did the background reading and studying that enabled me to crystallize my notions of what I wanted in an elementary work on probability. Secondly, I thank my students at that university who struggled along with me through preliminary editions in the form of mimeographed notes, for as a result of our experiences and their suggestions the initial presentation has been refined in numerous ways. My thanks go too to Arthur T. Thompson, Dean of the College of Engineering, Boston University, who encouraged me to continue my writing even though I was assuming new responsibilities under him, and to my colleagues in the College of Engineering, Professors Louis J. Flamand and Aubrey H. Payne, who did the computer programming necessary to carry out the calculations in Chapter 8 and to create the table of $-p \log_2 p$ used in Chapter 11.

Finally, heartfelt words of gratitude to Mrs. Diane Boteler of Georgetown University and Mrs. Margaret Hamilton of Boston University who typed the two sets of mimeographed notes and to the staff at Tech-Type, College Park, Maryland, especially Mrs. Barbara Athey, who typed the final version. Without their perserverance and skills this manuscript could not have gone to press.

ANNE E. SCHEERER

Omaha, Nebraska
September, 1969

Contents

Contents

Probability on Discrete
Sample Spaces
with applications

PART I

SOME BASIC TOOLS

chapter 1

Mathematical Induction

1-1. THE NATURE OF THE PROCESS

Many propositions in mathematics are about the integers. Two such statements are: (1) the sum of the first n positive integer is $\frac{1}{2}n\,(n\,+\,1)$ and (2) the sum of the interior angles of a polygon of n sides is equal to n–2 straight angles. The second of these statements is familiar to anyone who has studied elementary geometry, and the first is perhaps better understood if written in the form

$$1 + 2 + 3 + \cdots + n = \frac{1}{2}n\,(n\,+\,1)$$

where the three dots indicate inclusion of all the integers between 3 and n. We have

$$1 \qquad\qquad = 1 = \frac{1}{2}\,(1)(2)$$

$$1 + 2 \quad\ = 3 = \frac{1}{2}\,(2)(3)$$

$$1 + 2 + 3 = 6 = \frac{1}{2}\,(3)(4)$$

and if the statement is true for all n, the sum of the integers from one to a hundred is

$$1 + 2 + 3 + \cdots + 100 = \frac{1}{2}\,(100)(101) = 5050.$$

To prove a statement of this type we commonly show that it holds for a particular n (usually $n\,=\,1$) and that when it holds for $n\,=\,k$, it holds also for $n = k + 1$. We showed above that

$$1 + 2 = 3 + \cdots + n = \frac{1}{2}n\,(n\,+\,1)$$

holds for $n\,=\,1$, since

$$1 = \frac{1}{2}\,(1)\,(2).$$

Now suppose there is an integer k such that

$$1 + 2 + 3 + \cdots + k = \frac{1}{2} k(k+1).$$

We showed that 2 and 3 were such k's. Then consider:

$$1 + 2 + 3 + \cdots + k + (k+1) = (1 + 2 + 3 + \cdots + k) + (k+1)$$

$$= \frac{1}{2} k(k+1) + (k+1)$$

$$= (k+1)\left(\frac{1}{2}k + 1\right)$$

$$= (k+1)\left(\frac{k+2}{2}\right)$$

$$= \frac{1}{2}(k+1)(k+2).$$

Thus whenever k satisfies the statement, so does $k + 1$; and since it is true for one, it is true for two; and since it is true for two it is true for three; and so forth. Hence the statement holds for all positive integers.

A simple analogy might give some feeling for this method of proof. Consider a single column of identical toy soldiers.

Suppose that the distance between them is less than their height so that whenever one falls backward, it knocks down the one behind it. That is,

whenever we push over the first one, we knock down the entire column. To

knock over the column we must (1) push over the first and (2) the soldiers must be so spaced that if the kth one falls, it will knock down the one behind it (the $(k+1)$st). In our example concerning the sum of the first n integers, we pushed over the first soldier when we showed that the statement was true for $n = 1$, and we showed that if the kth fell so would the $(k + 1)$st when we demonstrated that if the proposition were true for $n = $ k, it is also true for $n = k + 1$. Usually the first soldier to fall is analogous to $n = $ 1, but this is not necessary. In the statement about a polygon with n sides enclosing $n - 2$ straight angles, the first soldier would be analogous to $n = 3$ (and then, if the rest of the proof goes through, the statement is true whenever it makes sense, i.e., for $n \geqslant 3$).

As a second example, let us apply this method to the proof of the proposition: if x is a positive number ($x \neq 1$), then for any positive integer n,

$$\frac{x^n - 1}{x - 1}$$

is also a positive integer. If $n = 1$,

$$\frac{x - 1}{x - 1} = 1$$

is a positive integer. Suppose there is an integer k such that

$$\frac{x^k - 1}{x - 1}$$

is a positive integer. Then consider

$$\frac{x^{k+1} - 1}{x - 1} = \frac{x^{k+1} - x^k + x^k - 1}{x - 1} \quad \begin{array}{l}\text{(We have added zero} \\ \text{in the form } x^k - x^k)\end{array}$$

$$= \frac{x^k(x - 1) + (x^k - 1)}{x - 1}$$

$$= x^k + \frac{x^k - 1}{x - 1}$$

Since x is a positive integer, so is x^k; and the sum of two integers, $x^k + \dfrac{x^k - 1}{x - 1}$, is also an integer. Hence this proposition is true for n $= k + 1$ when it is true for $n = k$, and so if x is a positive number ($x \neq 1$), for all integers $n \geqslant 1$

$$\frac{x^n - 1}{x - 1}$$

is a positive integer.

1-2. THE PRINCIPLE OF INDUCTION

The method of proof described above is known as *Mathematical Induction,* or simply, Induction. It is based on

1-2. *The Principle of Induction.* A proposition concerning the positive
integers is true for all positive integers $n \geqslant N$ if, and only if,
(1) it is true for $n = N$, and
(2) whenever it is true for $n = k$, it is also true for $n = k + 1$.

Let us emphasize again that there must be an initial N for which the proposition
is true (we have to be able to push over the first soldier) and the truth of the
proposition for one integer must imply its truth for the succeeding integer (each
soldier must knock down the one behind him). When we have shown that both
parts of the Principle of Induction are satisfied we say we have proved a statement
"by Induction."

Can we prove then $n^2 - n + 1$ is always a prime?* We see that

$$2^2 - 2 + 1 = 3$$

which is a prime, so the statement is true for $n = 2$. Suppose there is an integer
k such that

$$k^2 - k + 1$$

is prime. Then we consider

$$\begin{aligned}(k + 1)^2 - (k + 1) + 1 &= (k^2 + 2k + 1) - k \\ &= k^2 + k + 1 \\ &= (k^2 - k + 1) + 2k\end{aligned}$$

Now $k^2 - k + 1$ is a prime and $2k$ certainly is not (two is a factor); but this result
does not tell us a thing, for offhand we have no way of knowing whether the sum
$(k^2 - k + 1) + 2k$ is a prime or not. We have neither proved nor disproved the
statement. However, we observe that

$$3^2 - 3 + 1 = 7$$

is a prime and that

$$4^2 - 4 + 1 = 13$$

is a prime. But

$$5^2 - 5 + 1 = 21 = (7)(3)$$

is not a prime. In $n = 5$ we have found a *counterexample*—that is, we have
found a case for which the statement is not true. Hence since we had hoped
to prove it for all $n \geqslant 2$, we have now disproved the theorem. In most cases
when the second part of the induction does not yield a decisive result, one can
find a counterexample if he searches far enough.

*A prime number is a positive integer greater than one whose only divisors are itself and one.

EXERCISES 1-2

Prove or disprove the following statements.
1. The sum of the first n even positive integers is $n(n + 1)$.
2. The sum of the first n odd positive integers is n^2.
3. If $r \neq 1$, $\quad 1 + r + r^2 + \cdots + r^{n-1} = \dfrac{1 - r^n}{1 - r}$.
4. A man n years old measures $(50 + n)$ inches tall.
5. The product $n(n + 1)(n + 2)$ is divisible by 6.
6. The sum of the interior angles of a polygon of n sides is equal to $n - 2$ straight angles. (HINT: the sum of the interior angles of a triangle is one straight angle.)
7. $\dfrac{1}{1 \cdot 2} + \dfrac{1}{2 \cdot 3} + \dfrac{1}{3 \cdot 4} + \cdots + \dfrac{1}{n(n + 1)} = \dfrac{n}{n + 1}$. Can you prove this in a second way?

chapter 2

Sets and Their Arithmetic

2-1. SETS

a. What Is a Set?

When one starts to define a "set," or what is the same thing, when one tries to answer the question: "What is a set?" one usually says: "It is a collection of things" or "it is a group of objects." Unfortunately this leaves him right where he started, for he now must define the word "collection" or "group." Hence we shall leave the word "set" undefined and try to make its meaning clearly understood by examples. Some familiar nonmathematical sets are:

   A: the states that comprise the United States of America;
   B: the paintings that hang in the Denver Art Gallery;
   C: the ingredients called for in a given recipe;
   D: wild flowers indigenous to North America.

A few mathematical sets are:

   E: solutions of the equation: $x^2 - x - 12 = 0$;
   F: all numbers z such that $|z| < 3$;
   G: all ordered pairs (p,q) such that $p \leqslant 6$, $q \leqslant 6$ and $p + q > 7$.

These sets all have the property that given a certain object it is readily determined whether the object belongs to the set or not. For example, Pennsylvania belongs to A; Da Vinci's Mona Lisa is not in B (it is in the Louvre); $\frac{1}{2}$ is in F but -7 does not belong to F. Consider, however, the case of the small town baker who agrees to bake bread for those households, and only those households, which do not bake bread for themselves. Let S contain all households for whom the baker bakes bread. Is the baker's household a member of S? Since we can answer this question neither by "yes" nor by "no," we will not consider S a set.

The objects that make up a set will be called its *elements*. The elements of E are 4 and –3; violets and daisies are elements of D. It is convenient to notate the statement: "x is an element of the set S" by $x \in S$. The symbol "$\in$" can be read "is an element of," "is in," or "belongs to." Hence we will write $4 \in E$, $\frac{1}{2} \in F$, $(2,6) \in G$. Following common usage, the negation of $\in$ will be written $\notin$; and we have $-7 \notin F$, $(3,4) \notin G$.

Sets will be described in several ways. As we did with the sets A through G, we can describe them verbally; in many instances, especially when the elements of the set are numbers, we can shorten this by writing: $E = \{x | x^2 - x - 12 = 0\}$, $G = \{(p,q) | p + q > 7, p \leqslant 6, q \leqslant 6\}$. The vertical bar is read "such that" and the exact translations of the preceding are: E is the set of all x such that $x^2 - x - 12 = 0$, and G is the set of all ordered pairs (p,q) such that $p + q > 7$, $p \leqslant 6$ and $q \leqslant 6$. The curly brackets or braces are commonly used to enclose the description of a set. If there are only a finite number of elements in the set, as in C and E, we can list all the elements, for example: $E = \{4, -3\}$. (You will notice that already we have given three different ways of representing the set E.)

Often it is impossible to list all the elements of a set, but we can list enough of them that the reader can determine whether any other object belongs to the set. The set O of positive odd integers can be written as $O = \{1,3,5,7,\ldots\}$ where the ... is the mathematicians' favorite way of saying "and so on." Before employing the ... we must be very careful to ascertain that we have really indicated the elements of the set. $I^+ = \{1,2,3,4,5,\ldots\}$ is surely the set of all positive integers; $Q = \{2,4,6,8,\ldots\}$ is the set of even integers; $P = \{2,4,8,16,\ldots\}$ is the set of integral powers of two. But, what is $T = \{2,4,\ldots\}$?

We digress momentarily to state that henceforth the symbol I^+ will be used exclusively to denote the set of positive integers. That is,

2-1a₁ $$I^+ = \{1,2,3,4,\ldots\}.$$

This convention provides us with many notational short cuts; as examples we can write:

$$P = \{2,4,8,16,\ldots\} = \{2^n | n \in I^+\}$$
$$E = \{2,4,6,8,\ldots\} = \{2n | n \in I^+\}$$

We will find it useful to have a brief notation for the set which has no elements (such as the set α of six-legged dogs or the set β of real numbers x satisfying $x^2 + 1 = 0$). We set

2-1a₂ $$\varnothing = \text{the empty set (set with no elements)}$$

Thus $\alpha = \varnothing$ and $\beta = \varnothing$. The significance of $\varnothing$ will become clearer as our work progresses.

b. Subsets

2-1b_1 Definition. *Let A and B be sets; we say* **A is subset of B** *if and only if every element of A is an element of B.*

In symbols we write $A \subset B$ where the "$\subset$" is read either "is contained in" or "is a subset of." If $E = \{2n | n \in I^+\}$ and $P = \{2^n | n \in I^+\}$, $P \subset E$ since every positive integer power of two is a multiple of two. On the other hand, if $X = \{x | x^2 + x - 6 = 0\}$ and $Y = \{y | y^2 - 9 = 0\}$, $X \not\subset Y$ (X "is not contained in" Y or X "is not a subset of" Y) since $2 \in X$, but $2 \not\in Y$. Similarly $Y \not\subset X$ since $3 \in Y$ but $3 \not\in X$. From the definition we can immediately conclude that, for any set A, $A \subset A$ and $\varnothing \subset A$.

That the relation "is a subset of" is transitive is easily proved.

2-1b_2 *If $A \subset B$ and $B \subset C$, then $A \subset C$.*

Proof: Let

$$x \in A$$

then

$$x \in B \quad \text{(since } A \subset B, \text{ see 2-1}b_1)$$

also

$$x \in C \quad \text{(since } B \subset C, \text{ see 2-1}b_1)$$

Therefore

$$A \subset C \quad \text{(every element of } A \text{ is an element of } C).$$ ∎*

Let the set A have a finite number, say n, of elements; we can write

$$A = \{a_1, a_2, a_3, \ldots, a_n\}.$$

Suppose we want to determine the number of subsets A has. One way of doing this is by constructing a "tree"; we consider a_1 and we either put it in the subset or do not; having made that choice, we make a similar one with a_2, then with a_3 and so forth. This repeated branching forms the "tree." We readily see that the sequence of choices which always puts in the element produces the subset of A which is A itself, while that sequence which always discards the elements gives the subset $\varnothing$. Now since there are n elements in A and we treat each in either of two ways, there are 2^n subsets of A. We prove this last statement rigorously using the Principle of Induction.

---

* The symbol ∎ will be used to indicate the completion of a proof.

2-1b_3 Figure.

Tree for Determining Subsets of $A = \{a_1, a_2, a_3, a_4\}$ (a_i' indicates a_i is not included in subset)

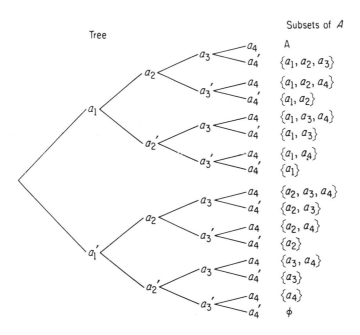

2-1b_4 Theorem.   *If a set A has n elements, it has 2^n subsets.*

Proof: Our proof will be by Induction. If A consists of one element, a_1, i.e., $A = \{a_1\}$, the subsets of A are A and $\varnothing$ and these are $2^1 = 2$ in number. Hence the theorem is true for $n = 1$. Suppose there is an integer k such that any set with k elements has 2^k subsets. Consider now the set

$$A = \{a_1, a_2, a_3, \dots, a_{k+1}\}$$

which has $k + 1$ elements. The k element subset $\{a_1, a_2, a_3, \dots, a_k\} \subset A$ and has 2^k subsets each of which is also a subset of A. These 2^k subsets are those subsets of A which do not contain the element a_{k+1}. Furthermore to each of these subsets we can append a_{k+1} and hence there are also 2^k subsets containing a_{k+1}. Every subset of A either contains a_{k+1} or it does not, hence A has $2(2^k) = 2^{k+1}$ subsets. Since the theorem is true for $n = 1$, and since whenever it is true for $n = k$ it is true for $n = k + 1$, by the Principle of Induction (1-2) the theorem is true for all integers. ■

Using our definition of subsets we can also define an equality of sets; specifically

2-1b_5   **Definition.**   *Let A and B be sets; **A equals B** if, and only if, $A \subset B$ and $B \subset A$.   We write $A = B$.*

We can phrase this definition in terms of elements, rather than subsets, by saying: $A = B$ if, and only if, every element of A is an element of B and every element of B is an element of A.   Then

$$\{x \,|\, x^3 - 8x^2 + 21x - 18 = 0\} = \{z \,|\, z^3 - 7z^2 + 16z - 12 = 0\}$$

for the elements of the set on the left are 2 and 3 $(x^3 - 8x^2 + 21x - 18 = = (x - 3)^2(x - 2))$ and the elements of the right hand set are also 2 and 3 $(z^3 - 7z^2 + 16z - 12 = (z - 2)^2(z - 3))$.   This example illustrates the tacit assumption that neither the order in which the elements of a set are listed nor the number of times a given element is listed affect the composition of the set.   We have

$$\{3,1,1,7,8\} = \{7,7,8,1,7,3\} = \{1,3,7,8\}.$$

EXERCISES 2-1b

1.   Let α be the set of state capitals in the United States.   Are the following cities elements of α?   Denver, Salem (Ore.), New York, Harrisburg, Richmond, Dallas, Fairbanks, Paris, Albany, Seattle.

2.   (a) Describe a set that contains Paris, London, and Rome and give three other elements of the set. (b) Describe a second set that also has Paris, London, and Rome as elements.

3.   Two dice are rolled.   What is the set of possible sums that can appear?

4.   In a medieval town the king decreed that the town's tailor must make clothes for all men and only those men who do not make their own.   Do the men for whom the tailor sews form a set?

5.   List five elements of each of the sets: $\{3^n \,|\, n \in I^+\}$ and $\{\frac{1}{4j} \,|\, j \in I^+\}$.

6.   For the following pairs of sets what is the set of elements common to them:
   (a) $\{2,4,8\}$ and $\{8,10,12\}$.
   (b) $\{x \,|\, x^2 - 4x - 12 = 0\}$ and $\{y \,|\, y^2 - 36 = 0\}$.
   (c)   Fords and Chevrolets.
   (d) $\{a,b,c,d\}$ and $\{a,b,c,\ldots,z\}$.

7.   In each of the pairs a,b,c,d of Prob. 6, what is the greatest set of elements belonging to at least one of them. For example: each element of $\{2,4,8,10,12\}$ belongs to at least one of the sets in (a).

8. Compare the following sets pairwise; i.e., for any two, say A and B, tell whether $A = B$ or $A \subset B$.
$A = \{2,4,6,8,10,\ldots\}$, $B = \{2^n | n \in I^+\}$, $C = \{4^k | k \in I^+\}$, and D: the set of positive even integers.

9. List the subsets of $\{a,b,c\}$.

10. How many subsets has the set consisting of the outcomes of rolling a die? Construct a tree diagram illustrating all these subsets.

11. For the following pairs of sets tell whether $A = B$, $A \subset B$, or $A \not\subset B$:
(a) $A = \{3^n | n \in I^+\}$ and B: the set of positive odd integers.
(b) $A = \{x | x^2 = 4\}$ and $B = \{2\}$.
(c) $A = \{x | (x^2 - 6x + 8)(x + 1) = 0\}$ and $B = \{y | (y^2 - 3y-4)(y - 2) = 0\}$.
(d) $A = \{3,3,8,9,8,7,8,7,1\}$ and $B = \{3,3,9,9,1,8,1,7\}$.

c. Classification of Sets

Sets are said to be *finite* or *infinite* depending upon whether or not they have a finite number of elements. Thus $\{x | x^2 - 1 = 0\} = \{1, -1\}$ is finite, it has two elements; the set of faces of a die is finite, and so is the set of taxpayers in the state of New York. If A is a finite set, we shall use $n(A)$ to denote the numbers of elements in A. Hence for $A = \{x | x^2 - 1 = 0\} = \{-1,1\}$, $n(A) = 2$, and for D, the set of faces of a die, $n(D) = 6$. On the other hand, I^+ and $T = \{t | 0 \leqslant t \leqslant 1\}$ are infinite.

The distinction between finite and infinite sets is always readily made; a further, and useful, subclassification of infinite sets is less easy. We begin by introducing:

2-1c_1 Definition. *A set A is in* **one-to-one correspondence** *with a set B if every element of A can be made to correspond to one and only one element of B, and conversely.*

If there are 30 desks in a classroom and 30 students in the class, the seating chart gives a one-to-one correspondence between the set of desks and the set of students. If $A = \{1,2,3,4\}$ and $B = \{-1,-2,-3,-4\}$, A and B are in one-to-one correspondence for we can pair 1 with -1, 2 with -2, 3 with -3, and 4 with -4. (It is illustrative to write $1, \longleftrightarrow -1$, $2 \longleftrightarrow -2$, $3 \longleftrightarrow -3$, $4 \longleftrightarrow -4$.) The correspondence is not unique for instead we could have set $1 \longleftrightarrow -4$, $2 \longleftrightarrow -3$, $3 \longleftrightarrow -2$, $4 \longleftrightarrow -1$. If $C = \{0,-1,-2,-3,-4\}$, A and C are not in one-to-one correspondence, for if we start by setting $1 \longleftrightarrow -1$, $2 \longleftrightarrow -2$, $3 \longleftrightarrow -3$, $4 \longleftrightarrow -4$, we have no element of A to pair with $0 \in C$. No matter how we try to pair the elements of A off with the elements of C, we will always have one element of C left over. We cannot, for example, let $1 \longleftrightarrow -1$ and $1 \longleftrightarrow 0$ for then $1 \in A$ corresponds to more than one element of C, contrary to definition. As a consequence of this discussion we see immediately how to prove:

2-1c₂ *Two finite sets can be put into one-to-one correspondence if, and only if, they have the same number of elements.*

An infinite set is said to be *countable** if it can be put into one-to-one correspondence with I^+. A set which is not countable is *uncountable.*[†] Establishing a one-to-one correspondence between an infinite set and I^+ is not always easy and we must keep in mind that we need only exhibit one possible correspondence. The set of all integers is countable, for we can pair its elements with those of I^+ in the following way:

$$0 \longleftrightarrow 1$$
$$1 \longleftrightarrow 2$$
$$-1 \longleftrightarrow 3$$
$$2 \longleftrightarrow 4$$
$$-2 \longleftrightarrow 5$$
$$3 \longleftrightarrow 6$$
$$-3 \longleftrightarrow 7$$

and for any $k > 0$,

$$k \longleftrightarrow 2k$$
$$-k \longleftrightarrow 2k + 1.$$

If instead we had started out by setting

$$0 \longleftrightarrow 1$$
$$1 \longleftrightarrow 2$$
$$2 \longleftrightarrow 3$$
$$\cdots$$
$$\cdots$$
$$\cdots$$

we would never have gotten to the point of considering the negative integers, and we could have drawn no conclusion.

We can show, however, that the set T of real numbers between zero and one, i.e., $T = \{t \mid 0 < t < 1\}$, is uncountable. We can express each element of T

---

*Enumerable and denumerable are frequently employed synonyms.
[†] Nonenumerable or nondenumerable.

uniquely as an infinite decimal. (Terminating decimals can be made infinite by decreasing the last digit by one and appending an infinite sequence of nines; for example, we shall write $.25 = .249999...$). If the set were countable we could list its elements sequentially:

$$t_1 = 0 \cdot t_{11}, t_{12}, t_{13}, t_{14}...$$
$$t_2 = 0 \cdot t_{21}, t_{22}, t_{23}, t_{24}...$$
$$t_3 = 0 \cdot t_{31}, t_{32}, t_{33}, t_{34}...$$
$$t_4 = 0 \cdot t_{41}, t_{42}, t_{43}, t_{44}...$$

. . . . .
. . . . .
. . . . .

where $t_i < t_{i+1}$ and for all i and j, t_{ij} is a digit. We have, in fact, made $t_i = 0.t_{i1}t_{i2}t_{i3}t_{i4}...$ correspond to $i \in I^+$. Consider now the number

$$0 \cdot t_{11}, t_{22}, t_{33}, t_{44}...$$

which is an element of T and form

$$N = 0 \cdot n_1, n_2, n_3, n_4...$$

where $n_j = t_{jj} + 1$ if $0 \leqslant t_{jj} < 9$ and $n_j = 2$ if $t_{jj} = 9$. N is an infinite decimal (it will not terminate in zeros); $N \in T$, and yet N is different from all those numbers which we claimed represented the elements of T (it differs from t_k in the kth digit). This contradicts our assumption that T is countable.

Since our concern in this work is exclusively with finite and countable sets, we need pursue this discussion no further. It will serve to give an initial insight into differences between countable and uncountable sets.

EXERCISES 2-1c

1. Which of the following sets are finite: (a) the hairs on your head, (b) the citizens of India, (c) the points in the circumference of a circle, (d) the prime numbers, (e) the rational numbers.

2. A class has 23 students. Can the teacher set up a one-to-one correspondence between them and the school's grading scale which assigns to each student an A, B, C, D, F, or I? Can the same teacher set up a one-to-one correspondence between these same students and the 25 lines in the official roll sheet?

3. Prove Statement 2-1c_2.

2-2.   SET ARITHMETIC

a.   The Universal Set

Every set is embedded in a *universe,* or *universal set.* The integers belong to the universe of the real numbers; $S = \{\frac{1}{2^n} | n \in I^+\}$ and $T = \{t | 0 \leqslant t < \frac{1}{2}\}$ are in the universe $U = \{u | 0 \leqslant u \leqslant 1\}$. We cannot say much about

$$B = \{b | b^2 + 1 = 0\}$$

until we know whether we are to consider B as belonging to the universe of real numbers, in which case $B = \varnothing$, or the universe of complex numbers, in which case $B = \{i, -i\}$. In many applications it is not necessary to give explicitly the universal set—it may be unimportant to the problem or obvious from the statement of the problem—but all set calculations are with respect to this universe.

In particular, with respect to its universe U every set X has a complement which we define as follows:

2-2a_1 Definition.   *Let X be a set in the universe U. The* **complement of X** *is the set of all elements of U which are not in X.*

We denote the complement of X by X'. If, again $U = \{u | 0 \leqslant u \leqslant 1\}$ and $T = \{t | 0 \leqslant t < \frac{1}{2}\}$, $T' = \{t | \frac{1}{2} \leqslant t < 1\}$, and if U is the set of faces on a die, $U = \{1,2,3,4,5,6\}$ and E is the set of even faces, $E = \{2,4,6\}$, $E' = \{1,3,5\}$, the set of odd faces. For any universe U, $U' = \varnothing$ and, conversely, $\varnothing' = U$.

b.   The Binary Operations

On sets, as on numbers, we define several binary operations* which give rise to other sets.

2-2b_1 Definition.   *The* **intersection** *or* **product of two sets** A *and* B *is the set of elements belonging to both A and B.*

Two common notations for this operation are AB and $A \cap B$. Let us look at a few examples.

(1)   Given $A = \{2,4,6\}$ and $B = \{1,2,3,4\}$, then $AB = \{2,4\}$.
(2)   Given $C = \{x | (x - 3)^2 = 0\}$ and $D = \{t | (t + 2)(t - 3) = 0\}$, then $C \cap D = \{3\}$.

---

* Recall that a binary operation is one which combines two elements in some way. If we wish to to combine three elements in that way, we combine two and with the result of that combination, combine the third. Thus, if a, b, and c are numbers, to obtain $a + b + c$ we add a and b and to their sum we add c.

(3) If $E = \{2n \,|\, n \in I^+\}$, then $EI^+ = E$. These last two examples illustrate that if $A \subset B$, $AB = A$.

(4) If $F = \{2k \,|\, k \in I^+\}$ and $G = \{2k - 1 \,|\, k \in I^+\}$, $F \cap G = \varnothing$.

This last example illustrates one very important reason for introducing the concept of an empty set; having $\varnothing$, we can say that the product of two sets is always a set. When the product of two sets is $\varnothing$, we say the sets are *disjoint*. Given a set of sets $\{A_j\}$, $j \in I^+$, finite or countable, we say the sets are *pairwise disjoint* or *mutually exclusive* if, whenever $i \ne j$, $A_i A_j = \varnothing$.

For any set A, $A\varnothing = \varnothing$. Notice that, in a certain sense, $\varnothing$, with respect to the set product, plays a role analogous to that of zero in the multiplication of numbers. However, looking at example 4 above we remark immediately that we have no counterpart to that property of real numbers which says that if a and b are real numbers, $ab = 0$ if and only if $a = 0$ or $b = 0$.

2-2b_2 **Definition.** *The **union of two sets** A and B is the set of all elements belonging to at least one of them.*

The most widely used notation for this operation is $A \cup B$; we shall also use $A + B$ whenever $AB = \varnothing$. (This indicates why forming the union is often referred to as set addition.) Looking again at the examples 1 through 4 which follow 2-2b_1; we compute

(1) $A \cup B = \{1,2,3,4,6\}$.

(2) $C \cup D = \{-2,3\}$.

(3) $E \cup I^+ = I^+$, illustrating that if $A \subset B$, $A \cup B = B$.

(4) $F + G = I^+$.

With respect to union, $\varnothing$ again plays a role similar to that of zero in the addition of numbers. For we have

$$A + \varnothing = A \cup \varnothing = A.$$

(We can write both $A + \varnothing$ and $A \cup \varnothing$ since we have $A\varnothing = \varnothing$ and $\varnothing \subset A$, and which we will use in a given instance will depend on other facets of the problem under consideration.)

2-2b_3 **Definition.** *The **difference of two sets** A and B is the set of elements belonging to A but not to B.*

We shall write simply $A - B$. Again returning to our previous examples, we have

(1) $A - B = \{6\} \ne \{1,3\} = B - A$.

(2) $C - D = \varnothing$, $D - C = \{-2\}$.

(3) $E - I^+ = \varnothing$, $I^+ - E = \{2n + 1 \,|\, n \in I^+\}$.

(4) $F - G = F$; $G - F = G$.

Directly from the definition, we have

$$A - A = \varnothing \text{ and } A - \varnothing = A.$$

Again $\varnothing$ acts somewhat like zero does in the subtraction of numbers, but the analogy is not complete for $A - B = \varnothing$ does not imply $A = B$ (see example 2 above). Whereas subtraction of numbers is an inverse operation to addition of numbers, i.e., if $a + b = c, c - b = a$, we do not have that the difference of sets is the inverse of the union of sets. To see that $A \cup B = C$ does not imply $C - B = A$, let $A = \{2,4,6\}$ and $B = \{2,4,8\}$.

A binary operation is said to be *closed* if whenever A and B belong to the same universe A operation B belongs to that universe. The set operations of union, intersection and difference are closed, for if $A \subset U$ and $B \subset U, A \cup B \subset U$, $AB \subset U$ and $A - B \subset U$. To prove that $A \cup B \subset U$, we observe that if $x \in A \cup B, x \in A$ or $x \in B$. If $x \in A, x \in U$ since $A \subset U$ (see 2-1b_1); if $x \in B, x \in U$, since $B \subset U$. In either case every element of $A \cup B$ is an element of U and so $A \cup B \subset U$. Examples of operations which are not closed are subtraction and division over the positive integers; the difference $a - b$ of two positive integers need not be positive and the quotient a/b of two positive integers is not necessarily an integer.

In 2-1b_5 we defined equality of sets; we said $A = B$ if, and only if, $A \subset B$ and $B \subset A$. We apply this definition to prove that two sets are equal. We start with $x \in A$ and show that $x \in B$. Then since every element of A is an element of B, $A \subset B$ by the definition of subset. Conversely we take any $y \in B$ and show that $y \in A$, and hence $B \subset A$, again by definition. Since $A \subset B$ and $B \subset A$, we have, by 2-1b_5, $A = B$. As an example let us prove

$$(A \cup B)' = A'B':$$

   (1)   Let $x \in (A \cup B)'$;
   (2)   then $x \notin A \cup B$ (def. of complement),
   (3)   then $x \notin A$ and $x \notin B$ (def. of union),
   (4)   then $x \in A'$ and $x \in B'$ (def. of complement),
   (5)   then $x \in A'B'$ (def. of intersection),
   (6)   and so $(A \cup B)' \subset A'B'$ (def. of subset).
   (7)   Now let $y \in A'B'$;
   (8)   then $y \in A'$ and $y \in B'$ (def. of intersection),
   (9)   then $y \notin A$ and $y \notin B$ (def. of complement),
 (10)   then $x \notin A \cup B$ (def. of union),
 (11)   then $x \in (A \cup B)'$ (def. of complement),
 (12)   and so $A'B' \subset (A \cup B)'$ (def. of subset).
 (13)   Finally, using steps 6 and 12 and the definition of equality,
            $(A \cup B)' = A'B'$.

EXERCISES 2-2b

1. $U = \{u \mid 0 \leqslant u \leqslant 1\}$ contains the following sets; with respect to U, what are their compliments?

$A = \{a \mid 0 \leqslant a \leqslant \frac{1}{2}\}$, $B = \{r \mid r$ a rational number$\}$, $C = \{\frac{1}{3}\}$

$D = \{d \mid 0 \leqslant d < \frac{1}{4}, \frac{3}{4} < d \leqslant 1\}$; $E = \{e \mid 0 < e < 1\}$

2. Let $\Omega = \{1,2,3,4,5,6\}$ be the set of possible outcomes when a die is rolled, and let $U = \{(r,g) \mid r \in \Omega, g \in \Omega\}$. Let $A = \{(r,g) \mid r + g \geqslant 9\}$, $B = \{(r,g) \mid r = 3\}$, $C = \{(r,g) \mid r + g$ is even$\}$ and $D = \{(r,g) \mid g = 6\}$. What are A', $A \cap B$, $A \cup B$, $A - B$, B', $B \cup A$, $B \cap C$, $B \cap D$, $B \cup D$, $C \cup D$, $C \cap D$, $B - A$, $D - C$?

3. A coin is tossed twice. Let H stand for heads and T for tails, and designate the set of possible otucomes by Ω. Let $A \subset \Omega$ be the set of pairs containing exactly one head and B the set containing at least one head. How many elements have A and B? What are A', B', $A - B$, $B - A$, $A \cup B$, and $A \cap B$?

·4. For any two nonempty sets C and D, such that $C \subset D$, what are:
(a) $C \cup D$ (c) $C - D$ (e) $C \cup C$ (g) $C \cup \varnothing$
(b) $C \cap D$ (d) $D - C$ (f) $D \cap D$ (h) $D \cap \varnothing$

5. Let $P = \{\delta \mid \delta$ is a letter in the word parallel$\}$.
(a) If $Q = \{\alpha \mid \alpha$ is a letter in the word real$\}$, is Q a subset of P?
(b) If $V = \{v \mid v$ is an English vowel$\}$, is V a subset of P?
(c) What are $Q \cap P$, $V \cup P$, $V - Q$?

The following exercises are more difficult and, in general, concern properties of set arithmetic.

6. Prove: (a) $AB = BA$ and (b) $A \cup B = B \cup A$ (The commutative property).

7. Prove: if $A \subset B$, (a) $AB = A$ and (b) $A \cup B = B$.

8. Prove: (a) $(A \cup B) \cup C = A \cup (B \cup C)$ and
(b) $(A \cap B) \cap C = A \cap (B \cap C)$ (the associative property).

9. Prove: $A \cap (B \cup C) = (A \cap B) \cup (A \cap C)$ and
$A \cup (B \cap C) = (A \cup B) \cap (A \cup C)$ (the distributive properties).

10. Show that the set of rational numbers is countable.

11. Prove: $(EF)' = E' \cup F'$.

12. If $A \subset U$ and $B \subset U$, prove (a) $AB \subset U$ and (b) $A - B \subset U$.

Earlier we introduced the notation $n(A)$ for the number of elements in A. Assume that if $AB = \varnothing$, $n(A + B) = n(A) + n(B)$ and do the remaining problems.

13. Prove: if $B \subset A$, $n(A - B) = n(A) - n(B)$.

14. Prove: $n(A \cup B) = n(A) + n(B) - n(AB)$.

15. Prove:
$n(A \cup B \cup C) = n(A) + n(B) + n(C) - n(AB) - n(AC) - n(BC) + n(ABC)$.

16. (a) If $AB = \varnothing$, whàt is $n(AB)$?

(b) If A, B, and C are pairwise disjoint, what is $n(A + B + C)$?

c.   The Cross-Product of Sets

The three operations already defined and the formation of complements are closed with respect to the universal set. We now define a binary operation on sets which is not closed and which will produce a new set that is fundamentally different than the contributing sets.

2-2c₁ **Definition.** *The* **cross-product** *or* **cartesian product** *of two sets A and B is the set of all ordered pairs (a,b) where $a \in A$ and $b \in B$.*

We write $A \times B$ and have

$$A \times B = \{(a,b) | a \in B \text{ and } b \in B\}.$$

Let us emphasize that the elements of $A \times B$ are ordered pairs; it matters which come first. Let $A = \{\alpha, \delta, \gamma\}$ and $B = \{\beta, \gamma, \delta\}$ then

$$A \times B = \{(\alpha,\beta), (\alpha,\gamma), (\alpha,\delta), (\delta,\beta), (\delta,\gamma), (\delta,\delta), (\gamma,\beta), (\gamma,\gamma), (\gamma,\delta)\}$$

and

$$B \times A = \{(\beta,\alpha), (\beta,\beta), (\beta,\gamma), (\gamma,\alpha), (\gamma,\beta), (\gamma,\gamma), (\delta,\alpha), (\delta,\beta), (\delta,\gamma)\};$$

we see immediately that, in general,

$$A \times B \neq B \times A.$$

The cross-product of a set with itself plays an important role in our work in probability. For example, if we designate the set of outcomes of tossing a coin by $\{H,T\}$, the set of outcomes of tossing the same coin twice is

$$\{(H,H), (H,T), (T,H), (T,T)\} = \{H,T\} \times \{H,T\}.$$

We extend the definitions to cross-products of a finite number of sets in a natural way:

$$A \times B \times C = \{(a,b,c) | a \in A, b \in B, c \in C\}.$$

2-2c₂ $A_1 \times A_2 \times \cdots \times A_n = \{(a_1, a_2, \ldots, a_n) | a_i \in A_i, i = 1,2,\ldots,n\}.$

We say that $A \times B \times C$ is a set of ordered triplets and that the elements of $A_1 \times A_2 \times \cdots \times A_n$ are ordered n-tuples.

We remark that the cross-product for $n > 2$ is not a binary operation, for

$$(A \times B) \times C = \{((a,b),c) | (a,b) \in A \times B, c \in C\} \neq A \times B \times C.$$

Thus the elements of $(A \times B) \times C$ are ordered pairs where the first member of a pair is itself an ordered pair. These are conceptually something different from the ordered triplets that make up the elements of $A \times B \times C$.

EXERCISES 2-2c

1. If $C = \{x|(x - 1)(x - 3) = 0\}$ and $D = \{y|(y + 2)(y - 3) = 0\}$ belong to the universe $U = \{-3,-2,-1,0,1,2,3\}$, what are $C \cup D$, CD, $C'D$, $D - C$ and $C \times D$?

2. If a coin is tossed and H stands for heads and T for tails, the set of possible outcomes is $\Omega = \{H,T\}$. Consider $U = \Omega \times \Omega \times \Omega$ and let $A \subset U$ be the set of triplets containing exactly one head and $B \subset U$ the set containing at least one head. What are $n(A)$ and $n(B)$? What are A', B', $A - B$, $B - A$, $A \cup B$ and AB?

3. A die is rolled twice. $D = \{1,2,3,4,5,6\}$ is the set of possible outcomes when a die is rolled once, so $D \times D$ is the set of outcomes when the die is rolled twice. What is the set of points whose sum is 7? If we call this set S, what is $n(S)$? If O is the subset of $D \times D$ whose elements have an odd sum, what are O and $n(O)$? If T_1 is the subset in which the first die turned up 3 and T_2 is the subset in which the second die turned up 3, what are T_1, T_2, $T_1 T_2$, $T_1 \cup T_2$, $n(T_1 \cup T_2)$ and $n(T_1 T_2)$? What is $T_1 \times T_2$?

chapter 3

A Mathematical Shorthand

3-1. THE SIGMA NOTATION

Operational symbols in mathematics have been familiar to us ever since we wrote our first plus sign. ·After that we added signs for subtraction, multiplication, division, and square root and perhaps others to our collection, and in the last chapter we introduced operational signs for union and intersection in set arithmetic. All the symbols mentioned here, except that for taking square root, apply to binary operations and we now extend them for use when we want to apply the operation to more than two numbers or two sets.

In the first chapter we proved that the sum of the first n integers is $\frac{1}{2}n(n + 1)$. Now if $n = 10$, we write this as

$$1 + 2 + 3 + \cdots + 10 = \frac{1}{2}(10)(11) = 55.$$

Another common way of writing this sum is

$$\sum_{k=1}^{10} k .$$

Here the large Greek sigma, Σ, stands for summation, telling us we are to add, the k to the right of the sigma says we are to add integers and the equation $k = 1$ below the sigma and the number 10 above it tell us to add the integers between one and ten. The letter k is referred to as an *index* or a *counter,* and the numbers 1 to 10 are called the lower and upper *limits* of the summation—they tell us where to start and where to stop our addition. With the introduction of the Σ notation,

$$1 + 2 + 3 + \cdots + 10 = \sum_{k=1}^{10} k$$

22

The notation

$$\sum_{k=1}^{5} k^2$$

tells us to add the squares of the integers from one to five; hence

$$\sum_{k=1}^{5} k^2 = 1^2 + 2^2 + 3^2 + 4^2 + 5^2.$$

The index need not be k.   If we write

$$\sum_{i=0}^{15} \left(\frac{1}{2}\right)^i$$

we want to add the integral powers of $\frac{1}{2}$ for the integers from zero to fifteen.
Thus.

$$\sum_{i=0}^{15} \left(\frac{1}{2}\right)^i = 1 + \frac{1}{2} + \left(\frac{1}{2}\right)^2 + \left(\frac{1}{2}\right)^3 + \left(\frac{1}{2}\right)^4 + \cdots + \left(\frac{1}{2}\right)^{15}.$$

(If to this sum we apply the result of Prob. 3 Exercises 1-2, in which we proved
for $r \neq 1$ that $1 + r + r^2 + \cdots + r^{n-1} = \dfrac{1 - r^n}{1 - r}$, we have

$$\sum_{i=0}^{15} \left(\frac{1}{2}\right)^i = \frac{1 - \left(\frac{1}{2}\right)^{16}}{1 - \left(\frac{1}{2}\right)} = 2 - \left(\frac{1}{2}\right)^{15}).$$

What letter we use to designate our index is not important, but it is important
that if more than one letter appear in the summation expression, we distinguish
between the index and any constant which is included.   For example, in

$$\sum_{j=1}^{20} j x^{j-1}$$

the index is j and x is a constant unchanged from term to term just as $\frac{1}{2}$ was a
constant in the previous example.   Thus

$$\sum_{j=1}^{20} j x^{j-1} = 1 + 2x + 3x^2 + \cdots + 20x^{19}.$$

In this sum, as in all the preceding ones, in the expression to the right of the Σ
we replace the index consecutively by the integers starting with the lower limit
and stopping with the upper one.   We then add all these terms together.

Any odd positive integer has the form $2n - 1$ where n itself is a positive integer. If we want to add together the first one hundred odd positive integers, we can write

$$1 + 3 + 5 + \cdots + 199,$$

or, more simply

$$\sum_{n=1}^{100} (2n - 1).$$

The sum of the first thousand positive integer powers of a variable r can be written

$$\sum_{v=1}^{1000} r^v$$

where v, the index, is the Greek letter *nu*.

In the examples we have used so far we have started our summation with either zero or one. This is not necessary; we can just as readily start at 3 or 10 or 150:

$$\sum_{k=40}^{185} \frac{k}{k+1} = \frac{40}{41} + \frac{41}{42} + \frac{43}{44} + \cdots + \frac{185}{186}.$$

Often the upper limit is variable. Returning again to our statement that for any positive integer n the sum of the first n positive integers is $\frac{1}{2} n(n + 1)$, using the sigma notation we write this as

$$\sum_{i=1}^{n} i = \frac{1}{2} n(n + 1).$$

Our index is i and we sum from one to n. The sum of the squares of the first N positive integers can be expressed as

$$\sum_{j=1}^{N} j^2$$

In general, if $A = \{a_1, a_2, a_3, \ldots, a_n\}$ is a set of numbers, the sum of the elements of A

$$a_1 + a_2 + a_3 + \cdots + a_n$$

can be written more compactly as

$$\sum_{i=1}^{n} a_i.$$

If B is a countable set of numbers, that is $B = \{b_1, b_2, b_3, \ldots\}$ and we want to indicate the sum of all of them, we write

$$\sum_{j=1} b_j.$$

As an example,

$$\sum_{i=0} \left(\frac{1}{2}\right)^i = 1 + \frac{1}{2} + \left(\frac{1}{2}\right)^2 + \left(\frac{1}{2}\right)^3 + \cdots$$

Whether this infinite sum can be evaluated is not our present concern; we are here interested only in the meaning of the notation. We point out that when we want to sum a countable number of terms, we give only the lower limit and we write this to the right of the Σ rather than below it:

$$\sum_{i=500} \frac{1}{i}$$

is the sum of the reciprocals of all integers greater than or equal to 500.

EXERCISES 3-1

1. Write in the Σ notation:
   (a) $2 + 4 + 6 + \cdots + 24$
   (b) $3 + 6 + 9 + \cdots + 57$
   (c) $4 + 7 + 10 + \cdots + 58$
   (d) $2 + 4 + 8 + \cdots + 128$
   (e) $2 + 6 + 18 + \cdots + 162$
   (f) $\dfrac{1}{1 \cdot 2} + \dfrac{1}{2 \cdot 3} + \dfrac{1}{3 \cdot 4} + \cdots + \dfrac{1}{n(n+1)}$
   (g) $1 + \dfrac{1}{2^2} + \dfrac{1}{3^3} + \cdots + \dfrac{1}{N^N}$
   (h) $\dfrac{1}{(j+1)^3} + \dfrac{2}{(j+2)^3} + \dfrac{3}{(j+3)^3} + \cdots + \dfrac{v}{(j+v)^3}$
   (i) $50y^{50} + 51y^{51} + 52y^{52} + \cdots + 99y^{99}$
   (j) $1 + r + r^2 + r^3 + \cdots$
   (k) $q + qp + qp^2 + qp^3 + \cdots$
   (l) $a^n + a^{n-1}b + a^{n-2}b^2 + \cdots + b^n.$

2. Write out the first three and the last term of the following summations, for example:

$$\sum_{k=7}^{29} (2k + 1) = 15 + 17 + 19 + \cdots + 59.$$

(a) $\displaystyle\sum_{k=0}^{14} (k^2 + 1)$ (e) $\displaystyle\sum_{i=1}^{20} \left(\frac{1}{i} - \frac{1}{i+1}\right)$

(b) $\displaystyle\sum_{j=1}^{30} \frac{x^j}{j^2}$ (f) $\displaystyle\sum_{n=0}^{10} ar^n$

(c) $\displaystyle\sum_{v=6}^{40} \frac{v^2}{v+1}$ (g) $\displaystyle\sum_{n=1}^{30} pq^{n-1}$

(d) $\displaystyle\sum_{n=0}^{15} \frac{(-1)^n}{n+1}$ (h) $\displaystyle\sum_{k=0}^{N} \frac{k}{2} a^{N-k} b^k$

3.   (a)   Write in Σ notation the sum of the cubes of the first N integers.
      (b)   Write in Σ notation the sum of the reciprocals of the first n integers.

4.   Express in words:

    (a) $\displaystyle\sum_{k=1} \frac{1}{k^2}$ (b) $\displaystyle\sum_{j=50} \frac{j}{j+1}$ (c) $\displaystyle\sum_{i=0} iy^{i+1}$

5.   (a)   Write in Σ notation the sum of the nonnegative integral powers of $\frac{1}{3}$.

      (b)   Write in Σ notation the sum of the reciprocals of all the integers greater than N.

3-2.   NOTATIONS FOR REPEATED SET OPERATIONS

Since the binary set operations are frequently applied to more than two sets, their symbols have been adapted for use similar to that of the sigma notation. Let $A_1, A_2, \ldots, A_n$ be a set of n sets, then the set

$$A_1 \cup A_2 \cup \ldots \cup A_n$$

can be written as

$$\bigcup_{i=1}^{n} A_i$$

where $\displaystyle\bigcup_{i=1}^{n}$ indicates the union of the sets over the index i between the limits 1 and n. Similarly, if we have a countable set of sets $B_1, B_2, B_3, \ldots$, we write their union as

$$\bigcup_{j=1} B_j.$$

As we did with the Σ sign, when our operation is applied to more than a finite number of sets we use only the lower limit and write that to the left of the operational sign. For the union of two disjoint sets we replaced $\cup$ by $+$ and similarly for the union of more than two we will replace the large $\cup$ by a Σ. Thus

$$\sum_{i=1}^{n} A_i$$

is the union of the n mutually exclusive sets $A_1, A_2, \ldots, A_n$ and

$$\sum_{j=1} B_j$$

is the union of the countable set of mutually exclusive sets $B_1, B_2, B_3, \ldots$ From the context it which it is used it will be clear whether a Σ sign refers to addition of numbers or to the union of disjoint sets.

The set

$$\bigcap_{v=1}^{n} C_v$$

is the intersection of the n sets $C_1, C_2, \ldots, C_n$ and the set

$$\bigcap_{\mu=1} D_\mu$$

(μ is the Greek letter *mu*) is the intersection of the countable set of sets $D_1, D_2, D_3, \ldots$

Although we shall make only occasional use of the notations $\bigcup_{i=1}^{n} A_i$ and $\bigcap_{v=1}^{n} C_v$, we shall repeatedly need the corresponding notation for the crossproduct of several sets:

$$\underset{i=1}{\overset{n}{\text{X}}} A_i = A_1 \times A_2 \times \ldots \times A_n.$$

From 2-2c_2,

$$\underset{i=1}{\overset{n}{\text{X}}} A_i = \{(a_1, a_2, \ldots, a_n) | a_i \in A_i, \ i = 1, 2, \ldots, n\}$$

is a set of ordered n-tuples. While we can indeed consider the cross-product of a countable number of sets $B_1, B_2, B_3, \ldots$,

$$\underset{j=1}{\text{X}} B_j$$

we shall have no reason to do so.

A special variation of the cross-product will later be of particular use in our work—it is the cross-product of a set n times with itself. The cross-product of a set A three times with itself is

$$A \times A \times A = \{(a,b,c)|a \in A, b \in A, c \in A\}.$$

Instead of writing

$$A \times A \times A$$

we shall find it convenient to write

$$(X\ A)_3$$

where the absence of the index indicates that we take the cross-product of the set A with itself and the subscript 3 tells us how many times. This is especially helpful if the number of times is great, for example the product of a set Ω with itself a hundred times will be written

$$(X\ \Omega)_{100}.$$

In general, for $n \in I^+$, we define:

3-2 $(X\ C)_n = C \times C \times \cdots \times C\ n$ times.

In the sets above,

$$(X\ \Omega)_{100} = \{(\omega_1, \omega_2, \ldots, \omega_{100})|\omega_i \in \Omega,\ i = 1,2,\ldots,100\}$$

and

$$(X\ C_n = \{(c_1, c_2, \ldots, c_n)|c_j \in C, j = 1,2,\ldots,n\}$$

so that $(X\ \Omega)_{100}$ is a set of ordered 100-tuples and an element of $(X\ C)_n$ is an ordered n-tuple. (Ω and ω are respectively the capital and small Greek letters *omega*). If $D = \{1,2,3,4,5,6\}$, the elements of $(X\ D)_{10}$ are ordered 10-tuples each member of which is an element of D. We have, for example,

$$(3,6,4,5,5,1,2,1,2,3) \in (X\ D)_{10} \text{ and } (6,6,6,2,4,4,1,5,6,6) \in (X\ D)_{10}.$$

EXERCISES 3-2

1. Given the n sets $C_1, C_2, \ldots, C_n$. Define $\displaystyle\bigcup_{i=1}^{n} C_i$ and $\displaystyle\bigcap_{i=1}^{n} C_i$.

2. If the N sets $C_1, C_2, \ldots, C_N$ are mutually exclusive, prove

$$n\left(\sum_{i=1}^{N} C_i\right) = \sum_{i=1}^{N} n(C_i).$$

(HINT: See Prob. 16 in Exercises 2-2b.)

3. Let $C_i = \{n|n \in I^+ \text{ and } n \geqslant i\}$—that is, C_i is the set of all positive integers greater than or equal to i.   Explain:

(a) $\displaystyle\bigcup_{i=1}^{30} C_i$,   (b) $\displaystyle\bigcap_{i=1}^{30} C_i$,   (c) $\displaystyle\bigcup_{i=1}^{N} C_i$ (d) $\displaystyle\bigcap_{i=1}^{N} C_i$

What is $\bigcup_{i=1} C_i$?   How would you define $\bigcap_{i=1} C_i$?

4. At the end of the section we spoke of $D = \{1,2,3,4,5,6\}$ and $(\times D)_{10}$. Give physical meaning to D and $(\times D)_{10}$.

5. Suppose that each time an experiment E is performed it yields one of 10 well-defined results.   We can write $E = \{r_1,r_2,r_3,\ldots,r_{10}\}$.   What are the elements of $(\times E)_n$ and what does any single element represent in terms of the experiment?

3-3.   VARIATIONS ON THE NOTATION

Certain variations of the notations will appear occasionally in our work. For the most part they will be self-explanatory, but we mention a few here to indicate what is possible.

If $U = \{u_1, u_2, u_3,\ldots\}$ is a finite or countable set of numbers and A is a finite or countable subset of U, then

$$\sum_{u_i \in A} u_i$$

is the sum of the elements of A.   Let $U = \{1,2,\ldots, 10\}$ and $E = \{2,4,6,8\}$, then

$$\sum_{i \in E} i = 2 + 4 + 6 + 8 = 20.$$

Again let $U = \{u_1, u_2, u_3,\ldots\}$ be a countable set and $A \subset U$; if

$$p(u_i) = \left(\frac{1}{2}\right)^i \text{ for every } i,$$

$$\sum_{u_i \in A} p(u_i) = \sum_{u_i \in A} \left(\frac{1}{2}\right)^i$$

is the sum of the numbers $\left(\frac{1}{2}\right)^i$ associated with the elements of A.   Specifically, if $A = \{u_1, u_2, u_3, u_4\}$,

$$\sum_{u_i \in A} p(u_i) = \frac{1}{2} + \left(\frac{1}{2}\right)^2 + \left(\frac{1}{2}\right)^3 + \left(\frac{1}{2}\right)^4$$

If $\{p_1, p_2, p_3,\ldots, p_{100}\}$ is a set of numbers,

$$\sum_{\substack{i=1 \\ i \neq 3}}^{10} p_i$$

is the sum of the first ten elements of the set except for p_3.

Given $A = \{a_1, a_2, \ldots, a_{10}\}$ and $B = \{b_1, b_2, \ldots, b_{20}\}$, let

$$p = \{p(a_i, b_j) | a_i \in A, b_j \in B\}$$

be a set of numbers associated with $A \times B$; then

$$\sum_{\substack{2 \leqslant i \leqslant 7 \\ 4 \leqslant j \leqslant 11}} p(a_i, b_j)$$

is the sum of those elements of p which corresponds to the ordered pairs (a_i, b_j) where i goes from 2 to 7 and j goes from 4 to 11.

chapter **4**

Counting Techniques

4-1. COUNTING THE ELEMENTS IN A SET

In our study of probability we shall frequently be concerned with the number of elements in a set. We shall need to answer questions such as how many ways can ten guests be seated around a dinner table and how many different bridge hands contain three kings and one queen. To enable us to do this we now introduce some common counting procedures.

a. The First Counting Principle

Suppose a certain coed has 3 skirts and 4 sweaters, and that any skirt can be worn with any sweater. How many outfits consisting of a skirt and a sweater has the girl to wear? Since with each of the 3 skirts she can pair any one of the 4 sweaters, she has in all $3 \cdot 4 = 12$ outfits as indicated in Figure 4-1a_1.

4-1a_1 **Figure**

Skirts: b g w

Sweaters: $B\ Y\ R\ W$ $B\ Y\ R\ W$ $B\ Y\ R\ W$

Skirts: b – black; g – grey; w – white
Sweaters: B – black; Y – yellow; R – red; W – white

A dancing class has 20 girls and 15 boys; the teacher would like to see every boy dance with every girl sometime during the course. How many couples will he have to check on? Each of the 20 girls can have as a partner any one of the 15 boys; hence there are $20 \cdot 15 = 300$ couples.

In a given lottery tickets numbered 1 to 500 are placed in a drum and, after

31

mixing, one ticket is drawn and the holder of that number wins the first prize. Without replacing the one already drawn a second ticket is drawn and a second prize awarded.   How many different pairs of winning numbers can be selected in this way?   The first ticket can be any one of the 500 in the drum and with it can be coupled any one of the 499 remaining tickets; hence there are $(500 \times 499) = 249,500$ possible winning pairs.

If the rules of the lottery were changed so that the first ticket was returned to the drum before the second drawing, then there are $(500)(500) = 250,000$ winning pairs (this method permits the same ticket holder to win both prizes).

The preceding examples illustrate our first counting principle:

4-1a_2   **Counting Principle I.**   *If there are N ways of making one choice and M ways of making another choice, there are NM ways of making both choices.*

Thus the freshman who enters a university offering five basic science courses and six foreign languages and who must satisfy the graduation requirement of one course in science and one in language can meet this requirement in $5 \cdot 6 = 30$ ways.   If A and B are sets containing respectively $n(A)$ and $n(B)$ elements; then the number of elements in $A \times B$ is $n(A)n(B)$; i.e., $n(A \times B) = n(A) \times n(B)$.

We have tried to show why Counting Principle I is reasonable, that it is in line with the physical world as we know it.   We make no attempt to prove it. However, with this as a postulate (a property assumed without proof) we prove all further statements concerning the number of elements in a set.

This principle can be easily extended to more than two sets.   Suppose a menu offers a choice of five appetizers, eight entrees, three salads and six desserts; consider the set of all possible dinners each consisting of one appetizer, one entree, one salad and one dessert.   That set contains $5 \cdot 8 \cdot 3 \cdot 6 = 720$ dinners, since with each of five appetizers we can select one of the eight entrees and to each of these forty pairs we can append one of the three salads, and, finally for each of these 120 triplets we can make a fourth choice, one of the six desserts. This example illustrates not only the direct extension of Counting Principle I but also indicates how we shall use Mathematical Induction to verify it.

4-1a_3   **Counting Principle I Extended.**   *If there are N choices to be made and A_1 ways of making the first choice, A_2 ways of making the second,..., A_N ways of making the Nth, then there are $A_1 A_2 \cdots A_N$ ways of making all N choices.*

Proof: Our proof will be by Induction.   Since by Counting Principle I ($4-1a_2$) there are $A_1 A_2$ ways of making two choices, the theorem is true for $n = 2$.

Let k be an integer for which the theorem holds, so that there are $A_1A_2\cdots A_k$ ways of making the first k choices and A_{k+1} ways of making the $(k+1)$st choice. By Counting Principle I there are $(A_1A_2\cdots A_k)A_{k+1} = A_1A_2\cdots A_{k+1}$ ways of first making k choices and then a $(k+1)$st choice, but this is the same as making $k+1$ choices. So whenever the theorem is true for k, it is true for $k + 1$ and our proof is complete. ∎

(Future references to Counting Principle I will include both the initial and extended forms.)

If a senior class in a high school has five home rooms having respectively 30, 31, 29, 30, 33 students and if the class council is made up of one representative from each home room, there are $30\cdot31\cdot29\cdot30\cdot33 = 26,700,300$ different councils that could be formed. If A_1, $A_2,\ldots,A_N$ are sets containing $n(A_1)$, $n(A_2),\ldots, n(A_N)$ elements then

$$n(A_1 \times A_2 \times \cdots \times A_N) = n(A_1)n(A_2)\cdots n(A_N).$$

b.   Permutations

Let us consider the senior class council of five members which we spoke of at the end of the last paragraph. If the council is responsible for the five major class projects; the prom, the graduation festivities, the show, the gift and the yearbook, how many ways can these projects be assigned to its members? On the assumption that everyone is equally qualified to fill every post, there are five ways the prom chairman can be chosen, then having chosen him, 4 ways to pick the graduation chairman, then 3 ways to select the show chairman, leaving leaving two choices for the gift chairman and only one for the yearbook post. In all there are $5\cdot4\cdot3\cdot2\cdot1 = 120$ ways of assigning the council members to the projects. These 120 assignments are the 120 permutations of the council members.

Consider the number of "words" that can be formed with the letters a, b, c, d. If we want the number of four letter "words," there are four choices for the first letter, three for the second, two for the third and one for the last. There are $4\cdot3\cdot2\cdot1 = 24$ "words," and they are

abcd	*bacd*	*cabd*	*dabc*
abdc	*badc*	*cadb*	*dacb*
acbd	*bcad*	*cbad*	*dbac*
acdb	*bcda*	*cbda*	*dbca*
adbc	*bdac*	*cdab*	*dcab*
adcb	*bdca*	*cdba*	*dcba*

These "words" are the permutations of the four letters a, b, c, d. We remark that instead of letters forming "words," we could be placing books on a shelf, digits in a number, or bands in a parade.

We have looked at several examples of what we have called "permutations"; let us now define the term.

4-1b_1 Definition. *Given n distinct objects. A* **permutation** *is an ordered array of these objects.*

Arrangement and rearrangement are descriptive synonyms for permutation. Given the five digits 0, 1, 2, 3, 4, (4 3 2 1 0), (0 1 2 3 4) and (2 1 0 4 3) are permutations of them. The permutations of a, b, c, d is the "word" list given above. If instead of dialing the number 5279364 you intended, you dialed 5279346, your wrong number is a permutation of the correct one. In order to count the number of permutations of n objects, it will be useful to recall some notation. Let $n \in I^+$; we call the product of the first n positive integers n *factorial,* and we write.

4-1b_2 $$n! = n(n - 1)(n - 2)\cdots 3\cdot2\cdot1.$$

Thus $3! = 3\cdot2\cdot1 = 6$, $7! = 7\cdot6\cdot5\cdot4\cdot3\cdot2\cdot1 = 5040$. In addition to our definition of $n!$ for $n \in I^+$, we shall find it useful to also define the symbol $0!$; we shall set

4-1b_3 $$0! = 1.$$

An immediate consequence of our definition is the property

4-1b_4 $$(n + 1)! = (n + 1)(n!).$$

Further properties will be explored in the exercises.

We said there were $5\cdot4\cdot3\cdot2\cdot1 = 5!$ permutations of the five class councilmen and $4\cdot3\cdot2\cdot1 = 4!$ "words" that could be formed from a, b, c, d; these are examples of our second counting principle.

4-1b_5 Counting Principle II. *A collection of n distinct elements has n! permutations.*

Proof: We use Induction. The statement is true for $n = 1$ since one object has itself as its only permutation. Let the principle be true for some $k \in I^+$, so that k objects have $k!$ permutations. Consider $k + 1$ objects. The first element of a permutation can be anyone of the $k + 1$ objects, the remaining k objects can be permuted in $k!$ ways (the principle is true for a set of k elements). Hence there are $(k + 1)(k!) = (k + 1)!$ permutations of $k + 1$ objects. By the Principle of Induction Counting Principle II is proved.   ∎

Often we are not interested in using all the elements of our original set. For example, the five member senior class council may only have three major projects to worry about, or we may be only concerned with the two letter "words" which can be formed from a, b, c, d. In the first case there would be five ways to fill the first job, four ways to fill the second and three ways to fill the last, or, $5 \cdot 4 \cdot 3 = 60$ ways in all. The two letter "words" in $\{a,b,c,d\}$ are $4 \cdot 3 = 12$ in number, they are:

$$
\begin{array}{cccc}
ab & ba & ca & da \\
ac & bc & cb & db \\
ad & bd & cd & dc
\end{array}
$$

This list of "words" is the set of permutations of four things taken two at a time.

4-1b_6 **Definition.** *Any ordered array of r of n distinct objects ($r \leqslant n$) is a* **permutation of n things taken r at a time.**

Given a, b, c, d, the set of permutations of these four things taken two at a time is the set of "words" that precede the definition. A salesman has a territory of ten cities and in any two-week period he can visit three of these cities. The different two-week trips he can plan are the permutations of ten things taken three at a time.

To count the number of permutations of n things r at a time we set, for $r \leqslant n$.

4-1b_7 $$(n)_r = n(n-1)(n-2)\cdots(n-r+1)$$

so that $(n)_r$ is the product of r consecutive integers starting with n and decreasing to $n - r + 1$. $(9)_5 = 9 \cdot 8 \cdot 7 \cdot 6 \cdot 5$ and $(18)_3 - 18 \cdot 17 \cdot 16$. $P(n,r)$ and P_n^r are other common notations for this product. We see that if $r = n$,

$$(n)_n = n(n-1)(n-2)\cdots(n-n+1) = n!$$

and if $r = 1$,

$$(n)_1 = n.$$

Other useful properties of $(n)_r$ that come directly from the definition are:

4-1b_8 $$(n)_{k+1} = n(n-1)\cdots(n-k+1)(n-k) = (n)_k(n-k).$$

4-1b_9 $$(n)_r(n-r)! = n!$$

4-1b_{10}   **Counting Principle II Extended.**   *The number of permutations of n things r at a time is* $(n)_r$.

Proof: Given a set of n objects, $O = \{o_1, o_2, \ldots, o_n\}$ and the set P of permutations of $r(r \leqslant n)$ of the elements of O, i.e.,

$$P = \{(p_1 p_2 \cdots p_r) | p_i \in O, \, p_i \neq p_k, \, k \neq i\}.$$

Now p_1 can be chosen in n ways, p_2 in $n - 1$ ways, p_3 in $n - 2$ ways and so on until p_r can be chosen in $n - r + 1$ ways. Using Counting Principle I, the number of permutations of n things r at a time is $n(n-1)(n-2)\cdots(n-r+1) = (n)_r$.   ∎

A man has two sets of books, one consisting of seven volumes, the other of four. If he keeps each set together, how many ways can he arrange the eleven books on his shelf? The seven volume set can be arranged in 7! orders and the four volume set in 4! orders. Then by Counting Principle I (4-1a_2) there are (7!)(4!) ways the books can be placed on the shelf if the seven volume set goes to the left of the other set; also there are (7!)(4!) number of ways the books can be placed on the shelf if the four volume set goes to the left of the seven volume set. Hence, altogether there are

$$2(7!)(4!) = 2(5040)(24) = 241,920$$

ways of arranging the eleven volumes on the shelf if each set is kept together.

A doll collector has 15 antique dolls, 20 Japanese dolls and 12 modern dolls (she classifies each doll so that it falls into only one category). She decides to enter three antique dolls, four Japanese dolls and three modern dolls in a hobby contest, and to display them by setting them in a row in a glass case. How many different exhibits can she form if she keeps each category together? She can display three of her 15 antique dolls in $(15)_3$ ways, four of her 20 Japanese dolls in $(20)_4$ ways and three of her 12 modern dolls in $(12)_3$ ways. The number of orderings of the three categories is 3! Thus the answer to the question, using Counting Principle I, is

$$3!(15)_3(20)_4(12)_3.$$

Whenever we want to count the number of different arrangements of the elements of a set which depend upon arrangement or order, we are concerned with permutations. A hostess making seating plans for her dinner guests is concerned with permutations, a coach testing a hitherto unknown squad tries various permutations of his recruits; on the other hand a poker player very much cares about which cards he is dealt and not at all about the order in which they are dealt him. His worry is not a permutation of the cards.

EXERCISES 4-1b

1. A man has 5 suits and 9 ties. How many different suit-tie combinations can he make?

2. The tenth, eleventh and twelfth grades of a certain high school have respectively 197, 213, and 160 students. How many different school councils can be formed if the council consists of one student from each grade?

3. Three dice are rolled—a red one, a blue one, and a green one. How many different ways can they come up?

4. (a) A local telephone number is a sequence of seven digits. If the first digit can be neither zero nor one, how many phone numbers are possible? (Of course, any digit may appear more than once.) (b) All telephone numbers are a series of ten digits in which the first is not zero and the fourth is neither zero nor one. How many such phone numbers can be issued?

5. Government Agencies in Washington are commonly referred to by their initials, such as the FCC or AID. How many different sets of initials are possible if any agency can use two, three, or four letters?

6. A student has five different subjects that he must prepare on a certain night. In how many different orders can he study his subjects?

7. A class of 21 lines up for a picture. (a) In how many different ways can they arrange themselves? (b) If they are to stand in two rows, 8 in the front and 13 in the back, in how many different ways can they be arranged?

8. (a) In how many different ways can a harried hostess seat herself and nine guests about a round table? (b) If there are five men and five women (including herself) and she wishes to alternate the sexes, how many seating arrangements are possible? (c) Would the answers be the same if the guests were to be seated on one side of a long table?

9. How many ten digit numbers can be made from the digits 0, 1, 2, 3, 4, 5, 6, 7, 8, 9, (a) if the first digit is not zero and no digit is repeated? (b) if the first digit is not zero but the digits can be repeated? (c) if the first digit is a two or a four, and no digit appears more than once? (d) if the last three digits are 5, 0, 0, the first digit is not zero but, obviously, repetition is permitted?

10. A college residence hall has 50 single rooms, 25 on each of two floors. (a) In how many ways can 47 men be assigned to the rooms? (b) If 20 women are put on one floor and 23 men on the other, in how many ways can the rooms be assigned?

11. How many four digit numbers greater than 5000 can be formed from the digits 2,4,5,6,7,8,9, (a) if no digit appears more than once? (b) if repetition of the digits is allowed?

12. The letters *a, b, c, d, e, f* are permuted to form six letter "words" and then are listed alphabetically. (a) How many words are in the list? (b) What is the 250th word? (c) the 425th?

13. How many ways can a man display 9 of the 14 ship models in his collection?

14. A test consists of a list of ten events in history and another list of ten men. The student is to match a name with each event. If the totally unprepared student matches them at random, how many different sets of answers are possible?

15. Consider the permutations of $\{1,2,3,4,5\}$. Let A_i, $i = 1, 2, 3, 4, 5$, be the set of permutations in which the digit i is the ith digit in the permutation. For example, $(1,4,5,3,2) \in A_1$, $(2,4,3,1,5) \in A_3 \cap A_5$. How many elements are there in each of the following sets: A_5, $A_1 \cap A_2$, $A_2 - A_3$, $A_3 \cup A_4$, $A_3 \cap A_4 \cap A_5$, $A_1 \cup A_3 \cup A_5$, A_5', $A_4 A_4'$?

16. Prove: $(n)_n = (n)_{n-1}$.

17. Prove: $\dfrac{n!}{(n-2)!} = n(n-1)$, $n \geqslant 2$.

18. For $r \leqslant n$, prove $(n)_r = \dfrac{n!}{(n-r)!}$.

19. For $k \leqslant n$, prove $(n)_k = n(n-1)_{k-1}$.

c.   **Combinations**

Like the poker player who cares only about what five cards he is dealt, the club president appointing a committee of ten from a total membership of 200 is not concerned with the order in which he appoints it members but only the final make-up of the group. We indicate how we can answer the questions how many poker hands are there and how many committees can the club president appoint by considering first a problem whose answer we can display completely.

Let $A = \{a,b,c,d\}$ and ask ourselves how many two-letter subsets does A have (as opposed to how many two letter "words" can be formed from the elements of A)? Thus $\{a,b\}$ is a two-letter subset, whereas we had the two "words" ab and ba. The two-letter subsets are

$$
\begin{array}{lll}
\{a,b\} & & \\
\{a,c\} & \{b,c\} & \\
\{a,d\} & \{b,d\} & \{c,d\}.
\end{array}
$$

Could we have calculated their number, namely 6, without listing them? Actually what we have done is to look at the various permutations and discard those whose membership has previously appeared; having included $\{a,b\}$ we do not include $\{b,a\}$ since they are the same set. Each set of two elements has 2!

permutations, so we must divide the total number of permutations of four letters two at a time by the number of permutations of two elements. In symbols,

$$\frac{(4)_2}{2!} = \frac{4 \cdot 3}{1 \cdot 2} = 6.$$

The total number of poker hands in an ordinary bridge deck is the number of five-element subsets in a set of 52 elements. There are $(52)_5$ ways the five element subset can be selected but each set of five elements appears $5!$ times in the totality of permutations of 52 cards taken five at a time. Hence the total number of poker hands is

$$\frac{(52)_5}{5!} = \frac{52 \cdot 51 \cdot 50 \cdot 49 \cdot 48}{1 \cdot 2 \cdot 3 \cdot 4 \cdot 5} = 2,598,960.$$

We call this a combination of 52 cards five at a time.

In the same way the club president can select his ten committeemen from the total membership of two hundred in $(200)_{10}$ ways but each group of ten members is included in these permutations $10!$ times. Hence, the total number of committees, the number of combinations of two hundred members ten at a time, is

$$\frac{(200)_{10}}{10!}.$$

4-1c_1 **Definition.** *Given a set of n elements. A **combination of n things taken r at a time** is an r element subset of the set of n elements.*

4-1c_2 **Counting Principle III.** *The number of combinations of n things r at a time is*

$$\frac{(n)_r}{r!},$$

Proof: The number of permutations of n things taken r at a time is $(n)_r$ by 4-1b_{10} (Counting Principle II Extended). But this set of permutations contains $r!$ permutations of any set of r elements. Thus the number of combinations (the number of distinct r element subsets) is

$$\frac{(n)_r}{r!}.$$ ■

For convenience we set

$$\frac{(n)_r}{r!} = \binom{n}{r}$$

and if we multiply numerator and denominator by $(n - r)!$ we have

4-1c$_3$
$$\binom{n}{r} = \frac{(n)_r (n - r)!}{r! \ (n - r)!} = \frac{n!}{r! \ (n - r)!}.$$

Direct consequences of the notation are

4-1c$_4$
$$\binom{n}{1} = n \quad \text{and} \quad \binom{n}{n} = 1.$$

and

4-1c$_5$
$$\binom{n}{j} = \binom{n}{n - j}.$$

The first of these is not surprising, since a set of n elements has n one-element subsets and one n-element subset. We shall find it convenient to set

4-1c$_6$
$$\binom{n}{0} = 1$$

and this is consistent with the fact that a set of n elements has one empty subset. We have already determined that there are $\binom{52}{5}$ poker hands. How many of these contain exactly three kings? To compute this we consider the deck of cards as the union of two disjoint sets, one consisting of the four kings and the other consisting of the 48 nonkings. There are $\binom{4}{3}$ combinations of the four kings three at a time and $\binom{48}{2}$ combinations of the 48 nonkings two at a time; hence using Counting Principle I, the number of poker hands containing exactly three kings is

$$\binom{4}{3}\binom{48}{2} = \left(\frac{4 \cdot 3 \cdot 2}{1 \cdot 2 \cdot 3}\right)\left(\frac{48 \cdot 47}{1 \cdot 2}\right) = 4512.$$

A citizens' committee is to be composed of representatives of each of four precincts having 2,297, 2,938, 1,946 and 3,215 registered voters. It is decided that these precincts will have respectively 2, 3, 2, and 3 representatives on the committee. If the names are "picked out of a hat," more formally "chosen at random," how many different committees can be formed? Using Counting Principles III and I there are

$$\binom{2297}{2}\binom{2938}{3}\binom{1946}{2}\binom{3215}{3}$$

possible committees.

d.   Collections In Which All the Members Are Not Distinct

In the preceding paragraphs we have repeatedly spoken of sets with n elements, and since we were working with sets we tacitly assumed that the n elements of the set were distinguishable.   If the set were a set of books, we could distinguish one from the other by their titles, the various cards in a deck are distinguishable by the suit (spade, heart, diamond, club) and the denomination (ace, 2, 3, 4, 5, 6, 7, 8, 9, 10, jack, queen, king) of each card.   However, we will from time to time be concerned with collections in which the elements are not distinguishable.

Suppose a collection contains four a's and three b's, that is $P = [a,a,a,a,b,b,b]$ where we use square brackets instead of braces to indicate we are interested in the entire collection, not just the set $\{a,b\}$. P might be a fruit bowl containing four oranges and three peaches or a purse containing three nickels and four dimes.   In another situation P could be a group of four girls and three boys and, although the girls and boys are surely distinguishable among themselves, we would be concerned only with their sexes.

One question often asked is how many distinguishable arrangements can be made from these four a's and three b's.   If instead of P we had

$$P' = \{a_1, a_2, a_3, a_4, b_1, b_2, b_3\}$$

there would be 7! permutations of the elements, but in any permutations of P the four a's and three b's can be permuted among themselves without changing the appearance of the arrangement; for example, the permutations of P'

$$a_1 a_2 b_1 a_3 b_2 a_4 b_3 \quad \text{and} \quad a_2 a_3 b_2 a_1 b_3 a_4 b_1$$

would give rise to the same arrangement of P, namely,

aabababb.

Each arrangement of P appears among the 7! permutations of P in $(4!)(3!)$ ways (the number of ways the a's can be permuted times the number of ways the b's can be permuted).   Thus the number of distinguishable arrangements of P is

$$\frac{7!}{(3!)(4!)}.$$

A child has twenty strips of colored paper.   Five are green, six red, four blue, and five yellow.   He is told to paste them side by side on a master sheet just large enough to accommodate all of them.   How many color patterns can he produce?   If the strips were each a different color, there would be 20! permutations of the colors; but since in any of the arrangements of the strips there are 5! ways the green, 6! ways the red, 4! ways the blue and 5! ways the yellow strips can be interchanged without alternating the pattern (i.e., we can rearrange two yellow strips or three red ones or do both without changing the appearance of the arrangement), there are only

$$\frac{20!}{(5!)(6!)(4!)(5!)}$$

different patterns.

This problem could also be attacked in a different way. We could rule the master sheet into twenty spaces so that each would receive one colored strip. We could then select five of the 20 spaces to cover with green strips in $\binom{20}{5}$ ways; six of the remaining 15 spaces for red in $\binom{15}{6}$ ways; four of those that are still left for the blue in $\binom{9}{4}$ ways and the yellow must go in the last five spaces in $\binom{5}{5} = 1$ ways. There are, then

$$\binom{20}{5}\binom{15}{6}\binom{9}{4}\binom{5}{5} = \frac{(20)_5(15)_6(9)_4(5)_5}{5!\,6!\,4!\,5!} = \frac{20!}{5!\,6!\,4!\,5!}$$

ways of arranging the strips.

In bridge the 52 cards in the deck are distributed evenly among four players. In how many ways can this be done? There are $\binom{52}{13}$ hands (a thirteen element subset of the deck) which can be dealt to the first player; then $\binom{39}{13}$ hands that can go to the second player; $\binom{26}{13}$ for the third and $\binom{13}{13}$ for the fourth. In all, the number of ways the cards can be distributed among the players is

$$\binom{52}{13}\binom{39}{13}\binom{26}{13}\binom{13}{13} = \frac{52!}{(13!)^4}.$$

We seem to have two problems here: (1) finding the number of arrangements of n elements not all of which are distinct, and (2) subdividing a set of n elements into k subsets. Actually they are the same problem. In finding the number of arrangements of n elements not all of which are distinct we can look at the n places to be filled and from them select one subset to be filled by elements of the first kind, another subset to be filled by elements of the second kind, and so on.

This was our second approach to solving the problem of how many color patterns we could make with five green, six red, four blue, and five yellow strips of paper. The problem is governed by Counting Principle IV.

4-1d_1 **Counting Principle IV.** *Let $r_1 + r_2 + \cdots + r_k = n$. The number of ways a collection of n elements can be subdivided into k parts containing respectively $r_1, r_2, \cdots r_k$ elements is*

$$\frac{n!}{r_1!r_2!\cdots r_k!}.$$

Proof: The first subset of r_1 elements can be selected in $\binom{n}{r_1}$ ways; the second is selected from the remaining $n - r_1$ elements in $\binom{n-r_1}{r_2}$ ways; the third from the still remaining $n - r_1 - r_2$ elements in $\binom{n-r_1-r_2}{r_3}$ ways; and so on, until the last subset is selected in $\binom{n-r_1-r_2-\cdots-r_{k-1}}{r_k} = \binom{r_k}{r_k}$ ways. Using Counting Principle I, the number of ways a set can be divided into k subsets containing $r_1, r_2, \ldots, r_k$ elements ($r_1 + r_2 + \cdots + r_k = n$) is

$$\binom{n}{r_1}\binom{n-r_1}{r_2}\cdots\binom{n-r_1-r_2-\cdots-r_{k-1}}{r_k}$$

$$= \left(\frac{n(n-1)\cdots(n-r_1+1)}{r_1!}\right)\left(\frac{(n-r_1)(n-r_1-1)\cdots(n-r_1-r_2+1)}{r_2!}\right)\cdots\left(\frac{r_k(r_{k-1})\ldots 1}{r_k!}\right)$$

$$= \frac{n!}{r_1!r_2!\cdots r_k!}. \qquad \blacksquare$$

As an extension of the notation for combinations of n things taken r at a time, we write

$$\frac{n!}{r_1!r_2!\cdots r_k!} = \binom{n}{r_1,r_2,\ldots,r_k}.$$

When $k = 2$, that is, when $r_1 + r_2 = n$ (an element is either put in or kept out of the subset), we have

$$\frac{n!}{r_1!r_2!} = \frac{n!}{r_1!(n-r_1)!} = \binom{n}{r_1}.$$

How many ways can five distinguishable balls be put into five cells such that three are put into one cell, one in each of two cells, and two cells left empty.

(Balls in cells may be replaced by patients referred to doctors, aircraft put into landing patterns, or students assigned to class sections.)   There are two sets here, the balls and the cells; and the number of arrangements that will put the balls into the cells in the prescribed manner involves both the number of ways we can group the balls into sets of size, three, one, and one, and the number of ways we can select the cells to receive these balls.   If the cells are numbered 1, 2, 3, 4, 5 and the balls bear the letters a, b, c, d, e we must first choose three balls to go into the same cell and the remaining two will go into different cells.   We thus separate the five balls into three classes where $r_1 = 3$, $r_2 = 1$, and $r_3 = 1$ and we can do this in $\binom{5}{3,1,1}$ ways.   Next we must select the one cell to contain the three balls, the two cells which will each contain one ball and the two cells which will remain empty.   The five cells have also been separated into three classes where $r_1 = 1$, $r_2 = 2$ and $r_3 = 2$ and this can be done in $\binom{5}{1,2,2}$ ways. Using the Counting Principle IV, there are

$$\binom{5}{3,1,1}$$

ways we can group the balls into one set of three and two sets of one; and there are

$$\binom{5}{2,2,1}$$

ways we can select the two cells to receive one ball each, the two to remain empty and the one to receive the three balls.   Applying Counting Principle I, there are

$$\binom{5}{3,1,1}\binom{5}{2,2,1} = \left(\frac{5!}{3!1!1!}\right)\left(\frac{5!}{2!2!1!}\right) = 600$$

ways we can put five balls into five cells in the prescribed fashion.   Two of the 600 possible arrangements are illustrated below; each of these uses the same cells but the placement of the balls is different.

4-1d_2 Figure

   Illustrations of the arrangements of five balls in five cells.

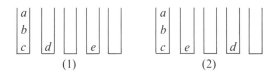

   (1)   (2)

EXERCISES 4-1d

1. A young woman has 15 different outfits but because of packing problems she can take only four on her vacation. How many different wardrobes can she choose?

2. Aunt Martha's Old Fashion Ice Cream Shoppe offers its customers a choice of ten sundae toppings, and for the basic price the customer is entitled to any three of the toppings on a scoop of vanilla ice cream. How many "basic" sundaes are available?

3. Eight candidates are running for the three seats on the city council. Each voter marks his ballot for the three of his choice. How many different choices can he make?

4. Given 15 points in the plane, no three of which are collinear. How many straight lines do they determine?

5. How many four-letter "words" can be made from the letters in *survey?*

6. A small child is asked to choose three out of five coins of different denominations. How many different amounts could he choose?

7. A certain class has 13 girls and 10 boys. How many different committees of 3 girls and 2 boys can be formed from members of the class?

8. A firm decides to close a plant employing 200 men and to divide them among 3 other plants so that 75 go to plant *A*, 100 to plant *B* and 25 to plant *C*. In how many ways can this be accomplished?

9. From a squad of 21 boys, how many baseball teams can a coach field, (a) if three boys must be included on the team? (b) if there are no restrictions?

10. Twenty dead batteries have gotten mixed with 80 good ones. Fifteen are chosen at random. In how many ways can you pick (a) all good ones? (b) five good and ten dead ones? (c) ten good and five dead ones?

11. In poker, a player holds five cards. How many poker hands contain: (a) Three of a kind (three of the same denomination and two other denominations)? (b) A flush (five cards in the same suit but not in sequence)? (c) A straight (five cards in sequence regardless of suit)? (d) Two pairs (two pairs plus one other card whose denomination is different from those of the pairs)? (e) A full house (one pair plus three of a kind)?

12. In bridge, a player holds 13 cards. How many bridge hands contain: (a) exactly three kings? (b) at most three kings? (c) exactly seven hearts? (d) no face cards (the face cards are J, Q, K, A)? (e) two queens and three aces?

13. (a) How many different eight-letter "words" can be made from the letters in *calculus?* (b) How many different ten-letter "words" from *initiation?*

14. How many different ways can the 52 cards in a deck be distributed evenly among 4 people such that each player has an ace?

15. How many bridge hands contain 4 hearts, 4 diamonds, and 5 spades?

16. In a certain restaurant, the price of a dinner includes one appetizer, one meat, two vegetables, one salad and one dessert. If on a certain day the menu offers 6 appetizers, 8 meats, 5 vegetables, 3 salads, and 10 desserts, how many different dinners can be ordered?

17. In Morse code, a letter of the English alphabet is a sequence of dots and dashes. If every letter is to be represented by at most n symbols, where n is as small as possible, what is n?

18. Four indistinguishable dice are rolled. How many different outcomes of the faces are possible?

19. A coin is tossed 16 times. (a) In the set of possible outcomes, how many elements have exactly four heads and twelve tails? (b) How many elements have at most two heads?

20. Consider the set of possible sex distributions in families with five children (we are interested not only in whether there are three girls and two boys but also in the order in which the sexes appear). Let C be the subset in which the first child is a boy and D the subset in which at least three of the children are boys. What are $n(C)$, $n(D)$, $n(CD)$, $n(CD')$, $n(C-D)$?

21. If $E = \{S,F\}$ and $\Omega = (\times E)_n$, Ω is a set of n-tuples in which each component is either S or F. (a) If $A \subset \Omega$ is the subset of n-tuples with exactly k S-components, what is $n(A)$? (b) If $B \subset \Omega$ is the subset of n-tuples with at least k S-components, what is $n(B)$?

22. In how many ways can the four players North, South, East, and West receive respectively N, S, E, W (N + S + E + W = 13) spades in a bridge game?

23. Evaluate: $\binom{5}{4}$; $\binom{14}{7}$; $\binom{100}{3}$; $\binom{200}{198}$.

24. Prove: $\binom{n}{2} \in I^+$.

4-2. MISCELLANY

a. Binomial Coefficients

By straightforward multiplication we can obtain the first several powers of the binomial, $a + b$. They are

$$(a + b)^0 = 1$$
$$(a + b)^1 = a + b$$
$$(a + b)^2 = a^2 + 2ab + b^2$$

$$(a + b)^3 = a^3 + 3a^2b + 3ab^2 + b^3$$
$$(a + b)^4 = a^4 + 4a^3b + 6a^2b^2 + 4ab^3 + b^4$$
$$(a + b)^5 = a^5 + 5a^4b + 10a^3b^2 + 10a^2b^3 + 5ab^4 + b^5$$

Is there any way of determining the nature of the terms in $(a + b)^n$ for any $n \in I^+$? If the cases above are any indication of general behavior, the exponents of a and b would appear in regular fashion; in fact, we would expect

$$(a + b)^n = c_0 a^n b^0 + c_1 a^{n-1} b + c_2 a^{n-2} b^2 + \cdots + c_{n-1} a b^{n-1} + c_n a^0 b^n$$

where the c_i's ($i = 0,1,\ldots,n$) stand for the numerical coefficients. Also our examples suggest that $c_0 = c_n = 1$, $c_1 = c_{n-1} = n$ and that perhaps $c_j = c_{n-j}$. Furthermore we can observe from the listing above that for $n = 1,2,3,4,5,$ and $0 \leqslant i \leqslant n$, i an integer,

$$c_i = (c_{i-1}) \frac{n - i + 1}{i} ,$$

for we have

$$c_0 = 1, \quad c_1 = (1)\frac{n}{1}, \quad c_2 = \left(1 \cdot \frac{n}{1}\right)\frac{n - 1}{2}, \quad c_3 = \left(1 \cdot \frac{n}{1} \cdot \frac{n-1}{2}\right)\frac{n - 2}{3} .$$

These coefficients are the numbers we earlier wrote as $\binom{n}{r}$ (see Counting Principle III and the remarks which follow it). In this notation

$$c_0 = \binom{n}{0}, \quad c_1 = \binom{n}{1}, \quad c_2 = \binom{n}{1}\frac{n - 1}{2} = \binom{n}{2}, \quad c_3 = \binom{n}{2}\frac{n - 2}{3} = \binom{n}{3}$$

Since $\binom{n}{j} = \binom{n}{n-j}$ (4-1c_5), we also have $c_j = c_{n-j}$.

We now "guess" that

$$(a + b)^n = \binom{n}{0} a^n b^0 + \binom{n}{1} a^{n-1} b + \binom{n}{2} a^{n-2} b^2 + \cdots + \binom{n}{n} a^0 b^n$$

where for $j = 0,1,\ldots,n$ the form of the jth term is

$$\binom{n}{j} a^{n-j} b^j$$

We shall use Mathematical Induction to prove that our guess is correct.*
    We prove first a property of the coefficients which we will use in the proof of the theorem.

*We remark here that this is the way in which statements concerning the integers are frequently discovered. In working with a given situation, for small integers the investigator sees a pattern emerging which he feels may hold for all $n \in I^+$. He then tries the second part of the proof by Induction; if it works, he knows the statement is true for all n; if it does not, he begins searching for a counter-example.

4-2a_1 **The Addition Property of the Coefficients.** *If n and j are nonnegative integers with $j \leqslant n$, then*

$$\binom{n}{j} + \binom{n}{j-1} = \binom{n+1}{j}.$$

Proof:

$$\binom{n}{j} + \binom{n}{j-1} = \frac{n!}{j!(n-j)!} + \frac{n!}{(j-1)!(n-j+1)!} \qquad (4\text{-}1c_3)$$

$$= \frac{(n-j+1)n! + (j)n!}{j!(n-j+1)!}$$

$$= \frac{(n!)((n-j+1)+j)}{j!(n-j+1)!} = \frac{(n!)(n+1)}{j!(n-j+1)!}$$

$$= \frac{(n+1)!}{j!(n+1-j)!}$$

$$= \binom{n+1}{j}. \qquad \blacksquare$$

4-2a_2 **Theorem.** *If $n \in I^+$,*

$$(a+b)^n = \binom{n}{0}a^n + \binom{n}{1}a^{n-1}b + \cdots + \binom{n}{j}a^{n-j}b^j + \cdots + \binom{n}{n}b^n.$$

Proof: The theorem is true for $n = 1$, since $(a+b)^1 = \binom{1}{0}a + \binom{1}{1}b$.

Let k be an integer for which the proposition holds, so that
$(a+b)^k$

$$= \binom{k}{0}a^k + \binom{k}{1}a^{k-1}b + \cdots + \binom{k}{j-1}a^{k-j+1}b^{j-1} + \binom{k}{j}a^{k-j}b^j + \cdots + \binom{k}{k}b^k.$$

We multiply $(a+b)^k$ by $(a+b)$ and obtain

$(a+b)^{k+1}$

$$= \binom{k}{0}a^{k+1} + \binom{k}{1}a^k b + \cdots + \binom{k}{j}a^{k-j+1}b^j + \cdots + \binom{k}{k}ab^k$$

$$+ \binom{k}{0}a^k b + \cdots + \binom{k}{j-1}a^{k-j+1}b^j + \cdots + \binom{k}{k-1}ab^k + \binom{k}{k}b^{k+1}.$$

$(a + b)^{k+1}$

$$= \binom{k}{0} a^{k+1} + \left[\binom{k}{1} + \binom{k}{0} \right] a^k b + \cdots + \left[\binom{k}{j} + \binom{k}{j-1} \right] a^{k+1-j} b^j$$

$$+ \cdots + \left[\binom{k}{k} + \binom{k}{k-1} \right] ab^k + \binom{k}{k} b^{k+1}$$

We know

$$\binom{k}{0} = \binom{k+1}{0} = 1; \binom{k}{k} = \binom{k+1}{k+1} = 1$$

and for $j = 1, 2, 3, \ldots, k$,

$$\binom{k}{j} + \binom{k}{j-1} = \binom{k+1}{j} \qquad (4\text{-}2a_1)$$

Hence

$(a + b)^{k+1}$

$$= \binom{k+1}{0} a^{k+1} + \binom{k+1}{1} a^k b + \cdots + \binom{k+1}{j} a^{k+1-j} b^j + \cdots + \binom{k+1}{k+1} b^{k+1}$$

and the proposition is true for $k + 1$. By the Principle of Induction the theorem is proved. ∎

Henceforth the numbers $\binom{n}{r}$ will be referred to as the *binomial coefficients* and so the number of combinations of n things r at a time is the binomial coefficient $\binom{n}{r}$.

The preceding discussion and proof make no demands on a and b other than those which are required to give the expression $(a + b)^n$ meaning. (If $a = -b$, we would have 0^n which does not make a particularly interesting binomial, nor does $(a + 0)^n = a^n$) If we let $a = b = 1$, we have

$$(1 + 1)^n = 1^n + \binom{n}{1} 1^{n-1}(1) + \binom{n}{2} 1^{n-2}(1)^2 + \cdots + \binom{n}{n}(1)^n$$

and hence

4-2a₃

$$\sum_{j=0}^{n} \binom{n}{j} = 2^n$$

In 2-1b_4 we proved every set of n elements has 2^n subsets and we have the same results here. A set of n elements has $\binom{n}{j}$ subsets with exactly j elements (see 4-1c_1 and 4-1c_2) and hence the total number of subsets is

$$\sum_{j=0}^{n}\binom{n}{j} = 2^n .$$

Of historical interest is Pascal's triangle, named after Blaise Pascal (1623-1662) who, although he was not the first to use the triangle, extensively investigated its properties. Two forms of the triangle whose elements are the binomial coefficients are shown in Figure 4-2a_4; from it Pascal was able to surmise the nature of the coefficients for any n and to show the addition property 4-2a_1.

4-2a_4 Figure

Two Forms of Pascal's Triangle.

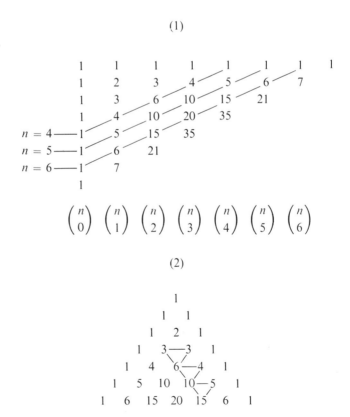

(1)

$$\binom{n}{0}\ \binom{n}{1}\ \binom{n}{2}\ \binom{n}{3}\ \binom{n}{4}\ \binom{n}{5}\ \binom{n}{6}$$

(2)

EXERCISES 4-2a

1. Compute the following sums without adding the individual terms:

(a) $\displaystyle\sum_{k=0}^{4} \binom{4}{k} (2^k)(1)^{4-k} = \sum_{k=0}^{4} \binom{4}{k} 2^k$

(b) $\displaystyle\sum_{j=0}^{6} \binom{6}{j} (\tfrac{1}{3})^j (\tfrac{2}{3})^{6-j}$

(c) $\displaystyle\sum_{n=0}^{5} \binom{5}{n} (3)^n (2)^{5-n}$

(d) $\displaystyle\sum_{e=1}^{6} \binom{5}{e-1} (2)^{e-1}(3)^{6-e}$

(e) $\displaystyle\sum_{j=0}^{14} (-1)^j \binom{14}{j}$

(f) $\displaystyle\sum_{r=0}^{30} \binom{30}{r} \left(\tfrac{1}{2}\right)^{30}$

(g) $\displaystyle\sum_{k=0}^{7} \binom{7}{k} (2)^k (-1)^{7-k}$

(h) $\displaystyle\sum_{k=0}^{300} \binom{300}{k} \frac{5^k}{6^{300}}$

2. Prove the identities:

(a) $k \dbinom{n}{k} = n \dbinom{n-1}{k-1}$

(b) $\dbinom{n}{k+1} = \dfrac{n-k}{k+1} \dbinom{n}{k}$

(c) $n \dbinom{n}{k} = (k+1) \dbinom{n}{k+1} + k \dbinom{n}{k}$

3. Prove the identities:

(a) $\dbinom{n}{0} - \dbinom{n}{1} + \dbinom{n}{2} - \dbinom{n}{3} + \cdots + (-1)^n \dbinom{n}{n} = 0$

(b) $\dbinom{n}{0} + \dbinom{n}{2} + \dbinom{n}{4} + \cdots + \dbinom{n}{n} = \dbinom{n}{1} + \dbinom{n}{3} + \cdots + \dbinom{n}{n-1} = 2^{n-1}$

if n is even. (HINT: Use 4-2a_1.)

(c) $\displaystyle\sum_{k=0}^{n} k \dbinom{n}{k} p^k q^{n-k} = np$ if $(p+q) = 1$

4. (a) Show that: $\dbinom{b+N}{p}$

$= \dbinom{N}{0}\dbinom{b}{p} + \dbinom{N}{1}\dbinom{b}{p-1} + \dbinom{N}{2}\dbinom{b}{p-2} + \cdots + \dbinom{N}{p}\dbinom{b}{0}$

(HINT: Show that for any b and any p, $\dbinom{b+1}{p} = \dbinom{1}{0}\dbinom{b}{p} + \dbinom{1}{1}\dbinom{b}{p-1}$

then use Induction on N and 4-2a_1 to complete the proof.)

(b) Use 4-2a_1 repeatedly to show that $\dbinom{n+1}{k} = \displaystyle\sum_{j=0}^{k} \dbinom{n-j}{k-j}$

5. Use Theorem 4-2a_2 to find approximations to three decimal places for:
   (a) $(1.01)^4 = (1 + .01)^4$
   (b) $(.99)^6$
   (c) $(1.002)^7$

6. Prove the identities

(a) $\dbinom{n}{k}\dbinom{k}{j} = \dfrac{n!}{(n-k)!j!(k-j)!} = \dbinom{n}{j,k-j,n-k}$

(b) $\dbinom{n}{j}\dbinom{n-j}{r_1, r_2, \ldots r_k} = \dbinom{n}{r_1, r_2, \ldots r_k, j}$

(c) $(p+q+r)^4 = ((p+q)+r)^4 = \displaystyle\sum\dbinom{4}{i,j,k}p^i q^j r^k$ where the summation

is over all triplets i,j,k of nonnegative integers such that $i + j + k = 4$.

(d) $(p+q+r)^n = \displaystyle\sum_{\substack{i+j+k=n \\ (i,j,k \geqslant 0)}} \dbinom{n}{i,j,k} p^i q^j r^k.$

(NOTE: Problems 7 and 8 are all that will formally be done with multi-nomials. Illustrative examples and a few problems using the chief result are inserted when they might add interest or further expláin a topic under discussion and these may always be omitted without loss of continuity. The only major use of them will be made in Chapter 10.)

7. For fixed Q, prove by induction on n that

$$(a_1 + a_2 + \cdots + a_n)^Q = \sum_{r_1 + r_2 + \cdots + r_n \,=\, Q} \binom{Q}{r_1, r_2, \ldots, r_n} a_1^{r_1} a_2^{r_2} \cdots a_n^{r_n}$$

(HINT: Use Prob. 6b above) and observe that

$$\sum_{j \,=\, 0}^{Q} \sum_{r_1 + r_2 + \cdots + r_k = Q - j} \binom{Q}{j}\binom{Q-1}{r_1, r_2, \ldots, r_k} a_1^{r_1} a_2^{r_2} \cdots a_k^{r_k} a_{k+1}^{j}$$

can be written

$$\sum_{r_1 + r_2 + \cdots + r_k + j = Q} \binom{Q}{r_1, r_2, \ldots, r_k, j} a_1^{r_1} a_2^{r_2} \cdots a_k^{r_k} a_{k+1}^{j}$$

8. (a) Prove: If $r_1 + r_2 + \cdots + r_k = n$

$$\binom{n-1}{r_1 - 1, r_2, \ldots, r_k} + \binom{n-1}{r_1, r_2 - 1, \ldots, r_k} + \cdots + \binom{n-1}{r_1, r_2, \ldots, r_k - 1} = \binom{n}{r_1, r_2, \ldots, r_k}$$

(This identity is analogous to the identity 4-2a_1.)

(b) For fixed q, prove by induction on n that

$$(a_1 + a_2 + \cdots + a_q)^n = \sum_{r_1 + r_2 + \cdots + r_q \,=\, n} \binom{n}{r_1, r_2, \ldots, r_q} a_1^{r_1} a_2^{r_2} \cdots a_q^{r_q}.$$

HINT:

$$(a_1 + a_2 + \cdots + a_q)^{k+1} = \sum_{r_1 + r_2 + \cdots + r_q \,=\, k} \binom{k}{r_1, r_2, \ldots, r_q} a_1^{r_1+1} a_2^{r_2} \cdots a_q^{r_q}$$

$$+ \sum_{r_1 + r_2 + \cdots + r_q \,=\, k} \binom{k}{r_1, r_2, \ldots, r_q} a_1^{r_1} a_2^{r_2+1} \cdots a_q^{r_q}$$

$$+ \sum_{r_1 + r_2 + \cdots + r_q \,=\, k} \binom{k}{r_1, r_2, \ldots, r_q} a_1^{r_1} a_2^{r_2} \cdots a_q^{r_q+1}$$

The sum of the exponents in each term is $k + 1$. Let $R_1 + R_2 + \cdots + R_q = k + 1$ where each R_i is a nonnegative integer. Reorder the terms as follows:

$$(a_1 + a_2 + \cdots + a_q)^{k+1} = \sum_{R_1 + R_2 + \cdots + R_q \,=\, k + 1} \left[\binom{k}{R_1-1, R_2, \ldots, R_q} \right.$$

$$\left. + \binom{k}{R_1, R_2-1, \ldots, R_q} + \cdots + \binom{k}{R_1, R_2, \ldots, R_q-1} \right] a_1^{R_1} a_2^{R_2} \cdots a_q^{R_q}$$

and then use part (a) to complete the Induction.

Together Probs. 7 and 8 prove that for any n and N,

$$(a_1 + a_2 + \cdots + a_N)^n = \sum_{r_1 + r_2 + \cdots + r_N \,=\, n} \binom{n}{r_1, r_2, \ldots, r_N} a_1^{r_1} a_2^{r_2} \cdots a_N^{r_N}.$$

b.   The Geometric Series

A set of numbers of the form $\{a,ar,ar^2,ar^3,\ldots\} = \{ar^{n-1}|n \in I^+\}$, where a and r are arbitrary nonzero numbers, is called a *geometric sequence* and its elements are called *terms*.   In particular the element ar^k is called the kth term of the sequence (a is the 0th term).

For $n = 5$, let us write

$$s_5 = a + ar + ar^2 + ar^3 + ar^4$$

then

$$rs_5 = \quad ar + ar^2 + ar^3 + ar^4 + ar^5$$

and subtracting,

$$s_5 - rs_5 = a - ar^5$$

so that

$$s_5 = \frac{a - ar^5}{1 - r} = a\,\frac{1 - r^5}{1 - r}.$$

This suggests that we have for any n,

$$s_n = \sum_{i=1}^{n} ar^{i-1} = a\,\frac{1 - r^n}{1 - r}$$

A straightforward application of the Principle of Mathematical Induction proves:

4-2b_1 Theorem.   *If s_n is the sum of the first n terms of a geometric sequence.*

$$s_n = \sum_{i=1}^{n} ar^{i-1} = a\,\frac{1 - r^n}{1 - r}.$$

Proof: The theorem is true for $n = 1$, since

$$s_1 = a = a\,\frac{1 - r}{1 - r}.$$

Let k be such that the statement holds, and consider

$$s_{k+1} = a + ar + \cdots + ar^k = s_k + ar^k$$

$$= a\,\frac{1 - r^k}{1 - r} + ar^k = a\left(\frac{1 - r^k + r^k - r^{k+1}}{1 - r}\right)$$

$$= a\,\frac{1 - r^{k+1}}{1 - r}$$

Hence the statement holds also for $k + 1$ and by the Principle of Induction the theorem is proved. ∎

For example,

$$3 + 6 + 12 + 24 + 48 = 3 + 3(2) + 3(2)^2 + 3(2)^3 + 3(2)^4$$

$$= 3 \frac{1-(2)^5}{1-2} = 3 \frac{-31}{-1} = 93.$$

The formula 4-2b_1 for the sum of the first n terms of a geometric sequence requires the first term a, the ratio r and the number of terms n. The form of the counter is not important. Thus

$$\sum_{i=1}^{6} \left(\frac{1}{8}\right)\left(\frac{1}{2}\right)^{i-1} = \sum_{j=0}^{5} \frac{1}{2^{j+3}} = \sum_{k=1}^{6} \frac{1}{2^{j+2}} = \frac{1}{8}\left(\frac{1-\left(\frac{1}{2}\right)^6}{1-\frac{1}{2}}\right)$$

Of particular interest to us will be those geometric sequences in which $0 < r < 1$. Let us look at

$$\left\{1, \frac{1}{2}, \frac{1}{4}, \frac{1}{8}, \cdots, \frac{1}{2^n}, \cdots\right\} = \left\{\frac{1}{2^{n-1}} \middle| n \in I^+\right\}$$

and consider s_n for various values of n.

$$s_5 = 1 + \frac{1}{2} + \frac{1}{4} + \frac{1}{8} + \frac{1}{16} = \sum_{i=1}^{5}\left(\frac{1}{2}\right)^{i-1} = \frac{1-\left(\frac{1}{2}\right)^5}{1-\frac{1}{2}} = 2 - \left(\frac{1}{2}\right)^4 = \frac{31}{16}$$

$$s_6 = \sum_{i=1}^{6}\left(\frac{1}{2}\right)^{i-1} = \frac{1-\left(\frac{1}{2}\right)^6}{1-\frac{1}{2}} = 2 - \left(\frac{1}{2}\right)^5 = \frac{63}{32}$$

$$s_{11} = \sum_{i=1}^{11}\left(\frac{1}{2}\right)^{i-1} = \frac{1-\left(\frac{1}{2}\right)^{11}}{1-\frac{1}{2}} = 2 - \left(\frac{1}{2}\right)^{10} = \frac{2055}{1028}$$

As we sum more and more terms, s_n gets closer and closer to 2. For any n,

$$s_n = \sum_{i=1}^{n}\left(\frac{1}{2}\right)^{i-1} = \frac{1-\left(\frac{1}{2}\right)^n}{1-\frac{1}{2}} = 2 - \left(\frac{1}{2}\right)^{n-1}$$

and

$$2 - s_n = \left(\frac{1}{2}\right)^{n-1}$$

We can force the difference between s_n and 2 to be as small as we want, simply by choosing n large enough. If ϵ is any positive number, no matter how small,

$$2 - s_n = \left(\frac{1}{2}\right)^{n-1} < \epsilon$$

if

$$(n - 1) \log \left(\frac{1}{2}\right) < \log \epsilon$$

which, when we solve for n, has the form

$$n > \frac{\log \epsilon}{\log \left(\frac{1}{2}\right)} + 1.$$

(For $\epsilon < 1$, $\log \epsilon < 0$, as is $\log \left(\frac{1}{2}\right)$, so that the right-hand side of the inequality is positive.)

Let us move to the general case. We have just proved that

$$s_n = \sum_{i=1}^{n} ar^{i-1} = a \frac{1 - r^n}{1 - r} = \frac{a}{1 - r} - a \frac{r^n}{1 - r}$$

and we have

$$\frac{a}{1 - r} - s_n = a \frac{r^n}{1 - r}.$$

We can show further that, if $0 < r < 1$ and ϵ is an arbitrary positive number, no matter how small, we can find an n so large that

$$0 < \frac{a}{1 - r} - s_n = a \frac{r^n}{1 - r} < \epsilon. \qquad [1]$$

Thus by taking n large enough we can make the difference

$$\frac{a}{1 - r} - \sum_{i=1}^{n} ar^{i-1}$$

as small as we please. We use this as a basis of our definition of the sum of a countable number of terms of a geometric sequence in which $0 < r < 1$.

[1] Numbers in square brackets refer to explanatory notes in the Appendix.

4-2b_2 Definition. *If* $0 < r < 1$, $\sum_{i=1}^{\infty} ar^{i-1} = \dfrac{a}{1-r}$.

Using this definition

$$\sum_{i=1}^{\infty} \left(\frac{1}{2}\right)^{i-1} = \frac{1}{1-\frac{1}{2}} = 2$$

and

$$\sum_{i=1}^{\infty} (.3)(.1)^{i-1} = \frac{.3}{1-.1} = \frac{1}{3}.$$

We remark that

$$\sum_{j=0}^{\infty} ar^{j} = \sum_{i=1}^{\infty} ar^{i-1} = \frac{a}{1-r},$$

since in summing a countable number of geometric terms we need only to know the first term a and the ratio r.

EXERCISES 4-2b

1. Evaluate:

   (a) $\displaystyle\sum_{i=1}^{5} 4(2)^{i-1}$

   (b) $2 - 4 + 8 - 16 + 32 - 64 + 128$

   (c) $\displaystyle\sum_{j=1}^{7} 5\left(\frac{1}{3}\right)^{i-1}$

   (d) $\displaystyle\sum_{k=0}^{8} \left(\frac{2}{5}\right)^{k}$

2. Evaluate:

   (a) $\displaystyle\sum_{j=0}^{\infty} \left(\frac{1}{9}\right)\left(\frac{8}{9}\right)^{i-1}$

   (b) $\displaystyle\sum_{k=0}^{\infty} \left(\frac{1}{3}\right)^{k}$

(c) $\sum_{j=0} (.6)^j$

(d) .3333...

(e) $\sum_{n=0} \dfrac{1}{2^{n+3}}$

(f) .2474747...

PART II

PROBABILITY THEORY

chapter 5

Fundamental Concepts

5-1. EVENTS

a. Sample Spaces

From its unusual and perhaps suspect beginnings in the minds and writings of sixteenth-century gentlemen gamblers, probability has matured into a branch of mathematics of great theoretical interest and of wide application.   However, traces of the colorful and suggestive language of gaming have remained in the field to aid the student in understanding the development of the science as well as the problem at hand.

Problems of probability are formulated with respect to a universal set whose elements are the possible outcomes of an experiment.   The experiment might be tossing a coin, rolling a die, or recording a man's age at the time of his death. This universal set is called the *sample space* and its elements are called *elementary events*.   The Greek letter Ω is commonly used to designate the sample space.   If in tossing a coin, we let H and T stand respectively for "coin comes up heads" and "coin comes up tails," we have for this experiment

$$\Omega = \{H,T\}.$$

Someone might object and ask what if the coin had a wide edge and could come up on its side or what if it rolled out of sight and could not be retrieved.   For that experiment we would have

$$\Omega = \{H,T,X\}$$

where X represents the third event.   If there were additional alternatives, the sample space would have still more elementary events.

In constructing the sample space associated with an experiment we tend not to include those outcomes whose likelihood is negligible; in this sense we idealize the experiment.   Thus, although other results are conceivable, the sample space associated with tossing a coin is

$$\Omega = \{H,T\}$$

and that associated with rolling a die is

$$\Omega = \{1,2,3,4,5,6\}.$$

If we draw cards from an ordinary bridge deck and are interested only in the suit of the card drawn, then

$$\Omega = \{\text{heart, spade, diamond, club}\}.$$

If instead we were concerned with the denomination of the card, then

$$\Omega = \{A,2,3,4,5,6,7,8,9,10,J,Q,K\};$$

or if we were interested in the specific card drawn, then (letting S stand for spade, H for heart, D for diamond, and C for club)

$$\Omega = \{AS,2S,\ldots,KS,AH,2H,\ldots,KH,AD,\ldots,KD,AC,\ldots,KC\}.$$

The experiment of tossing a coin until a tail appears gives rise to a sample space with more than a finite number of elements, for here

$$\Omega = \{T,HT,HHT,HHHT,HHHHT,\ldots\}.$$

Three contestants X, Y, and Z enter a tournament where a two-person game which cannot end in a tie is played (cribbage is such a game). X and Y play the first game and in the second game Z plays the winner of the first game. At every step two contenders are playing the game and the third is waiting to play the victor in the next round. The tournament ends when one contestant wins two consecutive games. The structure of the sample space associated with the tournament is the collection of elementary events listed in Figure 5-1a_1.

5-1a_1 Figure

Sample Space for Tournament.

(The sequence of letters represents the order in which the players, X, Y, and Z, win the games.)

XX	*YY*
XZZ	*YZZ*
XZYY	*YZXX*
XZYXX	*YZXYY*
XZYXZZ	*YZXYZZ*
XZYXZYY	*YZXYZXX*
.	.
.	.
.	.

A coin is tossed twice, or two coins, say a nickel and a dime, are tossed once; both experiments give rise to the sample space

$$\Omega = \{(H,H), (H,T), (T,H), (T,T)\}$$

where for the first experiment the ordered pair represents the outcome on the first, then on the second toss; and in the second experiment the first member is the outcome on the nickel and the second the outcome on the dime. In either case, if $C = \{H,T\}$,

$$\Omega = C \times C.$$

Many useful experiments consist of repeating the same simple experiment a given number of times or, what is the same thing, of performing the same experiment with a fixed number of distinguishable objects. In such experiments the sample space can often most efficiently be expressed as the cross-product of two or more sets.

If a die is rolled four times, the sample space is the set of ordered 4-tuples where each component is an elementary event in $D = \{1,2,3,4,5,6\}$. Then the sample space associated with rolling a die four times is

$$\Omega = D \times D \times D \times D = (\times D)_4.$$

If B is the set of boys at a party and G the set of girls, then

$$\Omega = \{(b,g) \mid b \in B, g \in G\} = B \times G$$

is the set of ways the guests at the party can be paired off.

It is useful to remark here that a sociologist who, in making a study of the ages of married couples polls one hundred couples, records his findings as ordered pairs (h,w) where h is the husband's age and w the wife's obtains a sample space of ordered pairs which is not the cross-product of the set of husband's ages with the set of wife's ages. This points up the necessity of carefully describing the sample space associated with a given experiment.

The above sample spaces are all *discrete*—that is, they are at most countable and it is possible to distinguish between the points of the sample space. In this introduction to the theory of probability we shall limit our considerations to finite and countable sample spaces. However, very many simple experiments lead to nondiscrete—noncountable—sample spaces. Consider a spinner—the type that is often found in a child's game—and suppose a starting position is indicated (see accompanying figure). The experiment consists of activating the spinner and recording the angle the spinner makes with its initial position when it comes to stop. For this experiment

$$\Omega = \{\omega \mid 0 \leqslant \omega < 2\pi\}.$$

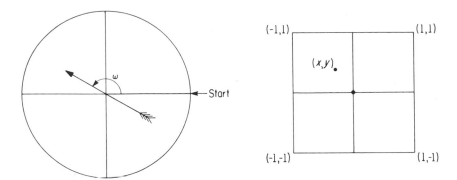

In another experiment a mouse is put down in the middle of a square pen and his position after a given time is recorded. If the center of the enclosure is taken as the origin of a rectangular coordinate system and the units on each axis are chosen so that the pen's corners have coordinates $(1,1)$, $(-1,1)$, $(-1,-1)$ and $(1,-1)$, then $\Omega = \{(x,y)|\ -1 < x < 1, -1 < y < 1\}$, where the elementary events in Ω are the ordered pairs (x,y).

EXERCISES 5-1a

Construct the sample spaces of the possible outcomes of the following experiments:

1.   Answers to a true-false question.
2.   Choosing a ball from an urn containing 3 red, 4 white, and 6 blue balls.
3.   Asking people the date of their birthdays.
4.   Color of a common traffic signal as a car reaches an intersection.
5.   The day of the week on which a traffic accident occurs.
6.   Choosing two balls simultaneously from the urn in Prob. 2.
7.   Choosing two balls consecutively from the urn in Prob. 2.   (Here we are interested in the order in which they are drawn.)
8.   Rolling a die and flipping a coin.
9.   (a) Position of an on-off transistor.   (b) Reading a row of 10 such transistors.   In large computing machines, such rows of transistors represent ten-digit binary numbers; so the set of possible positions represents the set of ten-digit binary numbers.   Really large machines operate with 32-digit numbers and larger.
10.   Rolling three distinguishable dice—say red, green, and blue—for the arrangement of their faces.
11.   Rolling three indistinguishable dice for the arrangement of their faces.
12.   Rolling three dice for the sum of their faces.

13. Observing sex distributions of children in families of five children. (Here we are interested not only in whether there are three girls and two boys but also the sequence in which the sexes appear.)

    14. (a) Putting three distinguishable balls into three cells.

          (b) Putting three nondistinguishable balls into three cells.

    15. Rolling a die until a 6 appears.

    16. Tossing a coin until a head appears twice.

    17. Tossing a coin until a head appears twice in succession.

    18. Three people, A, B, C enter an elevator on the main floor of a store and will get off on the first, second, or third floor.

          (a) We care who gets off at which floor.

          (b) We care only how many get off at each floor.

          (c) Compare (a) and (b) to Prob. 14a and 14b.

b.   Subsets of Ω

Along with Ω we consider its set of subsets which we shall denote by A_Ω. If $\Omega = \{H,T\}$, $A_\Omega = \{\emptyset, \{H\}, \{T\}, \{H,T\}\}$. From 2-1$b_4$ we know that if a sample space Ω has N points—that is, if $n(\Omega) = N$, then $n(A_\Omega) = 2^N$. However, if Ω is countable, A_Ω is uncountable.[2] The elements of A_Ω are called *events*.

Since throughout our work in probability we shall speak of a sample space instead of a universal set and events instead of elements and subsets, we shall now illustrate the use of these terms. Let an experiment consist of drawing a card from an ordinary bridge deck and noting its suit. Then

$$\Omega = \{S,H,D,C\}$$

and

$$A_\Omega = \{\emptyset, \{S\}, \{H\}, \{D\}, \{C\}, \{S,H\}, \{S,D\}, \{S,C\}, \{H,D\},$$
$$\{H,C\}, \{D,C\}, \{S,H,D\}, \{S,H,C\}, \{S,C,D\}, \{H,C,D\}, \Omega\}$$

The event the card drawn is red is $A_1 = \{H,D\} = \{H\} \cup \{D\}$; that is A_1 equals the union of the events a heart is drawn and a diamond is drawn. The event the card drawn is black is $A_2 = \{S,C\} = A_1'$; that is A_2 is the event the card drawn is not red. The event $\{S,C,D\}$ is the event the card drawn is not a heart and Ω, in this case, the event a card is drawn, is called the *sure event*.

A die is rolled; $\Omega = \{1,2,3,4,5,6\}$ is the set of faces that can come up. The event the die comes up with an even face is $A_1 = \{2,4,6\}$; that it comes up with a face greater than four is $A_2 = \{5,6\}$, and that it comes up with an even face greater than four is $A_1 \cap A_2 = \{6\}$ (both A_1 and A_2 had to occur to get an even face greater than four), $\{1,3,5\} = A_1'$ is the event the die does not come up even, or, what is the same thing, the event it comes up odd. $A_2 - A_1 = \{5\}$ is the event it comes up greater than four but not even. The event the die comes up four and odd is $\emptyset$, the *impossible event,* and if A_3 is the event the die comes up four or six, $A_3 = \{4,6\}$, $A_3 \subset A_1$—that is, the occurrence of A_3 guarantees the occurrence of A_1.

The preceding examples illustrate:

5-1b₁ **The probabilistic interpretations of set arithmetic** *If A_1 and A_2 are events in A_Ω;*

$A_1 \cup A_2$ *is the event that at least one of A_1 and A_2 occur;*

$A_1 \cap A_2$ *is the event that both A_1 and A_2 occur;*

$A_1 \quad A_2$ *is the event that A_1 but not A_2 occurs;*

A_1' *is the event that A_1 does not occur;*

$A_1 \subset A_2$ *means that the occurrence of A_1 implies the occurrence of A_2;*

Ω *is the* sure event, *it must occur and for all $A \in A_\Omega$, $A \subset \Omega$;*

$\emptyset$ *is the* impossible event, *it never occurs.*

If two distinguishable dice are rolled—say a red and a green one—or if a single die is rolled twice, $\Omega = D \times D$, where $D = \{1,2,3,4,5,6\}$. Elements of Ω are ordered pairs (r,g) where r is the face on the red die, or on the first roll, and g is the face on the green die, or on the second roll. $A_1 = \{(r,g)| r + g > 6\}$ is the event that the sum of the faces is greater than six. $A_2 = \{(r,g)| g = 3\}$ is the event that the green die, or second roll, comes up three. The event that both faces are greater than four is $A_3 = \{(r,g)| r > 4, g > 4\}$ and certainly if both faces are greater than four their sum is greater than six, so that $A_3 \subset A_1$. Since we cannot simultaneously have $g = 3$ and $g > 4$, $A_2 \cap A_3 = \emptyset$, and the event that the sum on the two faces is no more than six i s $A_1' = \{(r,g)| r + g \leqslant 6\}$.

EXERCISES 5-1b

1. What are the events
    (a) in $\Omega = \{H,T\}$?
    (b) in $\Omega = \{r\ b,g\}$?

2. Let $A \subset \Omega = \{(d,c)| d \in \{1,2,3,4,5,6\}, c \in \{H,T\}\}$ be the subset in which the first member of the ordered pair is 3. What are the elements of A?

3. From an urn containing 3 red, 4 blue, and 6 green balls, two balls are simultaneously drawn. What are the elements of $B \subset \Omega$, where B is the event at least one ball is green?

4. Three dice are rolled for the sum of the faces and C is the event the sum of the faces $\geqslant 17$. What are the elements of C?

   5. If three distinguishable dice are rolled for the arrangement of their faces, and C is the event in Prob. 4, what are the elements of C?
   6. A row of ten on-off transistors are read.
   (a) Let D be the event exactly three transistors are "on." How many elementary events are there in D?
   (b) If E is the event at least three transistors are "on," how many elementary events are there in E?
   7. The sex distributions of children in families of five children are observed. Let C be the event the first child is a boy and D the event at least three children are boys. What are the elements of C and D? What are $C \cup D$, $C \cap D$ and $C - D$?
   8. A coin is tossed until a head appears twice in succession. Let S be the event the second consecutive head appears before the sixth toss. What is S? Let T be the event the experiment ends at the first toss. What is T?
   9. Consider the set of permutations of 1, 2, 3, 4. Let A_i be the subset in which i appears in its natural place. What are the elements of A_3? What are $A_1 \cup A_3$, $A_2 \cap A_4$, A_2'?
   10. A coin is tossed 16 times. Let A be the event there are 4 heads and 12 tails. How many elements in A? If B is the event at most 2 heads, how many elements in B?
   11. A die is rolled 10 times. Let C be the event that exactly 3 aces and 4 fours appear. How many elements in C? How many elements in D, the event exactly 2 aces, 2 twos, and 3 fives?
   12. Four balls are distributed among four cells. Let A be the event two go into one cell and one into each of two cells (leaving one cell vacant). If the balls are distinguishable, how many elements are in A?
   13. Given a sample space Ω and the three events A, B, and C in Ω. In terms of A, B, C, express: (a) A alone occurs. (b) A and C but not B occur. (c) All three events occur. (d) At least one event occurs. (e) None occur. (f) Exactly two occur. (g) No more than two occur.

5-2. PROBABILITY

a. Functions

   5-2a₁ Definition. *A **function** is a set of ordered pairs no two of which have the same first member.*

The sets
$$f = \{(x,y) | y = 2x\}$$
$$g = \{(1,3),(2,4),(3,5),(5,5)\}$$
$$h = \{(x,y) | y = 3\}$$

are all functions, since whenever $x_1 = x_2$ we have $f(x_2) = f(x_2)$, $g(x_1) = g(x_2)$, and $h(x_1) = h(x_2)$. The set of ordered pairs

$$C = \{(x,y)|x^2 + y^2 = 25\}$$

is not a function, for if $x = 4$, $y = \pm 3$ so that $(4,3)$ and $(4,-3)$ both are elements of C.

All functions are not mathematical. The Weather Bureau keeps records consisting of ordered pairs (d,t), where d is the date and t is the mean temperature in that 24-hour period. The set of these pairs is a function. An endurance-test driver for an automobile manufacturer keeps a log in which at regular intervals he enters two numbers, s, the distance traveled since starting his run, and c, the amount of gasoline consumed. Such a log, the set of ordered pairs (s,c), is a function. The teacher hands his superior a function when he gives him a list of ordered pairs (n,g), where n is the student's name and g is his final grade.

Associated with every function f are two other sets; the set of the first members of its ordered pairs which is called the *domain* of f, and the set of the second members which is called the *range* of f. If we write respectively D_f and R_f, we have:

$$D_f = \{x|(x,y) \in f\}$$
$$R_f = \{y|(x,y) \in f\}.$$

We often speak of a function as being defined on its domain. Associated with $g = \{(1,3) , (2,4) , (3,5) , (5,5)\}$ are $D_g = \{1,2,3,5\}$ and $R_g = \{3,4,5\}$; with $h = \{(x,y)|y = 3\}$ are D_h the set of all real numbers, and $R_h = \{3\}$. The function $k = \{(x,y)|y = \sqrt{25 - x^2}\}$ is defined on $D_k = \{x|-5 \leqslant x \leqslant 5\}$ and has as its range, $R_k = \{y|0 \leqslant y \leqslant 5\}$. The Weather Bureau's mean temperature record is a function defined on the set of days from the establishment of the bureau; the teacher's grade list is a function defined on his class roll.

Knowing its domain, the description of the function will give a set which contains the range if not the range itself. The definitions of g, h, and k above enable us to determine their ranges exactly; in reviewing the Weather Bureau's record we know the range will be included between the lowest mean temperature recorded and the highest, but it may not include all intermediate values; the range of the teacher's grade record is contained in the set of symbols that comprise the school's grading scale.

A special group of functions which is fundamental in our work is the class of set functions.

5-2a_2 **Definition.** *A* **set function** *is a function whose domain is a nonempty set of sets and whose range is a subset of the real numbers.*

If $\Omega = \{H,T\}$ and f is defined on A_Ω, its set of subsets, by

$$f(\{H\}) = f(\Omega) = 1, f(\{T\}) = f(\varnothing) = 0$$

then f is a set function.   In terms of its ordered pairs,

$$f = \{(\{H\},1),(\Omega,1)(\{T\},0),(\varnothing,0)\}.$$

Let $\Omega = \{1,2,3,4,5,6\}$ and A_Ω be its set of subsets.   Define, for all $A \in A_\Omega$, $g(A)$ to be the sum of the points in A.   Then $g(A)$ is a real number for all $A \in A_\Omega$ and g is a set function.   We have $g(\varnothing) = 0$ and if $A = \{1,2,3\}$, $g(A) = 6$.

b.   Probability

5-2b_1 Definition.   *Given a sample space Ω and A_Ω its set of subsets.   A set function P defined on A_Ω is called a* **probability** *if, and only if, for all $A \in A_\Omega$.*

   (i) $P(A) \geqslant 0$;
   (ii)   *if $\{A_k | A_k \in A_\Omega, k \in I'\}$ is a finite or countable set of mutually disjoint sets, $P(\Sigma_k A_k) = \Sigma_k P(A_k)$;*
   (iii) $P(\Omega) = 1$.

$P(A)$ is read "the probability of the event A," or, more simply, "the probability of A."   The second condition on the set function P can be expressed by saying that, if the A_k are mutually disjoint sets, the probability of their sum is the sum of their probabilities.

An experiment consists of drawing one ball from an urn containing one red, two white, and three blue balls.   Set $\Omega = \{r,w,b\}$ and then $A_\Omega = \{\varnothing, \{r\}, \{w\}, \{b\}, \{r,w\}, \{r,b\}, \{w,b\}, \Omega\}$.   Define a function P on A_Ω as follows:

$$P(\varnothing) = 0 \qquad P(\{r,w\}) = \frac{1}{2}$$

$$P(\{r\}) = \frac{1}{6} \qquad P(\{r,b\}) = \frac{2}{3}$$

$$P(\{w\}) = \frac{1}{3} \qquad P(\{w,b\}) = \frac{5}{6}$$

$$P(\{b\}) = \frac{1}{2} \qquad P(\Omega) = 1$$

Is P a probability?   We see immediately that requirements (i) and (iii) of the definition are satisfied and we need only to check out (ii).   The condition requires that for all sums of mutually disjoint elements of A_Ω, P of the sum is the sum of the P's. Let us look first at sums of nonempty disjoint sets. If we take the sets two at a time we have:

$$P(\{r,w\}) = \frac{1}{2} = P(\{r\}) + P(\{w\}) = \frac{1}{6} + \frac{1}{3}$$

$$P(\{r,b\}) = \frac{2}{3} = P(\{r\}) + P(\{b\}) = \frac{1}{6} + \frac{1}{2}$$

$$P(\{w,b\}) = \frac{5}{6} = P(\{w\}) + P(\{b\}) = \frac{1}{3} + \frac{1}{2}$$

$$P(\{r,w,b\}) = P(\Omega) = 1 = P(\{r\}) + P(\{w,b\}) = \frac{1}{6} + \frac{5}{6}$$

$$P(\Omega) = 1 = P(\{w\}) + P(\{r,b\}) = \frac{1}{3} + \frac{2}{3}$$

$$P(\Omega) = 1 = P(\{b\}) + P(\{r,w\}) = \frac{1}{2} + \frac{1}{2}$$

All other pairs have nonempty intersections. If we take the nonempty sets three at a time, we have only one sum of disjoint sets; it is

$$P(\Omega) = 1 = P(\{r\}) + P(\{w\}) + P(\{b\}) = \frac{1}{6} + \frac{1}{3} + \frac{1}{2}.$$

There are no other sums of nonempty disjoint sets. Furthermore for all $A \in A_\Omega$, $A\varnothing = \varnothing$ and

$$P(A + \varnothing) = P(A) = P(A) + P(\varnothing).$$

This exhausts all sums and so P satisfies condition (ii) as well as (i) and (iii). Hence P is a probability.

Consider a second example. Let $\Omega = \{1,2,\ldots,10\}$, and for $A \in A_\Omega$, define

$$f(A) = \frac{2^{\sum\limits_{i \in A} i} - 1}{2^{55} - 1}$$

For $N \geqslant 0$, $2^N \geqslant 1$, hence for all A, $f(A) \geqslant 0$, so that (i) is satisfied. Also,

$$f(\Omega) = \frac{2^{\sum\limits_{i \in \Omega} i} - 1}{2^{55} - 1} = \frac{2^{\sum\limits_{i=1}^{10} i} - 1}{2^{55} - 1} = 1$$

and (iii) is satisfied.* If $A_1 = \{1\}$ and $A_2 = \{2\}$, $A_1 A_2 = \varnothing$,

$$f(A_1) = \frac{2^1 - 1}{2^{55} - 1}, \quad f(A_2) = \frac{2^2 - 1}{2^{55} - 1},$$

*In Sec. 1-1 we proved $\sum\limits_{i=1}^{n} i = \frac{1}{2} n(n + 1)$. Here $\sum\limits_{i=1}^{10} i = \frac{1}{2}(10)(11) = 55$.

and

$$f(A_1 + A_2) = f(\{1,2\}) = \frac{2^3 - 1}{2^{55} - 1} = \frac{7}{2^{55} - 1} \neq \frac{1}{2^{55} - 1} + \frac{3}{2^{55} - 1} = f(A_1) + f(A_2).$$

Hence f is not a probability. The set function f illustrates the importance of condition (ii) in differentiating set functions which are probabilities from those which are not.

Let us go back to urn containing one red, two white and three blue balls, but now let the experiment consist of drawing a ball, noting its color, returning it to the urn, and doing this three times. If we set $U = \{r,w,b\}$, $\Omega = (X \ U)_3 = \{(u_1,u_2,u_3)|u_i \in U\}$. Since there are three choices for each u_i, $n(\Omega) = 3^3 = 27$ (Counting Principle I) and $n(A_\Omega) = 2^{27}$. It would be quite a job to design a probability P by associating a real number with each of the 2^{27} events of A_Ω and making sure, at the same time, that (ii) is satisfied. Can we develop some easier way of assigning a probability to a sample space?

The need for a simpler method than the definition for ascertaining whether a function on a sample space is a probability is even more evident if we keep drawing from and replacing balls in the urn until we get the red one. Let x indicate that a white or blue ball is drawn, then $\Omega = \{r, xr, xxr, xxxr,....\}$. Ω is countable and so A_Ω is noncountable. If we are given a function f on A_Ω, we have to have some means of deciding whether f is a probability.

We digress a moment to remark that in the examples we have looked at so far, we started with an experiment that gave rise to a sample space Ω. Then on the events of Ω—that is, on A_Ω,—we define a set function P which is a probability if it satisfies the three requirements of Definition 5-2b_1. We start with the domain A_Ω and assign values to its elements, values which must lie between zero and one. How different from our earlier experiences with algebraic functions! When we work with algebraic functions, we invariably start with the set of ordered pairs and investigate properties of its domain and its range. In our work with probabilities we shall start with the domain and range and then somehow produce the ordered pairs.

In looking for a method of creating ordered pairs we observe first of all that if Ω is a finite or countable sample space, $\Omega = \{\omega_1,\omega_2,\omega_3,...\}$ and if P is a probability on A_Ω, for each one element subset $\{\omega_i\} \in A_\Omega$, $P(\{\omega_i\}) = P(\omega_i)$ is defined. (For simplicity we write $P(\omega_i)$ instead of $P(\{\omega_i\})$.) Furthermore if $A \in A_\Omega$ is a nonempty, finite or countable event, $A = \{\omega_{a_1}, \omega_{a_2}, \omega_{a_3},...\}$ where $\omega_{a_i} \in \Omega$ and we can write A as the disjoint union of its elementary events; that is, we can set

$$A = \sum_i \{\omega_{a_i}\} = \sum_{\omega_j \in A} \{\omega_j\}$$

(Each $\omega_{a_i} \in A$ is some $\omega_j \in \Omega$ and when we sum the elementary events in A, we sum those elementary events in Ω which are in A. Also, of course, if $i \neq j$, $\{\omega_i\} \cap \{\omega_j\} = \emptyset$.) Since P is a probability, we have

5-2b_2
$$P(A) = P\left(\sum_{\omega_i \in A} \{\omega_i\} \right) = \sum_{\omega_i \in A} P(\omega_i).$$

Here we have written the probability of an event as the sum of the probabilities of the elementary events which comprise the event. In the second example discussed above, the one in which $U = \{r,w,b\}$ and $\Omega = (\times U)_3$, the event that the second ball is the only white ball is

$$A = \{(r,w,r), (r,w,b), (b,w,r), (b,w,b)\},$$

and if P were a probability on A_Ω, we would have from 5-2b_2

$$P(A) = P\big((r,w,r)\big) + P\big((r,w,b)\big) + P\big((b,w,r)\big) + P\big((b,w,b)\big)$$

Why then, not go the other way; that is, why not explicitly define P only for the elementary events and then for all $A \in A_\Omega$ set $P(A) = \sum_{\omega_i \in A} P(\omega_i)$? We proceed to prove that subject to certain obvious restrictions we can design probabilities in this way.

5-2b_3   **Theorem.** *Let $\Omega = \{\omega_1, \omega_2, \omega_3, \ldots\}$ be a finite or countable sample space. Define a set function P on A_Ω satisfying:*

(i) *for all elementary events $\{\omega_i\} \in A_\Omega$, $P(\omega_i) \geqslant 0$;*

(ii) $\sum_{\omega_i \in \Omega} P(\omega_i) = 1,$

(iii) *for $A \in A_\Omega$, $P(A) = \sum_{\omega_i \in A} P(\omega_i)$.*

Then P is a probability.

Proof. Since by hypothesis $P(\omega_i) \geqslant 0$ for every $\omega_i \in \Omega$ (for all $\{\omega_i\} \in A_\Omega$), then for $A \in A_\Omega$, $P(A) = \sum_{\omega_i \in A} P(\omega_i) \geqslant 0$.

Let $A = \sum_k A_k$, where $A_i A_j = \emptyset$ if $i \neq j$, be finite or countable. If $\omega_i \in A$, $\omega_i \in A_k$ for one and only one k and we can reorder the terms in

$$P(A) = \sum_{\omega_i \in A} P(\omega_i)$$

by grouping together for every k those $P(\omega_i)$ such that $\omega_i \in A_k$. We then have

$$P(A) = \sum_{\omega_i \in A} P(\omega_i) = \sum_k \left(\sum_{\omega_i \in A_k} P(\omega_i) \right) = \sum_k P(A_k)^{[3]}.$$

Finally $\Omega = \sum_i \{\omega_i\}$ and $P(\Omega) = \sum_{\omega_i \in \Omega} P(\omega_i) = 1$.

Hence P is a probability since it satisfies the three conditions of Definition 5-2b_1. ∎

We hasten to remark that this theorem gives us only one way of designing probabilities; it is certainly not the only way. Granted however, it is a very useful method and is the one we will use repeatedly in our work.

In our initial example where we drew a ball from an urn containing one red, two white and three blue balls, we set $\Omega = \{r,w,b\}$ and constructed a probability by assigning an appropriate real number to each of the eight elements of A_Ω. Now using the theorem we need only to say: let $P(r) = \dfrac{1}{6}$, $P(w) = \dfrac{1}{3}$, $P(b) = \dfrac{1}{2}$ and for $A \in A_\Omega$, $P(A) = \sum_{\omega \in A} P(\omega)$. This set function P is a probability.

Let us look again at the experiment in which we draw and replace a ball in the urn three times. If $U = \{r,w,b\}$, $\Omega = (XU)_3 = \{(u_1,u_2,u_3) | u_i \in U\}$. In any ordered triplet $\omega \in \Omega$ there are i-r's, j-w's and k-b's where i, j, k are nonnegative integers and $i + j + k = 3$. The triplet $(b,b,w) \in \Omega$ and here $i = 0, j = 1$, $k = 2$. For $\{\omega\} \in A_\Omega$, set $P(\omega) = \left(\dfrac{1}{6}\right)^i \left(\dfrac{2}{6}\right)^j \left(\dfrac{3}{6}\right)^k$ and for $A \in A_\Omega$, set

$$P(A) = \sum_{\omega \in A} \left(\frac{1}{6}\right)^i \left(\frac{2}{6}\right)^j \left(\frac{3}{6}\right)^k.$$

The set $\{(b,b,w)\} \in A_\Omega$ and we set

$$P((b,b,w)) = \left(\frac{1}{6}\right)^0 \left(\frac{2}{6}\right) \left(\frac{3}{6}\right)^2$$

and the probability of the event A there are two b's and one w is

$$P(A) = P((b,b,w)) + P((b,w,b)) + P((w,b,b)) = \binom{3}{0,1,2}\left(\frac{1}{6}\right)^0 \left(\frac{2}{6}\right)\left(\frac{3}{6}\right)^2.$$

Obviously $P(\omega) \geqslant 0$; if we can show $\sum_{\omega \in \Omega} P(\omega) = 1$, P will be a probability for it will satisfy the conditions of the theorem. Three cells, the three places in the triplet, can be filled with i-r's, j-w's and k-b's in $\binom{3}{i,j,k}$ ways (Counting Principle IV) and each such point ω has $P(\omega) = \left(\dfrac{1}{6}\right)^i \left(\dfrac{2}{6}\right)^j \left(\dfrac{3}{6}\right)^k$. Hence

$$\sum_{\omega \in \Omega} P(\omega) = \sum_{i+j+k=3} \binom{3}{i,j,k}\left(\frac{1}{6}\right)^i \left(\frac{2}{6}\right)^j \left(\frac{3}{6}\right)^k = \left(\frac{1}{6} + \frac{2}{6} + \frac{3}{6}\right)^3 = 1$$

(see Prob. 6d, Exercise 4-2a).   So P is a probability.   Set aside just temporarily

any question concerning why we chose $P(\omega) = \left(\dfrac{1}{6}\right)^i \left(\dfrac{2}{6}\right)^j \left(\dfrac{3}{6}\right)^k$ The question

is an important one, but let us continue to focus our attention on the fact that, no matter where it came from, it does satisfy the requirements for a set function to be a probability.

One more example.   Let us look again at the case where we draw from and replace balls in the urn until we draw a red one.   We said

$$\Omega = \{r,\ xr,\ xxr,\ xxxr, \ldots\}$$

where x signifies the drawing of a nonred ball.   Let ω_i stand for the element in Ω which represents i draws; thus $\omega_3 = xxr$ and $\omega_6 = xxxxxr$.   For

$\{\omega_i\} \in A_\Omega$, set $P(\omega_i) = \dfrac{1}{6}\left(\dfrac{5}{6}\right)^{i-1}$ and for $A \in A_\Omega$ set $P(A) = \sum\limits_{\omega_i \in A} P(\omega_i)$.   In this

scheme $P(\omega_6) = \dfrac{1}{6}\left(\dfrac{5}{6}\right)^5$ and if A is the event it requires more than three

drawings $P(A) = \sum\limits_{\omega_i \in A} P(\omega_i) = \sum\limits_{i=4} \dfrac{1}{6}\left(\dfrac{5}{6}\right)^{i-1}$ $(A = \{\omega_4, \omega_5, \omega_6, \ldots\})$.   Is P a pro-

bability?   Since

$$\sum\limits_{\omega_i \in A} P(\omega_i) = \sum\limits_{i=1} \left(\dfrac{1}{6}\right)\left(\dfrac{5}{6}\right)^{i-1} = \dfrac{\dfrac{1}{6}}{1 - \dfrac{5}{6}} = 1$$

(see 4-2b_2), P is a probability.

EXERCISES 5-2b

1.   Which of the following sets are functions?
    (a) $\{(a,b)|a$ is a citizen of Washington, D.C.; b is his birthday.$\}$
    (b) $\{(c,u)|c$ is the color of the car; u is the color of the upholstery.$\}$
    (c) $\{(t,s)|s = \dfrac{1}{2}\ gt^2.\}$
    (d) $\{(x,y)|x + y^3 = 1,\ x$ and y are real numbers.$\}$
    (e) $\{(x,y)|x + y^3 = 1,\ x$ and y are complex numbers.$\}$
    (f) $\{(w,x)|w$ is face of a die, $x = 1$ if w even, $x = -1$ if w is odd.$\}$
    (g) $\{(p,n)|p,$ ticket-holder for the theater, $n,$ the seat number.$\}$
    (h) $\{(b,p)|b,$ brand of canned peaches; $p,$ price of no. 2 can.$\}$
2.   What are the domain and range of the function in Prob. 1?
3.   Let $\Omega = \{a,b,c,d\}.$ Set $f(a) = f(b) = f(c) = f(d) = \dfrac{1}{4};$

$$f(\{a,b\}) = f(\{a,c\}) = f(\{a,d\}) = f(\{b,c\}) = f(\{b,d\}) = f(\{c,d\}) = \frac{1}{2};$$

$$f(\{a,b,c\}) = f(\{a,b,d\}) = f(\{a,c,d\}) = f(\{b,c,d\}) = \frac{3}{4};$$

$f(\Omega) = 1$. Verify that f is a probability.

4. If in Prob. 3, a,b,c,d are the outcomes of tossing a coin twice, i.e., $a = HH$, $b = HT$, $c = TH$, $d = TT$, and f is the probability on A_Ω, what is the probability that (a) both outcomes are the same? (b) at least one outcome is a head? (c) a tail on the second toss? (d) two heads?

5. Let a coin be tossed until a head appears. $\Omega = \{H,TH,TTH,TTTH,\ldots\}$. Let $\omega_i \in \Omega$ be the point which requires i tosses. Define a function p on A_Ω by setting for $\{\omega_i\} \in A_\Omega$, $p(\omega_i) = \left(\frac{1}{2}\right)^i$ and for $A \in A_\Omega$, $p(A) = \sum_{\omega_i \in A} p(\omega_i)$. Is p a probability?

6. Let a die be rolled until a three or a six appears. $\Omega = \{S,FS,FFS, FFFS,\ldots\}$ where S means a three or six appeared and F means one did not. Let $\omega_i \in \Omega$ be the point which requires i rolls. Define a function on A_Ω by setting for $\{\omega_i\} \in A_\Omega$, $f(\omega_i) = \left(\frac{1}{3}\right)^i$ and for $A \in A_\Omega$, $f(A) = \sum_{\omega_i \in A} f(\omega_i)$ Is f a probability?

7. A die is rolled four times and as in Prob. 6 let S mean a three or six appears and F mean otherwise. If we set

$$D = \{S,F\}, \Omega = (X\,D)_4 = \{(d_1,d_2,d_3,d_4,)| d_i \in D\}.$$

Define a function g on A_Ω by setting for

$$\{\omega\} \in A_\Omega, \; g(\omega) = \left(\frac{1}{3}\right)^i \left(\frac{2}{3}\right)^{4-i}$$

where i is the number of S components in ω and for $A \in A_\Omega$, $g(A) = \sum_{\omega \in A} g(\omega)$. Is g a probability?

8. From a class of 8 boys and 7 girls, three students are selected at random to form a committee. Let Ω be the set of committees and for $\{\omega\} \in A_\Omega$, set

$$h(\omega) = \binom{8}{j}\binom{7}{3-j}\binom{15}{3}^{-1}$$

where j is the number of boys on the committee, and for $A \in A_\Omega$, set $h(A) = \sum_{\omega \in A} h(\omega)$. Is h a probability?

9. (Corollary to Theorem 5-2b_3) Given a finite sample space, with $n(\Omega) = N$; that is $\Omega = \{\omega_1,\omega_2,\ldots,\omega_N\}$. Define a set function P on A_Ω such that if $A \in A_\Omega$, $P(A) = \frac{n(A)}{N}$. Prove P is a probability.

5-3. PROBABILITY SPACES

a.   Choosing the Probability Function

Our discussion of probability has centered around three fundamental concepts: those of a sample space Ω, of A_Ω, the set of events in Ω (the set of subsets of Ω), and of a probability P, a set function on A_Ω.

5-3a_1 Definition.   *The ordered triplet $\{\Omega, A_\Omega, P\}$ is called a* **probability space.**

Every problem concerning probabilities is embedded in a probability space and until it is described no problem can be solved.

For the majority of beginning students of probability the choice P is the most puzzling aspect of the procedure.   If we consider the simple coin-tossing experiment giving rise to $\Omega = \{H, T\}$ and if we define P on A_Ω such that $P(H) = P(T) = \frac{1}{2}$, there will be no objection; but if we define P by $P(H) = \frac{1}{3}$, $P(T) = \frac{2}{3}$, someone is bound to cry "foul"; yet this choice of a probability satisfies our axioms.   For years—for centuries, even—people have been tossing coins and have observed that heads appear about half of the time.   In any set of N tosses, the number of heads is usually not exactly $\frac{N}{2}$ (or $\frac{N}{2} \pm \frac{1}{2}$, if N is odd), but if the experiment is continued, eventually the relative frequency of heads and tails will tend to be the same.   And so man's experience and his reason—why should not a coin tossed so it will turn over several times before it hits the ground come up heads as often as it comes up tails—lead him to set $P(H) = P(T) = \frac{1}{2}$.   Were we to toss a coin 10,000 times and obtain only 3,500 heads, we would suspect that the coin was biased in some way and we would begin to search for an explanation of this departure from the "norm." Could we find none, we would fully expect that if we gave the coin another 10,000 or 20,000 tosses, we would even the score.

The same is true of a die; we fully expect that if a die is rolled a large number of times, each of the faces will appear approximately one-sixth of the total number of rolls.   If in 100 rolls of a die, the ace came up 50 times, we would say the die was "loaded," and we would examine it carefully to see if it were sufficiently misshapened or weighted on one face to cause this unusual behavior.

We know there are

$$\binom{52}{13}\binom{39}{13}\binom{26}{13}\binom{13}{13} = \frac{52!}{(13!)^4} = 6 \times 10^{28}$$

different ways the cards in a bridge deck can be distributed (dealt) among the four players in a game.* One man dealing one round every second night and day for a year could only deal 3×10^7 games, hence experiments testing the frequency of occurrence of the games are not likely to be undertaken. But, we ask ourselves, if the cards are properly shuffled why would one combination of the cards occur more often than another, and consequently, on the space of outcomes of dealing bridge games we assign to each point in the sample space the same probability. (That is, if

$$N = \binom{52}{13}\binom{39}{13}\binom{26}{13}\binom{13}{13}$$

and $\Omega = \{\omega_1, \omega_2, \omega_3, \ldots, \omega_N\}$, we define the probability P on A_Ω by $P(\omega_i) = \dfrac{1}{N}$ for all ω_i) This is what your favorite bridge columnist is doing when he says such-and-such a combination occurs only once in so many games. Here, our choice of P is based on the decision that no other "reasonable" alternative is present; our probability is *a priori*.

Another way of assigning a probability to a sample space is illustrated by the manufacturer of plastic spoons who buys a new machine, checks the first thousand spoons produced by that machine and discovers ten of these do not meet his standard. He then figures one percent of the machine's output will be defective and he adjusts his production schedule accordingly. This is tantamount to classifying his output as a-acceptable and d-defective and embedding $\Omega = \{a,d\}$ in $\{\Omega, A_\Omega, P\}$ where $P(a) = .99$, $P(d) = .01$. If after several months the manufacturer again checks the output of the machine and finds that now 10 percent of his output is defective, he will look into the possible causes—the composition of the raw materials may have been altered or a screw may have worked loose—and finding one he will correct it and check again; or not finding one, he will continue testing his output and revise his estimate if this seems warranted. In this case the choice of the probability on the sample space is based on the experimental evidence before us; our probability is *a posteriori*.

Another very common problem in probability is that of distributing r balls into n cells. We shall speak of balls and cells, but this is a convenience and could represent elevator riders (balls) being discharged at various floors (cells), automobiles (balls) being shunted into repair berths (cells), or any number of situations. Rather than explore the problem for arbitrary r and n, let us look specifically at the case of distributing three balls into three cells, since this will illustrate fully the decisions that must be made. There are $3 \cdot 3 \cdot 3 = 27$ points in our sample space Ω: they are listed in Figure 5-3a_2.

*Based on calculations made from Herbert E., Salzer, *Tables of N! and T(n + 1/2) for the First Thousand Values of n*, NBS Applied Science Series, No. 16, 1951, Washington, D.C.

5-3a₂   **Figure**

Sample space associated with putting three distinguishable balls into three cells.

1. $x \mid y \mid z$	10. $yz \mid x \mid -$	19. $z \mid xy \mid -$
2. $x \mid z \mid y$	11. $yz \mid - \mid x$	20. $z \mid - \mid xy$
3. $y \mid x \mid z$	12. $- \mid yz \mid x$	21. $- \mid z \mid xy$
4. $y \mid z \mid x$	13. $y \mid xz \mid -$	22. $xy \mid z \mid -$
5. $z \mid x \mid y$	14. $y \mid - \mid xz$	23. $xy \mid - \mid z$
6. $z \mid y \mid x$	15. $- \mid y \mid xz$	24. $- \mid xy \mid z$
7. $x \mid yz \mid -$	16. $xz \mid y \mid -$	25. $xyz \mid - \mid -$
8. $x \mid - \mid yz$	17. $xz \mid - \mid y$	26. $- \mid xyz \mid -$
9. $- \mid x \mid yz$	18. $- \mid xz \mid y$	27. $- \mid - \mid xyz$

If for $A \in A_\Omega$ we define P by $P(A) = \dfrac{n(A)}{27}$, then $\{\Omega, A_\Omega, P\}$ is a probability space (see Prob. 9, Exercises 5-2b), and we can answer such questions as: (a) What is the probability all balls fall into the same cell? (b) What is the probability that at least two balls fall into the same cell? and (c) What is the probability that the x-ball falls into the first cell?

Suppose the balls are indistinguishable, or can be considered as such; for example, we may be interested only in how many elevator riders get off at the second floor rather than which ones. In this case Ω is the collection in Figure 5-3a₃.

5-3a₃   **Figure**

Sample space associated with putting three indistinguishable balls into three cells.

1. $b \mid b \mid b$	4. $b \mid - \mid bb$	7. $- \mid bb \mid b$
2. $bb \mid b \mid -$	5. $bb \mid - \mid b$	8. $bbb \mid - \mid -$
3. $b \mid bb \mid -$	6. $- \mid b \mid bb$	9. $- \mid bbb \mid -$
	10. $- \mid - \mid bbb$	

We can define P_1 on A_Ω by setting

$$P_1(1) = \frac{2}{9}, \ P_1(2) = P_1(3) = P_1(4) = P_1(5) = P_1(6) = P_1(7) = \frac{1}{9}$$

$$P_1(8) = P_1(9) = P_1(10) = \frac{1}{27}$$

and for $A \in A_\Omega$,

$$P_1(A) = \sum_{\omega_i \in A} P_1(\omega_i),$$

then $\{\Omega, A_{\Omega}, P_1\}$ is a probability space. However, it would be equally valid to set

$$P_2(1) = P_2(2) = \cdots = P_2(10) = \frac{1}{10}$$

and for $A \in A_{\Omega}$,

$$P_2(A) = \sum_{\omega_i \in A} P_2(\omega_i)$$

then $\{\Omega, A_{\Omega}, P_2\}$ is also a probability space. If A and B are the questions posed in the preceding paragraph, $P_1(A) = \frac{1}{9}$, $P_1(B) = \frac{7}{9}$ and $P_2(A) = \frac{3}{10}$, $P_2(B) = \frac{9}{10}$, where A and B are used not only to specify the questions but also to represent the events in A_{Ω} whose points satisfy the questions. Which sample space does one choose? Which one is "right"? These questions do not have a unique answer. In the classical theory of gases physicists believed that the molecules (balls) distribute themselves into volumes of space (cells) as described by $\{\Omega, A_{\Omega}, P_1\}$. However, experimentation has shown that certain types of particles behave in accordance with $\{\Omega, A_{\Omega}, P_2\}$ and that still others act like neither of these.

Consider rolling two dice for the sum of their faces. What is the probability that the sum is no more than four? Here $\Omega = \{2,3,4,\ldots,12\}$. For $A \in A_{\Omega}$, set $P_1(A) = \frac{n(A)}{11}$, then $\{\Omega, A_{\Omega}, P_1\}$ is a probability space. If F is the event, the sum is no more than four, $P_1(F) = \frac{3}{11}$. We may feel uneasy about this choice of P_1 since experience tells us that some sums are more likely to appear than others. However, P_1 satisfies the requirements for a set function to be a probability. Hence we can set up the probability space and in that space answer the question.

We can attack this problem differently. We still have $\Omega = \{2,3,4,\ldots,12\}$. However since we are rolling dice, let $D = \{1,2,3,4,5,6\}$ and $\Delta = (X\ D)_2 = \{(a,b) | a \in D,\ b \in D\}$. Now if the dice are fair we expect the 36 ordered pairs in Δ to occur with equal frequency. We then define a function P_2 on A_{Ω} in the following way: since $\{2\} \in A_{\Omega}$ results from $(1, 1) \in \Delta$, set $P_2(2) = \frac{1}{36}$; since $\{3\} \in A_{\Omega}$ results from $(1,2) \in \Delta$ and $(2,1) \in \Delta$, set $P_2(3) = \frac{2}{36}$; since $\{4\} \in A_{\Omega}$ results from $(1,3)$, $(2,2)$ and $(3,1)$ all in Δ, set $P_2(4) = \frac{3}{36}$; and so on. In this way we set

$$P_2(2) = P_2(12) = \frac{1}{36}$$

$$P_2(3) = P_2(11) = \frac{2}{36}$$

$$P_2(4) = P_2(10) = \frac{3}{36}$$

$$P_2(5) = P_2(9) = \frac{4}{36}$$

$$P_2(6) = P_2(8) = \frac{5}{36}$$

$$P_2(7) = \frac{6}{36}$$

and for $A \in A_\Omega$, $P_2(A) = \sum_{\omega \in A} P_2(\omega)$. Since $\sum_{\omega \in \Omega} P_2(\omega) = 1$, P_2 is a probability

and $\{\Omega, A_\Omega, P_2\}$ a probability space. In this space the probability of the event F

that the sum is no more than four is $P_2(F) = \sum_{\omega \in A} P_2(2) + P_2(3) + P_2(4) = \frac{6}{36}$,

and for most of us, this is a more acceptable value for the probability of that event than the first one.

In this last example we have a sample space Ω, its set of events A_Ω and two probabilities P_1 and P_2. The answer to a question concerning the probability of an event $F \in A_\Omega$ depends upon whether we work in $\{\Omega, A_\Omega, P_1\}$ or $\{\Omega, A_\Omega, P_2\}$. While we must avoid classifying one space as "right" and the other as "wrong," for they both satisfy the definition of a probability space, we shall indeed favor that choice of a probability which will most closely reflect the physical world as we know it. Since we study probability with the intent of applying it to scientific, economic and other problems, the "right" answer to any exercise is in that probability space which is in accord with physical reality.

To recapitulate, we have two major ways of assigning a probability function to a sample space. If we have, or can have, no experimental evidence to suggest the frequency with which the events in our space will occur, we look for factors that will influence their occurrence and "reason out" an appropriate probability. In most of our illustrative examples we shall use such an *a priori* probability. Thus in games of chance where we know of no external factors that will alter the play, we assume that no one of the elementary events will occur more than arother and we assign an equiprobability function (one in which every elementary event has the same probability) to the sample space. The alternative is to perform an experiment a great number of times and then to observe the frequencies of the possible outcomes and use these results to design an appropriate probability function. Manufacturer's claims for machine dependability, certain scientific hypothesis, future institutional needs are usually based upon *a posteriori* probabilities. Often the results of experimentation are tempered by the experimenter's knowledge of unusual conditions present during some phase of the work and he will adjust his *a posteriori* probability to reflect this. Also it is not unusual to "check out" an *a priori* decision by performing the experiment and observing the frequencies with which the several outcomes appear. Furthermore, although throughout our work we shall be concerned with determining the probabilities to be associated with experiments, this will not be our only concern. We are also interested in the theory of probability as a branch of mathematics and the application of this theory to a variety of problems.

b.   Some Basic Properties of a Probability

Several elementary properties of probabilities follow directly from Definition 5-2b_1 and Theorem 5-2b_3. Since they are frequently useful in the solution of problems, we introduce them here.

If P is a probability and $A \in A_\Omega$, we pointed out (in 5-2b_2) that

$$P(A) = P\left(\sum_{\omega_i \in A} \{\omega_i\} \right) = \sum_{\omega_i \in A} P(\omega_i).$$

Since $\Omega \in A_\Omega$ and $\Omega = \Sigma_i\{\omega_i\}$ we have as a special case

$$P(\Omega) = P(\sum_i \{\omega_i\}) = \sum_i P(\omega_i)$$

and along with condition (i) of the definition, this leads to

5-3b_1 $$0 \leqslant P(\omega_i) \leqslant 1$$

for all elementary events $\{\omega_i\} \in A_\Omega$.

Since $A + A' = \Omega$, $AA' = \varnothing$, from (ii) and (iii) of the definition,

$$P(\Omega) = P(A + A') = P(A) + P(A') = 1$$

and hence

5-3b_2 $$P(A') = 1 - P(A)$$

In particular, since $\Omega' = \varnothing$,

$$P(\varnothing) = 1 - P(\Omega)$$

and

5-3b_3 $$P(\varnothing) = 0.$$

In the definition we set $P(\Omega) = 1$ and in 5-3b_3 we have $P(\varnothing) = 0$; thus for any sample space the probability of the sure event is one and that of the impossible event is zero.

If $B \subset A$, $A = B + (A - B)$; from the definition

$$P(A) = P(B) + P(A - B)$$

and consequently

5-3b_4 whenever $B \subset A$, we have

$$P(B) \leqslant P(A)$$

and

$$P(A - B) = P(A) - P(B).$$

If A_1 and A_2 are any two events in A_Ω, $A_1 \cup A_2 = A_1 + (A_2 - A_1 A_2)$. From the definition, and 5-3b_4.

5-3b_5 $$P(A_1 \cup A_2) = P(A_1) + P(A_2) - P(A_1 A_2)$$

and

$$P(A_1 \cup A_2) \leqslant P(A_1) + P(A_2).$$

Equality holds in this last statement if, and only, if, $A_1 A_2 = \varnothing$. By use of the Principle of Mathematical Induction, we can prove that if $n \in I^+$.

5-3b_6

$$P\left(\bigcup_{i=1}^{n} A_i\right) \leqslant \sum_{i=1}^{n} P(A_i).$$

EXERCISES 5-3b

(These exercises extend the formulas developed above.)

1. Let P be a probability defined on the subsets of a sample space Ω. If A and B are any two events in A_Ω, find in terms of $P(A)$, $P(B)$ and $P(AB)$ formulas for: (a) $P(AB')$, (b) $P(A'B')$, (c) $P((AB)')$, (d) $P(A \cup B')$

2. Let P be a probability defined on the events in a sample space and let A, B, and C be events in A_Ω.

(a) Prove

$$P(A \cup B \cup C) = P(A) + P(B) + P(C) - P(AB) - P(AC) - P(BC) + P(ABC).$$

(b) If A, B, and C are mutually exclusive in pairs, prove

$$P(A \cup B \cup C) = P(A) + P(B) + P(C).$$

(c) A coin is tossed twice. If A is the event head on first toss, B the event head on both tosses, and C the event tail on first toss, $ABC = \varnothing$. Is the formula in part (b) valid in this case?

3. Let P be a probability defined on the events of a sample space Ω, and let A_k, $k = 1, 2, \ldots, n$ be events in A_Ω such that $A_i A_j = \emptyset$, whenever $i \neq j$.

(a) Prove

$$P(A_1 \cup A_2 \cup \ldots \cup A_n) = P(A_1) + P(A_2) + \cdots + P(A_n).$$

(b) If we do not require $A_i A_j = \emptyset$ for $i \neq j$, prove

$$P(A_1 \cup A_2 \cup \ldots \cup A_n) \leqslant P(A_1) + P(A_2) + \cdots + P(A_n).$$

c.   Additional Examples

The experiment of tossing a coin until a tail appears gives rise to the sample space

$$\Omega - \{T, HT, HHT, HHHT, HHHHT, \ldots\}.$$

Let ω_j be the element of Ω corresponding to a tail on the jth toss; we have

$$\omega_1 = T, \quad \omega_2 = HT, \quad \omega_3 = HHT,$$

and so on.   If the coin is true and we toss it once we expect to have equal chances of getting a head or a tail so we set

$$P(\omega_1) = P(T) = \frac{1}{2}.$$

Now, if we toss the coin twice, there are four possible and equally likely outcomes, HH, HT, TH, and TT; so we set $P(\omega_2) = P(HT) = \frac{1}{4} = \frac{1}{2^2}$.   Similarly if we toss a coin j times, we have 2^j possible and equally likely outcomes, one of which will be a run of $j - 1$ heads followed by a tail; so for $\{\omega_j\} \in A_\Omega$ we set $P(\omega_j) = \frac{1}{2^j}$.   By 4-2$b_2$,

$$\sum_{\omega_j \in \Omega} P(\omega_j) = \sum_{j=1} \frac{1}{2^j} = 1,$$

and if for $A \in A_\Omega$, we set

$$P(A) = \sum_{\omega_j \in A} P(\omega_i)$$

P is a probability.   Hence the triplet $\{\Omega, A_\Omega, P\}$ is a probability space and we can now calculate specific probabilities associated with this experiment.   Let A be the event it takes five or more tosses to produce a tail, then $A = \{\omega_5, \omega_6, \omega_7, \ldots\}$ and

$$P(A) = \sum_{j=5} \frac{1}{2^j} = \frac{1}{2^5} \left(\frac{1}{1 - \frac{1}{2}} \right) = \frac{1}{16}. \tag{4-2b_2}$$

Another way to calculate $P(A)$ is to note that

$$A = \Omega - \{\omega_1,\omega_2,\omega_3,\omega_4\} = \Omega - A'$$

and

$$P(A) = 1 - P(A')$$

$$= 1 - \sum_{j=1}^{4} \frac{1}{2^j} = 1 - \frac{15}{16} = \frac{1}{16}$$

The probability that the tail appears on even numbered toss is $P(C)$ where $C = \{\omega_2,\omega_4,\omega_6,\ldots\}$, and

$$P(C) = \sum_{k=1}^{\infty} \frac{1}{2^{2k}} = \frac{1}{4} \cdot \frac{1}{1-\frac{1}{4}} = \frac{1}{3}.$$

In the section on sample spaces, we described an experiment in which three people X, Y, and Z played repeatedly a two-person game which cannot end in a tie according to the following rules:

(i) X and Y play the first game; the second game is played by Z and the winner of the first game;

(ii) in subsequent games two players are competing and the third is waiting to play the victor of that game in the next round;

(iii) the play ends when one contestant wins two consecutive games.

The sample space for this experiment is given in Figure 5-1a_1, page 62.

Let us assume the players are equally matched and on this basis we construct P on A_Ω. If two games are played, there are four equally likely outcomes, XX, XZ, YY, YZ, and so we set $P(XX) = P(YY) = \frac{1}{4} = \frac{1}{2^2}$. The chance that the tournament goes on to three games is $\frac{1}{2}$ (1 minus the chance it is over in two games). If three games are played there are again four equally likely outcomes, XZZ, XZY, YZZ, YZX; two of these are events in A_Ω and we set

$$P(XZZ) = P(YZZ) = \frac{1}{2} \cdot \frac{1}{4} = \frac{1}{2^3}.$$

The tournament has a $\frac{1}{4}$ chance of continuing on to four games; if it does there are again four equally likely ways the four rounds can end. Two of these are in A_Ω and we set $P(XZYY) = P(YZXX) = \frac{1}{4} \cdot \frac{1}{4} = \frac{1}{2^4}$. Similarly the tournament will have a $\frac{1}{2^{k-2}}$ chance of continuing to k games; there will be four equally likely outcomes of the kth games; two of these will give rise to points in A_Ω and to these points we assign the probability $\frac{1}{2^{k-2}} \cdot \frac{1}{4} = \frac{1}{2^k}$.

$$\sum_{\omega \in \Omega} P(\omega) = 2 \sum_{j=2} \left(\frac{1}{2}\right)^{j} = 2 \sum_{i=0} \left(\frac{1}{2}\right)^{2+i} = 2 \cdot \frac{1}{4} \frac{1}{1-\frac{1}{2}} = 1$$

(see 4-2b_2). For $A \in A_\Omega$, set $P(A) = \sum_{\omega \in A} P(\omega)$; then $\{\Omega, A_\Omega, P\}$ is a probability

space. The player X can win in 2,4,5,7,8,10,11,... games; then if A_x is the event X wins the tournament,

$$P(A_x) = \left(\frac{1}{2}\right)^2 + \left(\frac{1}{2}\right)^4 + \left(\frac{1}{2}\right)^5 + \left(\frac{1}{2}\right)^7 + \left(\frac{1}{2}\right)^{10} + \cdots$$

$$= \left(\left(\frac{1}{2}\right)^2 + \left(\frac{1}{2}\right)^5 + \left(\frac{1}{2}\right)^8 + \cdots\right) + \left(\left(\frac{1}{2}\right)^4 + \left(\frac{1}{2}\right)^7 + \left(\frac{1}{2}\right)^{10} + \cdots\right) \quad {}^{[3]}$$

$$= \frac{1}{4} \frac{1}{1-\frac{1}{8}} + \frac{1}{16} \frac{1}{1-\frac{1}{8}}$$

$$= \frac{2}{7} + \frac{1}{14} = \frac{5}{14}$$

Similarly we can compute $P(A_y)$ and $P(A_z)$ where A_y and A_z are the events Y wins and Z wins respectively; we obtain $P(A_y) = \frac{5}{14}$, $P(A_z) = \frac{4}{14}$.

Since $A_x + A_y + A = \mathsf{C}\,\Omega$,

$$P((A_x + A_y + A_z)') = P(\Omega - (A_x + A_y + A_z)) = P(\Omega) - P(A_x + A_y + A_z)$$

$$= P(\Omega) - \left(P(A_x) + P(A_y) + P(A_z)\right) = 1 - 1 = 0.$$

From this we conclude that the event neither X nor Y nor Z wins is the impossible event and so the tournament will end after a finite number of games.

Consider the space Ω of all bridge hands. There are $\binom{52}{13}$ elements in Ω, and if the cards are properly shuffled and dealt they should appear with equal likelihood. Hence we define P on A_Ω so that for $A \in A_\Omega$,

$$P(A) = n(A) \binom{52}{13}^{-1}$$

P is a probability (see Prob. 9, Exercise 5-2b) and $\{\Omega, A_\Omega, P\}$ a probability space. Let $A \in A_\Omega$ be the event that a hand contains two aces; then A has $\binom{4}{2}\binom{48}{11}$ elements since the aces can be chosen in $\binom{4}{2}$ ways and the remaining eleven cards in $\binom{48}{11}$ ways. $P(A) = \binom{4}{2}\binom{48}{11}\binom{52}{13}^{-1}$. Let B be the

event the hand contains all the aces; then $P(B) = \binom{48}{9}\binom{52}{13}^{-1}$ since B has

$\binom{4}{4}\binom{48}{9}$ elements. The event C, that a hand contains exactly one run of

10, J, Q, K, A in one suit, has $\binom{4}{1}\left\{\binom{47}{8} - \binom{3}{1}\binom{42}{3}\right\}$ elements. To

arrive at this figure we observe that if we specified the 10, J, Q, K, A of hearts,

there would be $\binom{47}{8}$ ways of selecting the other eight cards to make up the hand;

but among these ways would be included those selections that contained another

10, J, Q, K, A run. There are $\binom{3}{1}\binom{42}{3}$ sets of eight cards that contain

another sequence 10, J, Q, K, A of one suit and so there are $\binom{47}{8} - \binom{3}{1}\binom{42}{3}$

hands that contain only the heart sequence. But the initial event C, was to con-

tain exactly one run, suit unspecified; therefore, any suit will satisfy and C has

$\binom{4}{1}\left\{\binom{47}{8} - \binom{3}{1}\binom{42}{3}\right\}$ elements.

$$P(C) = \binom{4}{1}\left\{\binom{47}{8} - \binom{3}{1}\binom{42}{3}\right\}\binom{52}{13}^{-1}.$$

EXERCISES 5-3c

In the problems an experiment is described. Set up a probability space associated with the experiment and with respect to the probability space, answer the questions.

1. A ball is drawn from an urn containing 3 red, 2 white, and 4 blue balls. What is the probability (a) the ball is white? (b) the ball is red or blue? (c) the ball is not blue?

2. A card is drawn from a standard deck. What is the probability (a) the card is a heart? (b) the card is an ace? (c) the card is black? (d) the card is a red-face card?

3. Two dice are rolled and the sum of their faces recorded. (a) What is the probability the sum is three? (b) What is the probability the sum is at least two? (c) What is the most probable sum? (d) What is the probability of getting no more than a ten?

4. Ten disks each bearing one of the digits 0,1,2,...,9 are put into an urn.

One is drawn, its number noted and it is returned to the urn. This is done five times so that a five digit number is constructed. A is the event, no digit is repeated; B the event no four appears; C the event the first digit is a three and D the event the number is greater than 6,000. What are $P(A)$, $P(B)$, $P(C)$, $P(D)$, $P(AB)$, $P(A \cup B)$, $P(CD)$?

5. From a class of 13 boys and 10 girls, a committee of five is selected. Find $P(A)$, $P(B)$ and $P(C)$ where A is the event the committee consists of 3 girls and 2 boys; B the event there is at least one boy; and C the event an even number of girls.

6. A bridge hand is dealt. What is the probability the hand contains (a) exactly three kings? (b) at least three kings? (c) at most one ace? (d) two aces, two kings, two queens? (e) the A, K, Q, in each of two suits and no other A, K, or Q? (f) two red cards?

7. There are n people at a party and each is asked the month of his birthday. (a) If $n = 12$, what is the probability each birthday falls in a different month? (b) If $n = 8$, what is the probability the birthdays fall in exactly three months? (c) If $n = 15$, find $P(C)$, where C is the event 4 birthdays fall in each two months, two fall in each of three months, and one falls in one month.

8. A die is rolled until a six appears. (a) What is the probability it will take at least six tries? (b) Is the experiment more likely to end on an even or odd numbered trial? (c) What is the probability it will take no more than three tries?

9. In a certain school with 300 students two thirds of the students are boys. A committee of 20 students is selected at random. What is the probability that (a) all are boys? (b) half are girls? (c) there are at most 2 girls? (d) there is at least 1 girl?

10. A coin is tossed until the second head appears. (a) What is the probability the first toss is a head? (b) What is the probability the second head appears on the third toss? (c) What is the probability the second head appears on the kth toss?

5-4. REPEATED TRIALS

a. Success and Failure Experiments

A die is rolled four times. What is the probability exactly two threes appear? Let $D = \{T,N\}$ (T-three, N – nonthree) and define a probability p on the events of D by setting $p(T) = \dfrac{1}{6}$, $p(N) = \dfrac{5}{6}$. In terms of D the sample space associated with rolling the die four times is $\Omega = (\mathbf{X}\, D)_4 = \{(d_1,d_2,d_3,d_4)\,|\,d_i \in D\}$.

To help us construct a probability p on A_Ω we construct the tree diagram of Figure 5-4a_1 in which the probability of a branch yielding a T is $\frac{1}{6}$ and that of

5-4a_1 Figure

Tree diagram associated with rolling a die four times and counting the number of threes.

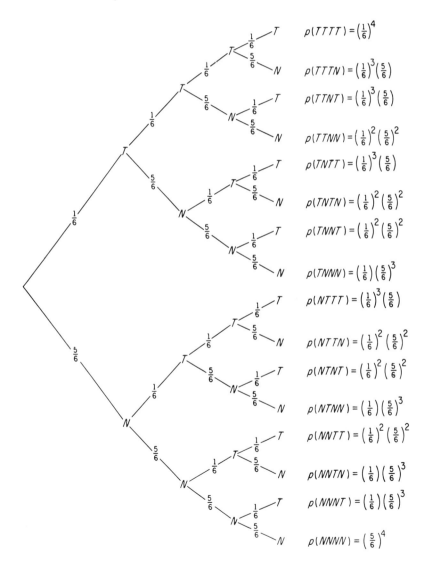

$$p(TTTT) = \left(\tfrac{1}{6}\right)^4$$

$$p(TTTN) = \left(\tfrac{1}{6}\right)^3\left(\tfrac{5}{6}\right)$$

$$p(TTNT) = \left(\tfrac{1}{6}\right)^3\left(\tfrac{5}{6}\right)$$

$$p(TTNN) = \left(\tfrac{1}{6}\right)^2\left(\tfrac{5}{6}\right)^2$$

$$p(TNTT) = \left(\tfrac{1}{6}\right)^3\left(\tfrac{5}{6}\right)$$

$$p(TNTN) = \left(\tfrac{1}{6}\right)^2\left(\tfrac{5}{6}\right)^2$$

$$p(TNNT) = \left(\tfrac{1}{6}\right)^2\left(\tfrac{5}{6}\right)^2$$

$$p(TNNN) = \left(\tfrac{1}{6}\right)\left(\tfrac{5}{6}\right)^3$$

$$p(NTTT) = \left(\tfrac{1}{6}\right)^3\left(\tfrac{5}{6}\right)$$

$$p(NTTN) = \left(\tfrac{1}{6}\right)^2\left(\tfrac{5}{6}\right)^2$$

$$p(NTNT) = \left(\tfrac{1}{6}\right)^2\left(\tfrac{5}{6}\right)^2$$

$$p(NTNN) = \left(\tfrac{1}{6}\right)\left(\tfrac{5}{6}\right)^3$$

$$p(NNTT) = \left(\tfrac{1}{6}\right)^2\left(\tfrac{5}{6}\right)^2$$

$$p(NNTN) = \left(\tfrac{1}{6}\right)\left(\tfrac{5}{6}\right)^3$$

$$p(NNNT) = \left(\tfrac{1}{6}\right)\left(\tfrac{5}{6}\right)^3$$

$$p(NNNN) = \left(\tfrac{5}{6}\right)^4$$

yielding an N is $\frac{5}{6}$. We argue that if there is a probability $\frac{1}{6}$ of getting a three in one roll and $\frac{5}{6}$ of not getting a three in a second roll, in two rolls there is $\left(\frac{1}{6}\right)\left(\frac{5}{6}\right)$ probability of getting one T and one N. In this way we define p on A_Ω as indicated in the figure, and if $\{\omega\} \in A_\Omega$, $p(\omega) = \left(\frac{1}{6}\right)^k \left(\frac{5}{6}\right)^{4-k}$ where k is the number of threes in ω. For $A \in A_\Omega$, set $p(A) = \sum_{\omega \in A} p(\omega)$. Is p a probability, and, if so, can we answer the question? Since $p(\omega) \geqslant 0$ for every $\omega \in \Omega$, p will be a probability if $\sum_{\omega \in \Omega} p(\omega) = 1$ (see 5-2b_3). In Ω there are $\binom{4}{k}$ elementary events that have k threes and $4-k$ nonthrees, hence

$$\sum_{\omega \in \Omega} p(\omega) = \sum_{k=0}^{4} \binom{4}{k} \left(\frac{1}{6}\right)^k \left(\frac{5}{6}\right)^{4-k} = \left(\frac{1}{6} + \frac{5}{6}\right)^4 = 1 \text{ (see 4-2}a_2)$$

and p is a probability. If B is the event two threes appear then

$$p(B) = \binom{4}{2}\left(\frac{1}{6}\right)^2\left(\frac{5}{6}\right)^2.$$

We now generalize. Consider an experiment that gives rise to a sample space of two elements, say $E = \{S,F\}$, which could represent heads or tails on a coin, six or not-six on a die, win or lose in a game, success or failure in any venture. On the events of E set $P(S) = p$ $(0 \leqslant p \leqslant 1)$, $P(F) = q = 1-p$. Repeat this experiment n times and so generate the sample space

$$\Omega = (X\, E)_n = \{(\omega_1,\omega_2,\ldots,\omega_n)|\omega \in E\}.$$

(The ith repetition of the original experiment is commonly referred to as the ith trial.) Ω has 2^n elements and of these 2^n elements $\binom{n}{k}$ of them have exactly k S's and $n - k$ F's (counting Principle IV). The following theorem gives one way of constructing a probability on Ω.

5-4a_2 **Theorem.** *Let $E = \{S,F\}$ and $\Omega = (X\,E)_n$. Define a probability p on the events of E by setting $\rho(S) = p$ $(0 \leqslant p \leqslant 1)$, $\rho(F) = q = 1-p$. On A_Ω define a function P by setting for $\{\omega\} \in A_\Omega$, $P(\omega) = p^k q^{n-k}$ where k is the number of S's in ω and for $A \in A_\Omega$ $P(A) = \sum_{\omega \in A} P(\omega)$. Then P is a probability and $\{\Omega, A_\Omega, P\}$ a probability space.*

Proof: We need only to show $\sum_{\omega \in \Omega} P(\omega) = 1$. With this established the theorem follows directly from Theorem 5-2b_3. Since in Ω there are exactly $\binom{n}{k}$ elementary events that have k S-components and $n - k$ F-components, and since k can assume all integral values from 0 to n inclusive, from 4-2a_2 and the hypotheses, we have

$$\sum_{\omega \in \Omega} P(\omega) = \sum_{k=0}^{n} \binom{n}{k} p^k q^{n-k} = (p + q)^n = 1.$$

Our theorem is proved. ∎

If $\{\Omega, A_\Omega, P\}$ is formed as in the theorem then the probability of A, the event that $\omega \in \Omega$ has exactly k S's, is

$$P(A) = \sum_{\omega \in A} P(\omega) = \binom{n}{k} p^k q^{n-k}$$

and B, the event there are at least j S's $(j \leqslant n)$, is

$$P(B) = \sum_{k=j}^{n} \binom{n}{k} p^k q^{n-k} = 1 - \sum_{k=0}^{j-1} \binom{n}{k} p^k q^{n-k} = 1 = P(B')$$

B' can also be interpreted as the event there are fewer than j S's.

Suppose a die is rolled five times (or five distinguishable dice are rolled simultaneously)* and we consider the appearance of a 6 as success and anything else as failure. This produces an Ω, the cross-product of $D = \{S, F\}$ five times with itself. On the events of D let us define a probability by $p(S) = \dfrac{1}{6}, p(F) = \dfrac{5}{6}$ and define P on A_Ω in the usual way, that is according to Theorem 5-4a_2. Then if $\omega_1 = (S, S, F, F, S)$, $P(\omega_1) = \left(\dfrac{1}{6}\right)^3 \left(\dfrac{5}{6}\right)^2$ and if A is the event there are exactly two sixes, $P(A) = \binom{5}{2} \left(\dfrac{1}{6}\right)^2 \left(\dfrac{5}{6}\right)^3$. The probability that there are at least two sixes (the probability of the event B that there are at least two sixes) is

$\sum_{k=2}^{5} \binom{5}{k} \left(\dfrac{1}{6}\right)^k \left(\dfrac{5}{6}\right)^{5-k}$ But this is more easily computed if we consider

---

*If a die is rolled five times, we can record the results in the order they occur and so create a point $(0_1, 0_2, 0_3, 0_4, 0_5)$. If the dice are distinguishable–perhaps by size or color–we can designate the individual dice as first, second, etc. and record the outcomes in that order, again creating an ordered 5-tuple.

instead the compliment of B, the event B' that there are fewer than two sixes, for then $B = \Omega - B'$ and

$$P(B) = 1 - P(B') = 1 - \left[\binom{5}{0}\left(\frac{1}{6}\right)^0 \left(\frac{5}{6}\right)^5 + \binom{5}{1}\left(\frac{1}{6}\right)\left(\frac{5}{6}\right)^5\right] = \frac{763}{3888}.$$

Let $C \in A_\Omega$ be the event the third component of ω is an S; C can be expressed as $C_0 + C_1 + C_2 + C_3 + C_4$, where C_j, $j = 0,1,2,3,4$ is the event that the third component of the 5-tuple is an S and j of the remaining four components are also S's. There are $\binom{4}{j}$ elements in C_j and each has probability $\left(\frac{1}{6}\right)^{j+1}\left(\frac{5}{6}\right)^{4-j}$

Hence

$$P(C_j) = \binom{4}{j}\left(\frac{1}{6}\right)^{j+1}\left(\frac{5}{6}\right)^{4-j}$$

$$P(C) = \sum_{j=0}^{4} P(C_j) = \sum_{j=0}^{4} \binom{4}{j}\left(\frac{1}{6}\right)^{j+1}\left(\frac{5}{6}\right)^{4-j}$$

$$= \frac{1}{6}\sum_{j=0}^{4} \binom{4}{j}\left(\frac{1}{6}\right)^{j}\left(\frac{5}{6}\right)^{4-j}$$

$$= \frac{1}{6}\left(\frac{1}{6} + \frac{5}{4}\right)^4 = \frac{1}{6}$$

(see 4-2a_2). We note here that $P(C)$ is equal to the probability of an S on the third roll; it is unaffected by what preceded or followed.

b. Multiple Outcome Experiments*

Suppose an urn contains two red, three white, and three blue balls. One ball is drawn from the urn, its color noted, and the ball returned to the urn; then a second ball is drawn and its color noted. What is the probability that both balls are red? To answer this question let us proceed as we did when there were but two alternatives. Let $B = \{r,w,b\}$ and on the events of B define a probability by setting $p(r) = \frac{2}{8}$, $p(w) = p(b) = \frac{3}{8}$. $\Omega = B \times B = \{(\alpha,\beta)|\alpha \in B, \beta \in B\}$ is the sample space for the experiment of drawing two balls consecutively. Using the same reasoning as in the earlier case, define a probability on A_Ω by setting for $\{\omega\} = \{(\alpha,\beta)\} \in A_\Omega$, $P(\omega) = p(\alpha)p(\beta)$ and for $A \in A_\Omega$, $P(A) = \sum_{\omega \in A} P(\omega)$.

*This section uses multinomials and can be omitted without loss of continuity.

Since

$$\sum_{\omega \in \Omega} P(\omega) = P((r,r)) + P((w,w)) + P((b,b)) + P((r,w)) + P((r,b))$$
$$+ P((w,r)) + P((w,b)) + P((b,r)) + P((b,w))$$
$$= p(r)p(r) + p(w)p(w) + p(b)p(b)$$
$$+ 2\,[p(r)p(w) + p(r)p(b) + p(w)p(b)]$$
$$= \left(\frac{2}{8}\right)^2 + \left(\frac{3}{8}\right)^2 + \left(\frac{3}{8}\right)^2 + 2\,\left[\frac{6}{8^2} + \frac{6}{8^2} + \frac{9}{8^2}\right]$$
$$= \left(\frac{2}{8} + \frac{3}{8} + \frac{3}{8}\right)^2 = 1,$$

P is a probability and $\{\Omega, A_\Omega, P\}$ a probability space. If R is the event both balls are red, $P(R) = \sum_{\omega \in R} P(\omega) = P((r,r)) = \frac{1}{16}$. If W is the event the first ball drawn is white, $P(W) = \sum_{\omega \in W} P(\omega) = P((w,r)) + P((w,w)) + P((w,b)) = \frac{21}{64}$.

If the second ball is returned and a third one drawn, what is the probability two of the balls are white? Let B and the probability on its events be unchanged. Now $\Omega = (XB)_3 = \{(b_1, b_2, b_3) | b_i \in B\}$, and define a probability on A_Ω by setting for $\{\omega\} \in A_\Omega$, $P(\omega) = \left(\frac{1}{8}\right)^r \left(\frac{3}{8}\right)^w \left(\frac{3}{8}\right)^b$ where r is the number of r's, w is the number of w's and b is the number of b's in the triplet ω, $(r + w + b = 3)$. For $A \in A_\Omega$ set $P(A) = \sum_{\omega \in A} P(\omega)$. Since there are $\binom{3}{r,w,b}$ points in Ω resulting from the drawing of exactly r red, w white, and b blue balls,

$$P(\omega) = \sum_{r+w+b=3} \binom{3}{r,w,b}\left(\frac{2}{8}\right)^r\left(\frac{3}{8}\right)^w\left(\frac{3}{8}\right)^b = \left(\frac{2}{8} + \frac{3}{8} + \frac{3}{8}\right)^3 = 1$$

and so P is a probability. The probability of the event C that exactly two of the balls are white is the sum of the events C_r and C_b where C_r is the event two white, one red, and C_b is the event two white, one blue. Therefore

$$P(C) = P(C_r) + P(C_b) = \binom{3}{1,2,0}\left(\frac{2}{8}\right)\left(\frac{3}{8}\right)^2 + \binom{3}{0,2,1}\left(\frac{3}{8}\right)^3$$

The two examples above illustrate the case of repeating an experiment in which more than two outcomes are possible. We prove a general theorem for assigning a probability to such sample space.

5-4b$_1$ Theorem.   *Let* $E = \{e_1, e_2, \ldots, e_n\}$ *and* $\Omega = (X\, E)_n$.   *On the events of*
E *define* $p(e_i) = p_i \geqslant 0$ *with* $\displaystyle\sum_{i=1}^{\eta} p_i = 1$, *and on* A_Ω *define* P *by setting*
for $\{\omega\} \in A_\Omega$, $P(\omega) = p_1^{i_1} \cdot p_2^{i_2} \cdots p_\eta^{i_\eta}$ *where* $i_1 + i_2 + \cdots + i_\eta = n$ *and* i_k *is*
the number of times e_k *appears among the components of* $\{\omega\} \in A_\Omega$.
For $A \in A_\Omega$ *define* $P(A) = \displaystyle\sum_{\omega \in A} P(\omega)$.   *Then* P *is a probability.*

Proof: Since $\displaystyle\binom{n}{i_1, i_2, \ldots, i_\eta}$ points of Ω have i_1, e_1s, i_2, e_2s, etc.,

$$\sum_{\omega \in \Omega} P(\omega) = \sum_{i_1 + i_2 + \cdots + i_\eta = n} \binom{n}{i_1, i_2, \ldots, i_\eta} p_1^{i_1} p_2^{i_2} \cdots p_\eta^{i_\eta}$$

$$= (p_1 + p_2 + \cdots + p_\eta)^n = 1 \qquad \text{(see Prob. 7 Exercises 4-2a)} \quad \blacksquare$$

EXERCISES 5-4b

1.   A coin is tossed 10 times.   Let A be the event, five heads appear; B the event an even number of tails appear; and C the event there are at most three tails. Compute $P(A)$, $P(B)$, $P(C)$, $P(BC)$, $P(AB)$, $P(B \cup C)$.

2.   Ninety-seven percent of the output of a certain production line is acceptable.   On a certain day an inspector tests 100 items of the output.   What is the probability   (a) all the tested items are acceptable?   (b) that there are five unacceptable items among those tested?

3.   Two dice are rolled and the outcome is considered a success if the sum of the dots is 6, 7, or 8.   This is done seven times.   What is the probability (a) there are two successes?   (b) there are more than three successes?   (c) there are an odd number of failures?

4.   In families of five children, what is the probability (a) there are three boys?   (b) there is at least one girl?   (c) the second and third children are boys?

5.   On a die one face is painted red, two white and three blue.   The die is rolled 10 times to observe the color which comes up.   What is the probability that: (a) it comes up white twice, red four times, and blue four times?   (b) it comes up blue all ten times?   (c) it comes up white three times, red five times and blue two times?   (d) exactly three white faces come up?   (e) nine or 10 blue faces appear?

6.   In a men's club 20 percent of the members are physicians, 30 percent businessmen, 25 percent government workers, 15 percent lawyers and 10 percent teachers. A committee of five is selected at random. What is the probability (a) that there is a committee member from each professional category? (b) that there are two physicians, two lawyers and one teacher? (c) that there are only businessmen and government workers?

chapter **6**

Random Variables

6-1. RANDOM VARIABLES

The gentlemen gamblers who initiated the study of probability were interested primarily in how likely they were to win or lose a given amount of money rather than how the dice came up or where the wheel stopped. A gambler who will be given as many dollars as the sum of the faces that come up when two dice are rolled does not care whether he gets a two and a three or a one and a four, only that he earns five dollars. The man who wins or loses a quarter at a carnival depending upon whether the wheel stops at an even or odd number does not regret a seven any more than a thirteen, in either case he has lost his quarter. In both of these situations the interest is not centered on the sample space associated with the given experiment but on returns that the occurence of events in the sample space yield. A poll taker, paid according to the number of responses collected, is not concerned with which of the people approached answer his queries, just with how many. In the theory of probability the rewards associated with elementary events are called *random variables,* a designation as misleading as it is entrenched.

a.   Definition

6-1a_1 Definition.   *A* **random variable** *is a real valued function defined on a sample space.*

This definition will serve as an explanation of the remark at the close of the last paragraph which said that the designation random variables is misleading; a *random variable* is a function. Let us emphasize that a random variable is a function whose domain is a sample space whereas a probability is a set function whose domain is the set of subsets of a sample space. Since in our study the sample space Ω is at most countable, the range of a random variable on Ω will be at most countable; these random variables are referred to as *discrete.*

The rewards of the gambler who wins or loses a dollar depending upon whether the roll of a die produces an even or odd face can be described as a random variable X on $D = \{1,2,3,4,5,6\}$ where $X(1) = X(3) = X(5) = -1$ and $X(2) = X(4) = X(6) = +1$. The earnings of a man who is given an amount equal to the sum of the faces on a roll of two dice are the range values of a random variable Y defined $\Omega = D \times D$ where if $(a,b) \in \Omega$, $Y((a,b)) = a + b$. The set of test scores is a random variable on the set of test papers; age in years is a random variable on the registered voters of Ardmore, Pennsylvania; the Dow-Jones average is a random variable on the business days of the stock market.

b.   Probability Distributions

Let $\{\Omega, A_\Omega, P\}$ be a probability space and let X be a random variable on Ω with $R_X = \{x_1, x_2, x_3, \ldots\}$ which may be finite or countable. The set

$$\{\omega \mid \omega \in \Omega, X(\omega) = x_i \in R_X\} \in A_\Omega,$$

and for simplicity we write

$$\{X = x_i\} = \{\omega \mid \omega \in \Omega,\ X(\omega) = x_i \in R_X\}.$$

Since $\{X = x_i\} \in A_\Omega$, it makes sense to talk about $P(X = x_i)$ and we introduce a new function.

6-1b_1   **Definition.**   *Given a probability space $\{\Omega, A_\Omega, P\}$ and a random variable X on Ω, the function*

$$P_X = \{(x_i, P(X = x_i)) \mid x_i \in R_X\}$$

is called the **probability distribution** *of X.*

The function P_X tells one how the probability is distributed over the range of the random variable. Again let $D = \{1,2,3,4,5,6\}$ and for $A \in A_D$ set $P(A) = \dfrac{n(A)}{6}$, then $\{D, A_D, P\}$ is a probability space. Define a random variable X on D as was done in the preceding paragraph: $X(1) = X(3) = X(5) = -1$ and $X(2) = X(4) = X(6) = 1$. Then $\{X = -1\} = \{1,3,5\}$, $\{X = 1\} = \{2,4,6\}$ and from 6-1b_1, $P_X = \{(-1, \frac{1}{2}), (1, \frac{1}{2})\}$. Now consider $\Omega = D \times D$ and for $A \in A_X$ set $P(A) = \dfrac{n(A)}{36}$. $\{\Omega, A_\Omega, P\}$ is a probability space and if Y is the random variable which is the sum of the members of the ordered pair, i.e., for

$$(a,b) \in \Omega,\ Y((a,b)) = a + b,\ P_Y = \{(j, P(Y = j)) \mid j = 2,3,\ldots,12\}.$$

We have, for example, $P_Y(5) = P(Y = 5) = P(\{(1,4),(2,3),(3,2),(4,1)\}) = \dfrac{4}{36}$.

Starting with a sample space Ω, we are now concerned with three functions. The first of these is the probability P defined on the events of Ω; P is a set function, its domain is A_Ω and it is so defined that its range is contained in the unit interval, that is, for all $A \in A_\Omega$, $0 \leqslant P(A) \leqslant 1$. The second function is the random variable which is defined on the elements of Ω; it is not a set function, its domain is a set of points, the elementary events in the sample space. The range of the random variable X, like that of P, is real valued. By definition of X each element of the range is a real number; but unlike P's, its range need not be limited to numbers between zero and one. Finally we have introduced probability distributions. By definition the domain of the probability distribution P_X is the range of a random variable and so the domain of P_X is a set of real numbers. We repeat, this is not true of the probability P and of the random variable X, the domain of the first is a set of sets and that of the second a set of points. The range values of P_X, being probabilities, are real numbers and so both the domain and range of P_X are contained in the set of real numbers. We say that P_X is a real valued function of a real variable; and, as such, P_X is subject to the well known rules of mathematical analysis. The use we make of this will become apparent as our work progresses.

Again consider a probability triple $\{\Omega, A_\Omega, P\}$ and a random variable X on Ω. For every $x_i \in R_X$, the set $\{X = x_i\} \in A_\Omega$ and

6-1b_2   if $x_i \neq x_j$, $\{X = x_i\} \cap \{X = x_j\} = \varnothing$.

This last statement is easy to verify. Assume $\{X = x_i\} \cap \{X = x_j\} \neq \varnothing$. This would imply that there exists an $\omega \in \Omega$ such that $X(\omega) = x_i$ and $X(\omega) = x_j$ which in turn implies that both the ordered pairs (ω, x_i) and (ω, x_j) belong to X. But this is a contradiction since X is a function. Hence the assumption is false and 6-1b_2 holds.

Let S_X designate the set of subsets of R_X and let $S \in S_X$; we have

$$S = \sum_{x_i \in S} \{X = x_i\}.$$

Since P is a probability on A_Ω,

$$P\left(\sum_{x_i \in S} \{X = x_i\} \right) = \sum_{x_i \in S} P(X = x_i) = \sum_{x_i \in S} P_X(x_i),$$

and we now extend the domain of P_X by setting

6-1b_3 $$P_X(S) = \sum_{x_i \in S} P_X(x_i) \text{ for all } S \in S_X.$$

With this extended domain P_X is a set function since it is now defined on the set of sets, S_X. It is clear that for $S \in S_X$, $P_X(S) \geqslant 0$, and it takes a calculation like that in the proof of 5-2b_3 to show that if $S_i S_j = \varnothing$, $i \neq j$, $\Sigma_j P_X(S_j) = P_X(\Sigma_j S_j)$. We have also

$$P_X(R_X) = \sum_{x_i \in R_X} P_X(x_i) = \sum_{x_i \in R_X} P(X = x_i) = P\left(\sum_{x_i \in R_X} \{X = x_i\}\right) = P(\Omega) = 1.$$

Thus the extended P_X is a probability on S_X and $\{R_X, S_X, P_X\}$ is a probability space.

We shall refer to $\{R_X, S_X, P_X\}$ as an *induced probability space*. Many problems in probability can be answered more readily in the induced space rather than in the original one. If a coin is tossed ten times, what is the probability there will be six or seven heads? On the basic probability space let Z be a random variable which counts the number of heads. Then

$$R_Z = \{0,1,2,\ldots,10\} \text{ and } P_Z = \left\{\left(i, \binom{10}{i} 2^{-10}\right) \middle| i \in R_Z\right\}.$$

If S_Z is the set of subsets of R_Z, $s = \{6,7\} \subset S_Z$, and since $s = \{6\} + \{7\}$,

$$P_Z(s) = P_Z(6) + P_Z(7)$$
$$= \binom{10}{6} 2^{-10} + \binom{10}{7} 2^{-10}$$
$$= \frac{165}{512}$$

answers the question. There was no need here to consider the 1,024 ten-tuples which comprise the sample space associated with the experiment; working the R_Z simplified the problem.

We needed to extend the domain of P_X in order to define the induced probability space $\{R_X, S_X, P_X\}$. Since, however, $P_X(S) = \sum_{x_i \in S} P_X(x_i)$ we shall always be working with the function defined on R_X (as we did in the example above), and so we shall continue to treat P_X as a real valued function of a real variable.

EXERCISES 6-1b

(NOTE: The first eight of these exercises will be used again in Exercises 6-3b.)

1. A ball is drawn from an urn containing 3 white, 2 red, and 4 blue balls. $\Omega = \{r,w,b\}$. Let X be a random variable on Ω where $X = \{(r,-1), (w,0), (b, 1)\}$. What are R_X and P_X?

2. A card is drawn from a standard deck. Let Y and Z be two random variables on the sample space Ω with $Y = \{(C, 1), (D, 2), (H, 3), (S, 4)\}$ (C, D, H, S are the suits) and $Z(j) = j$ where j is the denomination of the card ($j = 1, 2,\ldots,13$). What are R_Y, P_Y, R_Z, P_Z?

3. Nine disks each bearing one of the digits 1,2,...,9 are put into an urn. One is drawn out, its number noted and it is returned to the urn. This is done four times so that a four-digit number is generated. For $(a,b,c,d) \in \Omega$, let $X((a,b,c,d)) = a$. Describe the probability space induced by X. What is the probability (a) that the number generated lies between 6,000 and 7,000? (b) that is greater than 8,000? (c) that is less than or equal to 3,000?

4. At a certain camp there are 15 junior and 25 senior campers. Each day seven are chosen by lot for KP duty and Y counts the number of juniors in the group. Construct the probability space induced by Y. What is the probability (a) that $Y = 0$? (b) that two of the group are juniors? (c) that six or seven are juniors? (d) that all are juniors? If Z counts the number of seniors, how are P_Y and P_Z related?

5. Three indistinguishable dice are rolled and their faces recorded. Describe a probability space. Let X,Y,Z be random variables on Ω where if ω is the unordered triple (d_1,d_2,d_3), $X(\omega) = d_1 + d_2 + d_3$, $Y(\omega) = \max (d_1,d_2,d_3)$, $Z(\omega) = \min (d_1,d_2,d_3)$. [By max (a,b,c) is meant that value a, b, or c which is at least as large as the other two; min (a,b,c) is analogously defined.] What are R_X, P_X, R_Y, P_Y, R_Z, P_Z? In terms of the probability spaces induced by X, Y, and Z, how can the following questions be answered?

(a) What is the probability at least one of the dice comes up six?

(b) Describe the event every face is greater than three.

(c) Describe the event the sum of the faces is at least 5.

(d) What is the probability no face is greater than four?

(e) What is the probability all faces are greater than four?

(f) What is the probability the sum is less than five?

(g) Describe the event every face is less than three.

(h) Describe the event the smallest face is three and the sum is greater than nine.

(i) Describe the event all faces are the same.

(j) Describe the event a six or an ace appears.

6. A bridge hand is dealt. X counts the number of kings, Y the number of hearts and Z the number of black cards. What are the induced probability spaces? In terms of X,Y,Z and the induced probability spaces:

(a) What is the probability there are three kings?

(b) What is the event the hand contains at least one king?

(c) What is the probability the hand contains five hearts?

(d) What is the event the hand has no more than three hearts?

(e) What is the probability all cards are black?

(f) What is the event there are no kings and no hearts?

7. A die is rolled until an ace appears. Let X be the random variable that counts the number of rolls. Describe R_X and P_X. What is the probability (a) that an even number of rolls is necessary? (b) that the last trial will occur at a multiple of three? (c) that the ace will appear on or before the fifth roll? (Compute the results.)

8. A machine is known to be 95 percent reliable. An inspector tests 100 samples of its output. Let Y count the number of substandard items in the sample. What is P_Y? What is the probability (a) that no more than three items are substandard? (b) that more than 10 are substandard?

9. Let $\Omega = \{\omega_1, \omega_2, \omega_3, \ldots\}$ be countable and construct $\{\Omega, A_\Omega, P\}$ where $P(\omega_i) = p_i, (\Sigma_i p_i = 1)$. Let X be a random variable on Ω with $X(\omega_i) = i$. What is P_X? What is the probability (a) that X is even? (b) that X is a multiple of 8? (c) that X is no more than 5?

10. If X is a random variable on Ω in $\{\Omega, A_\Omega, P\}$, prove $\displaystyle\sum_{x_i \in R_X} \{X = x_i\} = \Omega$.

11. A coin is tossed until the second head appears and Y counts the number of tosses. What is P_Y? (See Prob. 10, Exercises 5-3c)

12. A card is drawn from a standard deck without replacement until a heart appears. Z counts the number of draws. What is P_Z? Can you show that $\displaystyle\sum_{j \in R_Z} P_Z(j) = 1$?

c. Joint Distribution

An urn contains four balls numbered 1, 2, 3, and 4; two are drawn. The points in the sample space Ω are the $\binom{4}{2} = 6$ combinations of the balls and if for $A \in A_\Omega$, $P(A) = \dfrac{n(A)}{6}$, $\{\Omega, A_\Omega, P\}$ is a probability space. Two random variables X and Y are defined on Ω where X is the sum of the two numbers drawn and Y is the larger of the numbers. $R_X = \{3,4,5,6,7\}$; $R_Y = \{2,3,4\}$; and each has a probability distribution. What is the probability we have both $X = i$ and $Y = j$? For example, what is the probability $X = 5$ and $Y = 4$? We have

$$\{X = 5\} = \{(1,4), (2,3)\} \text{ and } \{Y = 4\} = \{(1,4), (2,4), (3,4)\}$$

where in this problem the pairs (a,b) are unordered. The event

$$\{X = 5, Y = 4\} = \{X = 5\} \cap \{Y = 4\} = \{(1,4)\}$$

and

$$P(X = 5, Y = 4) = P((1,4)) = \frac{1}{6}.$$

We have found one value in the *joint distribution* of the random variables X and Y on Ω in $\{\Omega, A_\Omega, P\}$.

6-1c₁ **Definition.** *The* **joint distribution** *of two random variables X and Y on Ω in $\{\Omega, A_\Omega, P\}$ is the function*

$$\{((x_i, y_j), P(X = x_i, Y = y_j)) | x_i \in R_X, y_j \in R_Y\}.$$

The domain of the joint distribution is the set of ordered pairs (x_i, y_j) and its range is contained in $\{r | 0 \leqslant r \leqslant 1\}$.

The range values of the joint distribution of the random variable X and Y described at the beginning of this section are displayed in tabular form in Figure 6-1c₂, where the number in the ith column and jth row is the value of $P(X = i, Y = j)$. In the margins of the table, we obtain, by adding across the rows and down the columns, the probability distributions P_X and P_Y. Someone puzzled by the entries in this table should construct Ω, and, as was done on page 100, compute $P(X = i, Y = j)$ for all possible combinations of i and j.

6-1c₂ **Figure**

Joint Distribution Table of X and Y.

R_X \ R_Y	3	4	5	6	7	P_Y
2	$\frac{1}{6}$	0	0	0	0	$\frac{1}{6}$
3	0	$\frac{1}{6}$	$\frac{1}{6}$	0	0	$\frac{1}{3}$
4	0	0	$\frac{1}{6}$	$\frac{1}{6}$	$\frac{1}{6}$	$\frac{1}{2}$
P_X	$\frac{1}{6}$	$\frac{1}{6}$	$\frac{1}{3}$	$\frac{1}{6}$	$\frac{1}{6}$	1

The procedure above generalizes easily. Let $\{\Omega, A_\Omega, P\}$ be a probability space and let X and Y be two random variables, defined on Ω. Let their respective ranges be $R_X = \{x_1, x_2, x_3, \ldots\}$ and $R_Y = \{y_1, y_2, y_3, \ldots\}$ both of which are at most countable and, as usual, designate their probability distributions by P_X and P_Y. Set, for brevity,

$$P(X = x_i, Y = y_j) = p_{ij}$$

Then Figure 6-1c₃ is the table of the joint distribution of X and Y.

6-1c₃   Figure

Joint Distribution Table of Two Random Variables

R_X R_Y	x_1	x_2	x_3	$\cdots$	x_i	$\cdots$	P_Y
y_1	p_{11}	p_{21}	p_{31}	$\cdots$	p_{i1}	$\cdots$	$P_Y(y_1)$
y_2	p_{12}	p_{22}	p_{32}	$\cdots$	p_{i2}	$\cdots$	$P_Y(y_2)$
y_3	p_{13}	p_{23}	p_{33}	$\cdots$	p_{i3}	$\cdots$	$P_Y(y_3)$
.	.	.	.		.		.
.	.	.	.		.		.
.	.	.	.		.		.
y_j	p_{1j}	p_{2j}	p_{3j}	$\cdots$	p_{ij}	$\cdots$	$P_Y(y_j)$
.	.	.	.		.		.
.	.	.	.		.		.
.	.	.	.		.		.
P_X	$P_X(x_1)$	$P_X(x_2)$	$P_X(x_3)$	$\cdots$	$P_X(x_i)$	$\cdots$	1

When we add down any column, we obtain

$$p_{i1} + p_{i2} + \cdots + p_{ij} + \cdots = \sum_{j=1} P(X = x_i, Y = y_j)$$

$$= \sum_{j=1} P(\{X = x_i\} \{Y = y_j\})$$

$$= P\left(\sum_{j=1} \{X = x_i\} \{Y = y_j\} \right)$$

$$= P\left(\{X = x_i\} \sum_{j=1} \{Y = y_j\} \right)$$

(The distributive law
for sets see Prob. 9,
Exercises 2-2*b*)

$$= P(\{X = x_i\}\Omega) = P\{X = x_i\}$$

$$= P_X(x_i).$$

A similar argument yields the row sums $P_Y(y_j)$ for $j = 1,2,3,\ldots$ With respect
to the joint distribution, P_X and P_Y are referred to as marginal distributions.   We
remark that in Figure 6-1c₃ the array of ordered pairs (x_i, y_j) with $x_i \in R_X$,
$y_i \in R_Y$ form a sample space $\bar{\Omega}$.   Choosing $\bar{P}(x_i, y_j) = p_{ij}$ and for $A \in A_{\bar{\Omega}}$, setting

$$\bar{P}(A) = \sum_{(x_i + y_j) \in A} p_{ij}$$

$\{\bar{\Omega}, A_{\bar{\Omega}}, \bar{P}\}$ is a probability space.

Let a die be tossed three times generating $\Omega = D \times D \times D$ where $D = \{1,2,3,4,5,6\}$, and for $\omega \in \Omega$, set $p(\omega) = \left(\dfrac{1}{6}\right)^{i+j}\left(\dfrac{4}{6}\right)^{3-i-j}$ where i is the number of ones in the ordered triple ω and j is the number of twos. By Counting Principle IV (4-1d_1) there are $\left(\dfrac{3}{i,j,3-i-j}\right)$ points in Ω with i ones and j twos. On Ω let X be a random variable which counts the number of ones and Y the random variable which counts the number of twos. The joint distribution of X and Y is given in Figure 6-1c_4.

6-1c_4 **Figure**

Joint distribution of X and Y where X counts the ones and Y counts the twos in three rolls of a die

R_Y \ R_X	0	1	2	3	P_Y
0	$\dfrac{4^3}{6^3}$	$3\dfrac{4^2}{6^3}$	$3\dfrac{4}{6^3}$	$\dfrac{1}{6^3}$	
1	$3\dfrac{4^2}{6^3}$	$3\dfrac{4}{6^3}$	$\dfrac{1}{6^3}$	0	
2	$3\dfrac{4}{6^3}$	$\dfrac{1}{6^3}$	0	0	
3	$\dfrac{1}{6^3}$	0	0	0	
P_X					1

If we sum the entire set of p_{ij}'s, we have

$$\sum_{i,j} p_{ij} = \sum_{i,j}\left(\dfrac{3}{i,j,3-i-j}\right)\left(\dfrac{1}{6}\right)^{i+j}\left(\dfrac{4}{6}\right)^{3-i-j} = \left(\dfrac{1}{6}+\dfrac{1}{6}+\dfrac{4}{6}\right)^3 = 1$$

(see Exercises 4-2a, Prob. 7), and so the set of ordered pairs (i,j) is a sample space and $p_{ij} = P(X = i, Y = j)$ defines a probability on this sample space. To find the marginal distribution P_X we compute the column sums:

$$P_X(0) = \sum_{j=0}^{3} p_{0j} = \sum_{j=0}^{3}\left(\dfrac{3}{0,j,3-j}\right)\left(\dfrac{1}{6}\right)^{j}\left(\dfrac{4}{6}\right)^{3-j} = \left(\dfrac{1}{6}+\dfrac{4}{6}\right)^3 = \left(\dfrac{5}{6}\right)^3;$$

$$P_X(1) = \sum_{j=0}^{3} p_{1j} = \sum_{k=0}^{3} \binom{3}{1,j,2-j} \left(\frac{1}{6}\right)^{1+j} \left(\frac{4}{6}\right)^{2-j}$$

$$= 3 \left(\frac{1}{6}\right) \sum_{j=0}^{2} \binom{2}{j} \left(\frac{1}{6}\right)^{j} \left(\frac{4}{6}\right)^{2-j} = 3 \left(\frac{1}{6}\right) \left(\frac{5}{6}\right)^{2};$$

$$P_X(2) = \sum_{j=0}^{3} p_{2j} = \sum_{j=0}^{3} \binom{3}{2,j,1-j} \left(\frac{1}{6}\right)^{2+j} \left(\frac{4}{6}\right)^{1-j}$$

$$= 3 \left(\frac{1}{6}\right)^{2} \sum_{j=0}^{1} \binom{1}{j} \left(\frac{1}{6}\right)^{j} \left(\frac{4}{6}\right)^{1-j} = 3 \left(\frac{1}{6}\right)^{2} \left(\frac{5}{6}\right);$$

$$P_X(3) = \sum_{j=0}^{3} p_{3j} = \sum_{j=0}^{3} \binom{3}{3,0,0} \left(\frac{1}{6}\right)^{3} \left(\frac{4}{6}\right)^{0} = \left(\frac{1}{6}\right)^{3}$$

Furthermore, we have

$$\sum_{i=0}^{3} P_X(i) = \sum_{i=0}^{3} \binom{3}{i} \left(\frac{1}{6}\right)^{i} \left(\frac{5}{6}\right)^{3-i} = \left(\frac{1}{6} + \frac{5}{6}\right)^{3} = 1,$$

and that the probability distribution of X is given by

$$P_X = \left\{ \left(i, \binom{3}{i} \left(\frac{1}{6}\right)^{i} \left(\frac{5}{6}\right)^{3-i} \right) \Big| i = 0,1,2,3 \right\}.$$

We obtained P_X, not directly, but from the joint distribution of the two random variables X and Y. Summing across rows would give us the distribution P_Y.

d.*   Joint Distribution Of More Than Two Random Variables

The number of random variables that one can define on a sample space is not limited to two; however, the graphical representation of their joint distributions becomes increasingly difficult. If three random Y, Y, and Z are defined on Ω in the probability space $\{\Omega, A_\Omega, P\}$, to display the

$$p_{ijk} = P(X = x_i, Y = y_j, Z = z_k), \; x_i \in R_X, \; y_j \in R_Y, \; z_k \in R_Z$$

would require a parallelepiped ruled into cubes, each of which would contain an entry p_{ijk}. The various faces of this solid would contain the marginal distributions p_{ij}, p_{jk}, p_{ki}, and the common edges P_X, P_Y, P_Z (these would be analogous to the X, Y, Z axes in Euclidean three-space). An example might clarify the situation. Let a die be rolled four times; the sample space

---

*This paragraph can be omitted without loss of continuity.

is $\Omega = D \times D \times D \times D$ where $D = \{1,2,3,4,5,6\}$. For $\omega \in \Omega$, let $P(\omega) = 6^{-4}$ and for $A \in A_\Omega$, set $P(A) = \sum_{\omega \in A} P(\omega)$, then $\{\Omega, A_\Omega, P\}$ is a probability space.

Three random variables are defined on Ω: X, the number of twos; Y, the number of fours; and Z, the number of sixes. $R_X = R_Y = R_Z = \{0,1,2,3,4\}$. By Counting Principle IV the number of points in Ω containing i twos, j fours and k sixes is $\binom{4}{i,j,k,4-i-j-k}$. The probability of each such point is $\left(\frac{1}{6}\right)^{i+j+k}\left(\frac{3}{6}\right)^{4-i-j-k}$; and then

$$p_{ijk} = P(X = i, Y = j, Z = k) = \frac{4!}{i!j!k!(4-i-j-k)!}\left(\frac{1}{6}\right)^{i+j+k}\left(\frac{1}{2}\right)^{4-i-j-k}$$

Holding i and j fixed and summing over k we obtain the marginal probability p_{ij}:

$$p_{ij} = \sum_{k=0}^{4} p_{ijk} = \sum_{k=0}^{4} \frac{4!}{i!j!k!(4-i-j-k)!}\left(\frac{1}{6}\right)^{i+j+k}\left(\frac{1}{2}\right)^{4-i-j-k}$$

$$= \frac{4!}{i!j!(4-i-j)!}\left(\frac{1}{6}\right)^{i+j}\sum_{k=0}^{4}\frac{(4-i-j)!}{(4-i-j-k)!k!}\left(\frac{1}{6}\right)^{k}\left(\frac{1}{2}\right)^{4-i-j-k}$$

$$= \frac{4!}{i!j!(4-i-j)!}\left(\frac{1}{6}\right)^{i+j}\left(\frac{1}{6}+\frac{1}{2}\right)^{4-i-j}$$

where the sum over k is made up of the terms of the binomial expansion of $\left(\frac{1}{6}+\frac{1}{2}\right)^{4-i-j}$. Now, holding i fixed, we obtain

$$P_X(i) = \sum_{i=0}^{4} p_{ij} = \frac{4!}{i!(4-i)!}\left(\frac{1}{6}\right)^{i}\sum_{j=0}^{4}\frac{(4-i)!}{j!(4-i-j)!}\left(\frac{1}{6}\right)^{i}\left(\frac{4}{6}\right)^{4-i-j}$$

$$= \binom{4}{i}\left(\frac{1}{6}\right)^{i}\left(\frac{5}{6}\right)^{4-i},$$

and this is in agreement with earlier calculations when we computed the probability that in four rolls of a die we get exactly i twos.

EXERCISES 6-1d

1. Prepare the joint distribution table for the random variables Y and Z in Prob. 2 of Exercises 6-1b.

2. Prepare the joint distribution table for the random variables Y and Z in Prob. 4 of Exercises 6-1b.

3.   Three distinguishable balls are distributed among three cells. (This sample space is given in Figure 5-3a_2 of the text.)   Let X be the number of balls in the second cell, Y the number in the third cell, and Z the number of occupied cells.   Prepare joint distribution tables for X and Y and for X and Z.

4.   A class of 15 boys and 12 girls.   Five are chosen at random to form a class council.   Let X count the number of boys, Y the number of girls on the council.   What is the probability space induced by X?   What is the joint distribution of X and Y?

5.   Prepare joint distribution tables for X and Y, X and for Z and Y and Z in Prob. 5 of Exercises 6-1b.

   (a) What is the probability the sum is 12 and the maximum is 8?
   (b) What is the probability the sum is 9 and the minimum is 3?
   (c) What is the probability the maximum is 5 and the minimum is 2?
   (d) Prepare the joint distribution table of X, Y and Z.

6.   If X and Y are random variables with $R_X = \{0,1,2\}$ and $R_Y = \{0,1,2,3\}$ and $P(X = x,\ Y = y) = \dfrac{1}{54}(x + y^2)$, show that $P_X(x) = \dfrac{1}{27}(7 + 2x)$ and find P_Y.

7.   Compute the marginal distribution P_Y in Figure 6-1c_4.

e.   Special Probability Distributions

There are many experiments which result in two alternatives, such as heads and tails, success and failure, odds and evens.   It is convenient to set $\Omega = \{S,F\}$; and if $P(S) = p$ and $P(F) = 1-p$, $\{\Omega,A_\Omega,P\}$ is a probability space.

6-1e$_1$ *Let X be a random variable on Ω with $X(S) = 1$, $X(F) = 0$, then*

$$P_X = \{(1,p),(0,1-p)\}$$

is called a **Bernoulli distribution,***

or is said to obey the Bernoulli probability law.   The quantity p which varies from experiment to experiment is called a *parameter.*   The tossing of a coin has a Bernoulli distribution with parameter $\dfrac{1}{2}$; if in rolling a die, the appearance of a 6 is taken as S, the experiment is distributed according to the Bernoulli law with parameter $\dfrac{1}{6}$; if in drawing a ball from an urn containing two white and three black balls, drawing a black one is taken as S, then the experiment has a Bernoulli distribution with parameter $\dfrac{3}{5}$.

If the experiment above is repeated n times, the sample space Ω is a set of

*After James Bernoulli (1654-1704).

ordered n-tuples where each component is either S or F. If, further, $\omega \in \Omega$ has i S's and $n-i$ F's and $P(\omega) = p^i(1-p)^{n-i}$, $\{\Omega, A_\Omega, P\}$ is a probability space.

6-1e$_2$ *Let X on Ω be a random variable which counts the number of S's, then*

$$P_X = \left\{ \left(i, \binom{n}{i} p^i (1-p)^{n-i} \right) \Big| i \leqslant n \right\}$$

is called a **binomial distribution,**

or is said to obey the binomial law of probability. This distribution has two parameters n and p. If a coin is tossed ten times and X counts the number of heads, X has a binomial distribution with parameters $n = 10$, and $p = \frac{1}{2}$; if a die is rolled 20 times and Y counts the number of sixes, Y obeys the binomial law with parameters $n = 20$ and $p = \frac{1}{6}$.

Let $\Omega = \{\omega_1, \omega_2, \omega_3, \ldots\}$ be countable and let $X(\omega_i) = i$.

6-1e$_3$ $P_X = \{(n, p(1-p)^{n-1}) \mid n = 1, 2, 3, \ldots, 0 < p < 1\}$ *is called a* **geometric distribution.**

It has the single parameter p. A random variable with this distribution is the one which counts the number of rolls of a die until the first six appears; the parameter in this case is $\frac{1}{6}$. $P_X(1) = \frac{1}{6}$, $P_X(2) = \frac{5}{6}\frac{1}{6}$, $P_X(3) = \left(\frac{5}{6}\right)^2 \frac{1}{6}, \ldots$ A less trivial example is found in the life span of certain radioactive atoms. In any given time interval an atom decays with probability p and does not decay with probability $1 - p$. (In a certain sense these atoms do not age, whether they decay in a given time interval is independent of the length of their existence.) If T records the interval of decay of a given atom, T obeys the geometric law, i.e.,

$$P_T(k) = (1-p)^{k-1}p.$$

$P_T(3)$ is the probability of decay in the third time interval.

We spoke earlier of the induced probability space $\{R_X, S_X, P_X\}$ and we said that many problems could be solved more easily in terms of the induced space. This is particularly true if we are interested not in the sample space itself but rather in a random variable which has one of these special distributions. Let us illustrate. The chances for success in a certain military situation can be approximated by putting seven white and five black balls in an urn; one is drawn out, its color noted and the ball replaced. The balls are then stirred and the process repeated. This is done six times. The outcome is judged successful if at least five of the balls drawn are white. If X counts the number of white balls, X is a random variable distributed binomially with parameters $n = 6$ and $p = \frac{7}{12}$.

Then

$$P(X = 5 \text{ or } X = 6) = P_X(5) + P_X(6) = \binom{6}{5}\left(\frac{7}{12}\right)^5\left(\frac{5}{12}\right) + \binom{6}{6}\left(\frac{7}{12}\right)^6\left(\frac{5}{12}\right)^0$$

$$P_Y(5) + P_Y(6) = \left(\frac{7}{12}\right)^5\left(6\cdot\frac{5}{12} + \frac{7}{12}\right) = \left(\frac{7}{12}\right)^5\left(\frac{37}{12}\right)$$

Here there was no need to describe the sample space; we had only to recognize that the problem could be solved in terms of a random variable obeying the binomial law with known parameters.

There are other important probability distributions—in fact, two of the most important, the Poisson and the Normal distributions, we cannot discuss here because of computational difficulties—but the three given will serve to illustrate the concept and to indicate how they can be used in the solution of problems.

EXERCISES 6-1e

1. Let a card be drawn from a standard deck and let X be a random variable on the sample space where $X = 1$ if the card is a heart and $X = 0$ otherwise. What is the nature of P_X?

2. A card is drawn from a standard deck, its suit noted, the card replaced and the deck shuffled. This is done five times. If Y counts the number of hearts, what is the nature of P_Y? What are $P_Y(3)$? $P_Y\left(\frac{5}{2}\right)$?

3. Suppose the experiment in Prob. 2 is repeated 10 times, and Z counts the number of hearts. How does P_Z differ from P_Y of Prob. 2?

4. Let X count the number of heads when a coin is tossed 6 times. Let Y count the number of boys in families of 6 children. And let Z count the number of even faces which appear in 6 rolls of a die. Compare the probability distributions X, Y, and Z. What are $P_X(4)$, $P_Y(4)$ and $P_Z(4)$?

5. A die is rolled until a six appears. X counts the number of rolls. What is P_X?

6-2. DISTRIBUTION FUNCTIONS*

a. Definition

In the probability space $\{\Omega, A_\Omega, P\}$, the probability P is a set function on Ω and we have discussed how, if a random variable X is defined on Ω, we can work instead in the induced probability space $\{R_X, S_X, P_X\}$. As we pointed out earlier, the use of P_X rather than P has certain advantages for it is a numerical function. The advantage is pressed even further if, instead of P_X, we introduce F_X, the distribution function of X.

*This section can be omitted without loss of continuity.

6-2a₁ **Definition.** *The* **distribution function** *of a random variable X is the function*

$$F_X = \{(x, P(X < x)) \mid x \text{ is a real number}\}^{[4]}.$$

We see immediately that the domain of F_X is the set of real numbers and that the range of F_X is contained in $\{x \mid 0 \le x \le 1\}$. Since the domain of F_X contains more than a countable number of points, it can be studied by more sophisticated techniques than can P_X. In fairness it should be said here that the role of distribution functions increases in importance with the study of continuous sample spaces; however, the concept is fundamental and an acquaintance with it will increase our understanding of probability.

Since our concern is limited to discrete random variables, let us apply this definition to such a random variable. Let X be a random variable with

$$R_X = \{x_1, x_2, x_3, \ldots\}$$

which is either finite or countable and whose elements are designated so that $x_1 < x_2 < x_3 < \ldots$ If $b \in R$, the set of real numbers, we have from the definition that

$$F_X(b) = P(X < b).$$

Now if $x_1 < x_2 < x_3 < b$, the set

$$\{X < b\} = \{X < x_1\} + \{x_1 \le X < x_2\} + \{x_2 \le X < x_3\} + \{x_3 \le X < b\}$$

and since they are disjoint,

$$P(X < b) = P(X < x_1) + P(X_1 \le X < x_2) + P(x_2 \le X < x_3) + P(x_3 \le X < b).$$

What we have done here is to take the set of real numbers less than b and separate it into three disjoint intervals. Graphically we have

where in each of the finite intervals the left-hand end point, but not the right, is included. Looking at these intervals in terms of X, we see

$$
\begin{aligned}
&\{X < x_1\} &&= \varnothing &&\text{so that} &&P(X < x_1) &&= 0 \\
&\{x_1 \le X < x_2\} &&= \{X = x_1\} &&\text{so that} &&P(x_1 \le X < x_2) &&= P(X = x_1) \\
&\{x_2 \le X < x_3\} &&= \{X = x_2\} &&\text{so that} &&P(x_2 \le X < x_3) &&= P(X = x_2) \\
&\{x_3 \le X < b\} &&= \{X = x_3\} &&\text{so that} &&P(x_3 \le X < b) &&= P(X = x_3)
\end{aligned}
$$

Hence

$$F_X(b) = P(X < b) = P(X = x_1) + P(X = x_2) + P(X = x_3).$$

By use of the Principle of Induction we can prove that if $x \in R$ and

$$x_1 < x_2 \ldots < x_n < x,$$

then

6-2a_2
$$F_X(x) = \sum_{x_i < x} P(X = x_i) = \sum_{x_i < x} P_X(x_i).$$

We have in 6-2a_2 an alternative definition of the distribution function of a discrete random variable.

Two elementary properties of distribution functions can be derived directly from the definition and hold for all random variables, not just discrete ones. To find $P(a \leqslant X < b)$ we write

$$\{a \leqslant X < b\} = \{X < b\} - \{X < a\}$$

and observe that $\{X < a\} \subset \{X < b\}$. Hence from 5-3$b_4$, we have

$$P(a \leqslant X < b) = P(X < b) - P(X < a)$$

and from 6-2a_1

6-2a_3
$$P(a \leqslant X < b) = F_X(b) - F_X(a).$$

A numerical function f is said to be *nondecreasing* if, whenever $x_1 < x_2$, $f(x_1) \leqslant f(x_2)$. Since, if $x_1 < x_2$,

$$F_X(x_2) - F_X(x_1) = P(X < x_2) - P(X < x_1) = P(x_1 \leqslant X < x_2) \geqslant 0,$$

we have

6-2a_4 *The distribution function, F_X, is nondecreasing.*

b. Examples

Let β be a random variable with Bernoulli distribution with parameter p. Then $P_\beta = \{(0, 1 - p), (1, p)\}$ and

$$F_\beta(x) = \sum_{x_i < x} P_\beta(x_i) = \begin{cases} 0 & x \leqslant 0 \\ 1 - p & 0 < x \leqslant 1 \\ 1 & 1 < x \end{cases}$$

If $x \leqslant 0$, no range value of β is less than x and

$$F_\beta(x) = \sum_{x_i < x} P_\beta(x_i) = 0$$

and if $0 < x \leqslant 1$, the only range value below x is zero, so

$$F_\beta(x) = \sum_{x_i < x} P_\beta(x_i) = P_\beta(0) = 1 - p,$$

and if $1 < x$, both range values of β are below x and so

$$F_\beta(x) = \sum_{x_i < x} P_\beta(x_i) = P_\beta(0) + P_\beta(1) = 1.$$

The graph of F_β appears in Figure 6-2b_1.

6-2b_1 Figure
    Graph of F, where β has a Bernoulli distribution with parameter p.

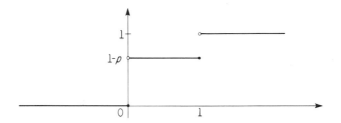

(In each interval the right-hand end point, but not the left, is included.)

For B a random variable obeying the binomial law with parameters n and p, its distribution function F_B is, from 6-2a_2,

$$F_B(x) = \sum_{k < x} \binom{n}{k} p^k (1 - p)^{n-k}.$$

If $n = 3$ and $p = \dfrac{1}{3}$ and

if $x \leqslant 0$,

$$F_B(x) = \sum_{k < 0} \binom{3}{k} \left(\frac{1}{3}\right)^k \left(\frac{2}{3}\right)^{3-k} = 0:$$

if $0 < x \leqslant 1$,

$$F_B(x) = \sum_{k < 1} \binom{3}{k} \left(\frac{1}{3}\right)^k \left(\frac{2}{3}\right)^{3-k} = \binom{3}{0}\left(\frac{1}{3}\right)^0\left(\frac{2}{3}\right)^3 = \frac{8}{27}:$$

if $1 < x \leqslant 2$,

$$F_B(x) = \sum_{k < 2} \binom{3}{k} \left(\frac{1}{3}\right)^k \left(\frac{2}{3}\right)^{3-k} = \binom{3}{0}\left(\frac{1}{3}\right)^0\left(\frac{2}{3}\right)^3 + \binom{3}{1}\left(\frac{1}{3}\right)\left(\frac{2}{3}\right)^2 = \frac{20}{27};$$

if $2 < x \leqslant 3$,

$$F_B(x) = \sum_{k < 3} \binom{3}{k}\left(\frac{1}{3}\right)^k\left(\frac{2}{3}\right)^{3-k} = \frac{20}{27} + \binom{3}{2}\left(\frac{1}{3}\right)^2\left(\frac{2}{3}\right) = \frac{26}{27};$$

if $3 < x$,

$$F_B(x) = \sum_{k \leqslant 3} \binom{3}{k}\left(\frac{1}{3}\right)^k\left(\frac{2}{3}\right)^{3-k} = \frac{26}{27} + \binom{3}{3}\left(\frac{1}{3}\right)^3\left(\frac{2}{3}\right)^0 = \frac{27}{27} = 1.$$

The graph of $F_B(x)$ for $n = 3$ and $p = \dfrac{1}{3}$ is given in Figure 6-2b_2.

6-2b_2 Figure

   Graph of F_B where B has a binomial distribution with parameters

   $n = 3$ *and* $p = \dfrac{1}{3}$.

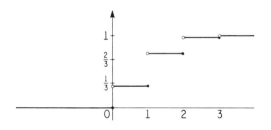

If a random variable G has a geometric probability distribution,

$$P_G(k) = p(1-p)^{k-1}$$

and

$$F_G(x) = \sum_{k < x} p(1-p)^{k-1} = p\,\frac{1-(1-p)^{\lfloor x \rfloor}}{1-(1-p)} = 1 - (1-p)^{\lfloor x \rfloor} \quad \text{(See 4-2}b_1\text{)}$$

where $\lfloor x \rfloor$ is the greatest integer less than or equal to x.   The graph of x with parameter $p = \dfrac{1}{2}$ is given in Figure 6-2b_3.

6-2b₃ **Figure**

Graph of F_G where G has a geometric distribution with parameter $\dfrac{1}{2}$.

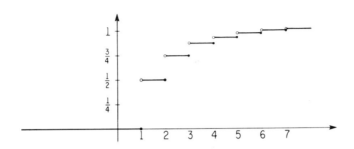

Our examples all illustrate the properties that if x is smaller than the minimum range value of X, $F_X(x) = 0$, and if x is larger than the maximum range value, $F_X(x) = 1$. Certain random variables, like G above, have no maximum range value.

EXERCISES 6-2b

1. Let a card be drawn from a standard deck and let X be a random variable on the sample space where $X = 1$ if the card is a heart and $X = 0$ otherwise. Graph F_X.

2. A card is drawn from a standard deck, its suit noted, the card replaced, and the deck shuffled. This is done five times. Y counts the number of hearts. Graph F_Y. What are $F_Y(-3)$? $F_Y\left(\dfrac{5}{2}\right)$? $F_Y\left(\dfrac{15}{4}\right)$? $F_Y(6)$?

3. Suppose the experiment in Prob. 2 is repeated ten times, and Z counts the number of hearts. How does F_Z differ from F_Y of Prob. 2?

4. Let X count the number of heads when a coin is tossed 6 times. Let Y count the number of boys in families of six children. And let Z count the number of even faces which appear in 6 rolls of a die. Compare the distribution functions of X, Y, and Z. What are $F_X\left(\dfrac{3}{2}\right)$, $F_Y\left(\dfrac{3}{2}\right)$, $F_Z\left(\dfrac{3}{2}\right)$?

5. A die is rolled until a six appears. X counts the number of rolls. What is F_X? Graph F_X. What is $F_X\left(\dfrac{7}{2}\right)$? $F_X\left(\dfrac{41}{2}\right)$?

6. A coin is tossed until the second head appears and Y counts the number of tosses. What is F_Y? (See Prob. 11, Exercises 6-1b.)

7. A card is drawn from a standard deck without replacement until a heart appears. Z counts the number of draws. What is F_Z? (See Prob. 12, Exercises 6-1b).

6-3   EXPECTATION

a.   Definition

It is frequently useful to be able to describe a random variable in terms of a typical value. Three such measures are commonly used; they are the mode, the median, and the mean. The mode of a ramdom variable X is that value $x_m \in R_X$ such that $P_X(x_m) \geqslant P_X(x_j)$ for all $x_j \in R_X$; it is the range value that has the greatest probability. If more than one x_i has the same maximum probability, the mode can be taken as any one of them. The median is that $x_M \in R_X$ such that $P(X < x_M) \leqslant \dfrac{1}{2}$ and $P(X > x_M) \leqslant \dfrac{1}{2}$; said differently, the median x_M is that range value of X which roughly has one-half the distribution on both sides of it.

The most widely used of the typical values and the one which lends itself most readily to mathematical analysis is the expectation or mean.

6-3a_1   **Definition.**   *Let X be a random variable with range $R_X = \{x_1, x_2, x_3, \ldots\}$ and probability distribution P_X. The **expectation** (or mean) of X is the number.*

$$E(X) = \sum_{x_i \in R_X} x_i P_X(x_i).$$

Another common notation for $E(X)$ is μ_X. If R_X has more than a finite number of elements, it is conceivable that $E(X)$ would not be finite and in such a case we simply say the expectation is not finite. Fortunately this situation does not arise in elementary problems; in fact, it does not arise in most applied problems, so that in all discussions and theorems which follow we shall tacitly assume that $E(X)$ is finite.

Consider the random variable X whose range and distribution are given in the following table:

R_X	0	1	2	3	4
P_X	$\dfrac{16}{81}$	$\dfrac{32}{81}$	$\dfrac{24}{81}$	$\dfrac{8}{81}$	$\dfrac{1}{81}$

The mode of X is obviously one since $P_Y(1) \geqslant P_X(i)$, $i \neq 1$; also, since

$$P(X < 1) < \frac{1}{2} \text{ and } P(X > 1) < \frac{1}{2},$$

the median is also one. Computing the expectation, we obtain

$$E(X) = \sum_{i=0}^{4} iP_X(i) = 0\left(\frac{16}{81}\right) + 1\left(\frac{32}{81}\right) + 2\left(\frac{24}{81}\right) + 3\left(\frac{8}{81}\right) + 4\left(\frac{1}{81}\right)$$

$$= \frac{1}{81}(32 + 48 + 24 + 4) = \frac{108}{81} = \frac{4}{3}.$$

Their definitions ensure that the mode and median are range values of X, while this example shows that this is not necessarily true of the expectation.

The expectation of a random variable β having a Bernoulli distribution (see 6-1e_1) is

6-3a_2 $$E(\beta) = 1 \cdot p + 0(1 - p) = p.$$

A random variable B obeying the binomial law with parameters n and p (6-1e_2) has expectation

$$E(B) = \sum_{i=0}^{n} i\binom{n}{i}p^i(1 - p)^{n-i} = \sum_{i=1}^{n} i\binom{n}{i}p^i(1 - p)^{n-i}$$

since when $i = 0$, the term is zero. We know $i\binom{n}{i} = n\binom{n-1}{i-1}$ (see Prob. 2a

Exercises 4-2a), and so

$$i\binom{n}{i}p^i(1 - p)^{n-i} = n\binom{n-1}{i-1}p^i(1 - p)^{n-i} = np\binom{n-1}{i-1}p^{i-1}(1 - p)^{n-i}$$

With this

$$E(B) = \sum_{i=1}^{n} np\binom{n-1}{i-1}p^{i-1}(1 - p)^{n-i} = np\sum_{i=1}^{n}\binom{n-1}{i-1}p^{i-1}(1 - p)^{n-i}$$

$$= np[p + (1 - p)]^{n-1} = np.$$

We have just shown that a random variable B with a binomial distribution and parameters n and p has

6-3a_3 $$E(B) = np.$$

Thus any binomially distributed random variable has its expectation equal to the product of its parameters. A gambler who rolls a die twenty times and gets a dollar for every ace he rolls can expect to earn $(20) \left(\frac{1}{6} \right) = \frac{10}{3}$ dollars, since the amount he earns is a random variable with a binomial distribution having parameters $n = 20$ and $p = \frac{1}{6}$. Obviously the gambler will win an integral number of dollars (he cannot roll three and a third aces).

It is perhaps worth noting that the probability the gambler wins exactly three dollars is small; it is $P_X(3) = \frac{(1140)5^{17}}{6^{20}} = .2379$. It is obvious here that the number of aces the gambler actually rolls may differ widely from the expected number—the probability of his rolling 15 aces, though small, is still positive. Whenever we work with expected values, we must keep in mind that actual values may vary greatly from the computed mean. The expectation is only one characteristic of a random variable and as such cannot describe it completely.

*A geometrically distributed random variable G with parameter p has its expectation equal to $\frac{1}{p}$. The proof hinges on being able to calculate the sum $\sum_{i=1} ipq^{i-1}$. Using the Principle of Induction, it can be shown that

$$\sum_{i=1}^{n} ipq^{i-1} = \frac{1}{p} - \frac{q^n}{p} (np + 1).$$

Since the difference

$$\frac{1}{p} - \sum_{i=1}^{n} ipq^{i-1} = \frac{q^n}{p} (np + 1) = nq^n + \frac{q^n}{p}$$

can be made arbitrarily small simply by choosing n sufficiently large,[5] we define

$$\sum_{i=1} ipq^{i-1} = \frac{1}{p}.$$

Hence

6-3a_4
$$E(G) = \sum_{i=1} ip(1-p)^{i-1} = \frac{1}{p}.$$

Thus if we toss a coin until a head appears and Y counts the number of tosses, since Y is a random variable with parameter $\frac{1}{2}$, the expected number of tosses is 2.

---

* This paragraph can be omitted without loss of continuity.

b. Some Applications of Expectation

A produce vendor buys tomatoes by the bushel and by watching his sales over a period of time he has calculated that on any one day his probability of selling a given quantity of tomatoes is shown in the following table:

Bushels sold	0	1	2	3	4	5
Probability	.05	.15	.20	.35	.20	.05

If a bushel cost him $1.00 and he can sell it for $2.50, how many bushels should he stock to make the maximum gross product if, at the end of the day, he must destroy those which are not sold? Let X_i be the random variable which records the profit made if the vendor purchases i bushels of tomatoes. The random variables, their probability distributions, and their expectations are:

X_0	0					
P_{X_0}	1					

$E(X_0) = 0$

X_1	−1.00	1.50				
P_{X_1}	.05	.95				

$E(X_1) = 1.37$

X_2	−2.00	.50	3.00			
P_{X_2}	.05	.15	.80			

$E(X_1) = 2.37$

X_3	−3.00	−.50	2.00	4.50		
P_{X_3}	.05	.15	.20	.60		

$E(X_3) = 2.88$

X_4	−4.00	−1.50	1.00	3.50	6.00	
P_{X_4}	.05	.15	.20	.35	.25	

$E(X_4) = 2.55$

X_5	−5.00	−2.50	0	2.50	5.00	7.50
P_{X_5}	.05	.15	.20	.35	.20	.05

$E(X_5) = 1.62$

Thus the vendor will maximize his espected gross profit if he buys three bushels of tomatoes. (He can make more money if he buys more than three bushels and sells them all, but he is also apt to lose more money. In the long run he can expect a maximum return if he buys three bushels each day.)

There is, of course, a direct relation bet veen this formally defined expectation of a probability distribution and the quantity commonly referred to as an average or mean. A teacher, for example, gives a test to 100 students and after grading them, constructs the following table:

Grade	50	55	60	65	70	75	80	85	90	95	100
Number of Students	3	2	10	10	15	20	20	10	7	2	1

He then computes the "average" or "mean" grade.

$$A = \frac{1}{100}[3(50) + 2(55) + 10(60) + 10(65) + 15(70) + 20(75) + 20(80)$$

$$+ 10(85) + 7(90) + 2(95) + (100)]$$

$$= 74.2.$$

There is no significant difference between obtaining an average in this way and defining a random variable X on the sample space of test papers where X is the test score and P_X is the *a posteriori* probability distribution $\left(\text{that is if } x_i \in R_X,\right.$ $P_X(x_i) = \dfrac{\text{no. of } x_i \text{ scores}}{\text{no. of students}}\left.\right)$ and computing the expected value of X.

In general, any collection of numbers recorded with a frequency of occurrence (test scores, recruit heights, gross receipts, professional salaries) can be regarded as the range of a random variable with a given probability distribution. Thus, if on a given sample space the quantities $z_1, z_2, z_3, \ldots, z_n$ occur with respective frequencies $f_1, f_2, f_3, \ldots, f_n$ where $f_1 + f_2 + \cdots + f_n = N$, we have defined on the sample space a random variable Z assuming values z_i with probability $\dfrac{f_i}{N}$.

Hence

$$E(Z) = \sum_{i=1}^{n} z_i \frac{f_i}{N} = \frac{1}{N} \sum_{i=1}^{n} z_i f_i.$$

A statistician does exactly this when he computes the average of a sample.

EXERCISES 6-3b

NOTE: The first eight of these problems are identical with those in Exercises 6-1b. There we were interested in the range and probability distribution of the random variables. Here we want their expectations.

1. A ball is drawn from an urn containing 3 white, 2 red, and 4 blue balls. $\Omega = \{r, w, b\}$. Let X be a random variable on Ω where $X = \{(r, -1), (w, 0), (b, 1)\}$ Compute $E(X)$.

2. A card is drawn from a standard deck. Let Y and Z be two random variables on the sample space Ω with $Y = \{(C, 1), (D, 2), (H, 3), (S, 4)\}$ (C, D, H, S are the suits) and $Z(j) = j$ where j is the denomination of the card ($j = 1, 2, \ldots, 13$). Compute $E(Y)$ and $E(Z)$.

3. Nine disks each bearing one of the digits $1,2,\ldots,9$ are put into an urn. One is drawn out, its number noted, and returned to the urn. This is done four times so that a four-digit number is generated. For $(a,b,c,d) \in \Omega$, let

$$X((a,b,c,d)) = a.$$

Compute $E(X)$.

4. At a certain camp there are 15 junior and 25 senior campers. Each day seven are chosen by lot for KP duty and Y counts the number of juniors in the group. Compute $E(Y)$.

5. Three indistinguishable dice are rolled and their faces recorded. Describe a probability space. Let X, Y, Z be random variables on Ω where if ω is the unordered triple (d_1, d_2, d_3), $X(\omega) = d_1 + d_2 + d_3$, $Y(\omega) = \max (d_1, d_2, d_3)$, $Z(\omega) = \min (d_1, d_2, d_3)$. [By max (a,b,c) is meant that value a, b, or c which is at least as large as the other two; min (a,b,c) is analogously defined.] Compute $E(X)$, $E(Y)$ and $E(Z)$.

6. A bridge hand is dealt. X counts the number of kings, Y the number of hearts, and Z the number of black cards. Compute $E(X)$, $E(Y)$, and $E(Z)$.

7. A die is rolled until an ace appears. Let X be the random variable that counts the number of rolls. Compute $E(X)$.

8. A machine is known to be 95 percent reliable. An inspector tests 100 samples of its output. Let Y count the number of substandard items in the sample. Compute $E(Y)$.

9. In 550 rolls of die the faces appeared with the following frequencies:

Face	1	2	3	4	5	6
Frequency	90	76	96	97	88	103

On $\Omega = \{1,2,3,4,5,6\}$ let $X(i) = i$ and $P_X(i) = $ frequency/550. Compute $E(X)$.

10. The usual roulette wheel has 38 numbered positions 00, $0,1,2,\ldots,36$. The person betting on the winning number has his bet returned plus 35 times as much more.

(a) What is the expected take of the house on a one-dollar bet on number 23?

(b) Of the 38 positions two are green, 18 red, and 18 black. A player can bet on red or black but not green. A man playing the winning color gets back his bet plus a matching amount. What is the value to the house of a one dollar bet on red?

(c) What are the expected earnings of a player who bets one dollar on 23 which is red and one dollar on black?

11. On entering a certain store customers wait a total of t minutes with probability $p(t)$ given in the following table:

Length of wait	0	1	2	3	4	5	6
Probability	.03	.12	.20	.35	.15	.10	.05

What is a customer's expected wait?

12. A submarine search cost the Navy \$5,000 a day. They have found that the length of times needed to locate an enemy submarine known to be off the coast of Florida varies according to the accompanying table:

Number of days out	1	2	3	4	5 (return without success)
Probability	.15	.20	.25	.15	.25

What is the expected cost of a search mission?

c.   Functions Of Random Variables

Since the range values of random variables are real numbers, it is meaningful to talk about algebraic functions of random variables. Two of particular interest in our work are the linear and quadratic functions of a random variable. Let us start with a random variable X on a sample space Ω. For any two real numbers a and b, we define a new random variable as

$$aX + b = \{(\omega, aX(\omega) + b) | \omega \in \Omega\}$$

which is a linear function of X and is such that if

$$(\omega, x) \in X, \ (\omega, ax + b) \in aX + b.$$

As a simple example let $\Omega = \{p, q, r, s\}$ and $X = \{(p,-2),(q,0),(r,1),(s,2)\}$. A linear function of X is the random variable

$$3X - 1 = \{(p,-7),(q,-1),(r,2),(s,5)\}$$

where, since $(p,-2) \in X$, $(p, 3(-2)-1) = (p,-7) \in 3X - 1$.

A second frequently used random variable derived from X is

$$X^2 = \{[\omega, (X(\omega))^2] | \omega \in \Omega\}.$$

Using the same Ω and X we defined above,

$$X^2 = \{(p,4),(q,0),(r,1),(s,4)\}$$

Here, since $(p,-2) \in X$, $(p,(-2)^2) = (p,4) \in X^2$.

More generally if f is any algebraic function and X is a random variable on a sample space Ω, we can create another random variable $f(X)$ on Ω, where

$$f(X) = \{|\omega, f(X(\omega))]| \omega \in \Omega\}$$

is such that if $(\omega, x) \in X$, $(\omega, f(x)) \in f(X)$ provided $f(X)$ is defined for all $x \in R_X$. Returning again to the random variable $X = \{(p,-2),(q,0),(r,1),(s,2)\}$ we cannot form a random variable $\dfrac{1}{X}$ since $\dfrac{1}{X(q)} = \dfrac{1}{0}$ is not defined; nor can we form $\sqrt{X}$ since $\sqrt{X(p)} = \sqrt{-2}$ is not defined. However, $\dfrac{1}{X}$ is defined for any random variable whose range does not contain zero and $\sqrt{X}$ is defined for any random variable whose range is nonnegative.

Since $aX + b$, X^2, and $f(X)$ are random variables we can talk about their expectations. If X is a random variable, by definition 6-3a_1,

$$E(X) = \sum_{x_i \in R_X} x_i\, P_X(x_i).$$

In order to suggest a theorem for $E(f(X))$, let us look at $E(X^2)$. The range and probability distribution of a random variable X are given by

R_X	-2	-1	0	1	2
P_X	$\dfrac{1}{4}$	$\dfrac{1}{2}$	$\dfrac{1}{12}$	$\dfrac{1}{12}$	$\dfrac{1}{12}$

Consider the sum

$$\sum_{x_j \in R_X} x_j^2 P_X(x_j) = 4\left(\frac{1}{4}\right) + 1\left(\frac{1}{2}\right) + 0\left(\frac{1}{2}\right) + 1\left(\frac{1}{12}\right) + 4\left(\frac{1}{12}\right) = \frac{23}{12}.$$

Look now at X^2:

R_{X^2}	0	1	4
P_{X^2}	$\dfrac{1}{12}$	$\dfrac{7}{12}$	$\dfrac{4}{12}$

$$E(X^2) = \sum_{\xi_k \in R_{X^2}} \xi_k P_{X^2}(\xi_k) = 0\left(\frac{1}{12}\right) + 1\left(\frac{7}{12}\right) + 4\left(\frac{4}{12}\right) = \frac{23}{12} = \sum_{x_j \in R_X} x_j^2 P_X(x_j).$$

Let us look at $E(X^2)$ more closely.

$$E(X^2) = \sum_{\xi_k \in R_X} \xi_k P_{X^2}(\xi_k)$$

$$= \sum_{\xi_k \in R_{X^2}} \xi_k P(X^2 = \xi_k)$$

$$E(X^2) = \sum_{\xi_k \in R_{X^2}} \xi_k P(X = \sqrt{\xi_k} \text{ or } X = -\sqrt{\xi_k})$$

$$= \sum_{\xi_k \in R_X} \xi_k \left[P(X = \sqrt{\xi_k}) + P(X = -\sqrt{\xi_k}) \right]$$

$$= \sum_{\xi_k \in R_{X^2}} \xi_k P_X(\sqrt{\xi_k}) + \sum_{\xi_k \in R_{X^2}} \xi_k P_X (-\sqrt{\xi_k}).$$

Now if $P_X(\sqrt{\xi_k}) \neq 0$, $\sqrt{\xi_k} = x_j \in R_X$ such that $x_j \geqslant 0$ and $x_j^2 = \xi_k$. Also if $P_X(-\sqrt{\xi_k}) \neq 0, -\sqrt{\xi_k} = x_i \in R_X$ such that $x_i < 0$ and $x_i^2 = \xi_k$. Hence

$$E(X^2) - \sum_{\genfrac{}{}{0pt}{}{x_j \in R_X}{x_j \geqslant 0}} x_j^2 P_X(x_i) + \sum_{\genfrac{}{}{0pt}{}{x_i \in R_X}{x_i < 0}} x_i^2 P_X(x_i)$$

$$= \sum_{x_k \in R_X} x_k^2 P_X(x_k).$$

This suggest that if f is an algebraic function and X is a random variable, for the expectation of the random variable $f(X)$ we would have

$$E(f(X)) = \sum_{x_j \in R_X} f(x_j) P_X(x_j).$$

In what follows we will always assume that the sum under consideration exists; it will do this if $f(x_j)$ is defined. We will consider no random variables $f(X)$ where one or more of the $f(x_j), x_j \in R_X$, are meaningless. Before stating and proving the theorem we remark that if $f_i \in R_{f(X)}$, then

$$P_{f(X)}(f_i) = P(f(X) = f_i)$$

$$= P\left(\sum_{f(x_j) = f_i} \{X = x_j\} \right)$$

$$= \sum_{f(x_j) = f_i} P_X(x_j).$$

We are ready to prove

6-3c$_1$ Theorem. *If f is an algebraic function and X is a random variable.*

$$E(f(X)) = \sum_{x_j \in R_X} f(x_j) P_X(x_j).$$

Proof: By definition,

$$E(f(X)) = \sum_{f_i \in R_{f(X)}} f_i P_{f(X)}(f_i)$$

$$= \sum_{f_r \in R_{f(X)}} f_i \left(\sum_{f(x_j) = f_i} P_X(x_j) \right)$$

$$= \sum_{f_i \in R_{f(X)}} \left(\sum_{f(x_j) = f_i} f(x_j) P_X(x_j) \right)$$

Now as we sum over all possible values f_i in $R_{f(X)}$, we sum over all $x_j \in R_X$, and so

the double summation can be written as $\sum\limits_{x_j \in R_X} f(x_j) \, P_X(x_j)$ and we have

$$E(f(X)) = \sum_{x_j \in R_X} f(x_j) \, P_X(x_j).$$

Our theorem is proved. ∎

From this we immediately obtain

6-3c₂ Corollary $E(aX + b) = aE(X) + b.$

Proof: $E(aX + b) = \sum\limits_{x_j \in R_X} (ax_j + b) \, P_X(x_j)$

$$= \sum_{x_j \in R_X} (ax_j \, P_X(x_j) + b \, P_X(x_j))$$

$$= \sum_{x_j \in R_X} ax_j \, P_X(x_j) + \sum_{x_j \in R_X} b P_X(x_j)^{[3]}$$

$$= a \sum_{x_j \in R_X} x_j P_X(x_j) + b \sum_{x_j \in R_X} P_X(x_j)$$

$$= aE(X) + b. \qquad \blacksquare$$

Using this corollary we have $E(X - E(X)) = E(X) - E(X) = 0$; that is the expected value of the difference between a random variable and its expectation is zero. The quantity $E(X - E(X))$ is called the *first moment* of X about its mean. Readers familiar with mechanics will readily see the analog between the mean of a random variable and the center of gravity of a unit mass distributed along the x-axis where the amount m_i is concentrated at the point x_i ($\sum_{i=1} m_i = 1$). $\mu_X = \sum_{i=1} x_i m_i$ is the center of gravity, and if a fulcrum is placed at μ_X the system will be in equilibrium, that is $\mu_{(X - \mu_X)} = \sum_{i=1} (x_i - \mu_X) m_i - 0$.

Higher moments are also defined. For any positive integer n the nth moments of X about the origin is the expectation of X^n, $E(X^n) = \sum\limits_{x_i \subset R_X} x_i^n P_X(x_i)$, and the nth moment of X about its mean is $E[(X - E(X))^n]$. Moments of all orders are important in the general theory, but in this introduction we will have use only for the second.

In addition to algebraic function of a single real variable we can consider algebraic combinations of two or more random variables defined on the same sample space. Let X and Y be random variables defined on a sample space Ω and let their ranges, probability distributions and joint distributions be those in Figure 6-3c₃. We can now define two other random variables on Ω, $X + Y$, and XY.

$$X + Y = \{(\omega, x_i + y_j) | X(\omega) = x_i, Y(\omega) = y_j\}$$

$$XY = \{(\omega, x_i y_j) | X(\omega) = x_i, Y(\omega) = y_j\}.$$

6-3c₃ **Figure**

Joint distribution table of the random variables X and Y.

Y \ X	0	1	2	3	P_Y
0	$\dfrac{1}{6}$	0	$\dfrac{1}{12}$	0	$\dfrac{3}{12}$
1	$\dfrac{1}{12}$	$\dfrac{1}{4}$	0	$\dfrac{1}{12}$	$\dfrac{5}{12}$
2	0	$\dfrac{3}{12}$	0	$\dfrac{1}{12}$	$\dfrac{4}{12}$
P_X	$\dfrac{3}{12}$	$\dfrac{6}{12}$	$\dfrac{1}{12}$	$\dfrac{2}{12}$	1

For X and Y in the figure, we have

$$R_{X+Y} = \{0,1,2,3,4,5\}$$

$$P_{X+Y} = \left\{ \left(0,\frac{2}{12}\right), \left(1,\frac{1}{12}\right), \left(2,\frac{4}{12}\right), \left(3,\frac{3}{12}\right), \left(4,\frac{1}{12}\right), \left(5,\frac{1}{12}\right) \right\}$$

where, for example,

$$P_{X+Y}(2) = P(X + Y = 2) = P(X = 2, Y = 0) + P(X = 1, Y = 1) + P(X = 0, Y = 2)$$

$$= \frac{1}{12} + \frac{1}{4} + 0 = \frac{4}{12}.$$

Also,

$$R_{XY} = \{0,1,2,3,4,6\}$$

$$P_{XY} = \left\{ \left(0,\frac{4}{12}\right), \left(1,\frac{3}{12}\right), \left(2,\frac{3}{12}\right), \left(3,\frac{1}{12}\right), \left(4,0\right), \left(6,\frac{1}{12}\right) \right\}$$

where, for example

$$P_{XY}(2) = P(XY = 2) = P(X = 1, Y = 2) + P(X = 2, Y = 1) = \frac{3}{12} + 0 = \frac{3}{12}.$$

Since X + Y and XY are random variables we can compute their expectations:

$$E(X + Y) = 0 \cdot \frac{2}{12} + 1 \cdot \frac{1}{12} + 2 \cdot \frac{4}{12} + 3 \cdot \frac{3}{12} + 4 \cdot \frac{1}{12} + 5 \cdot \frac{1}{12} = \frac{27}{12}$$

and
$$E(XY) = 0 \cdot \frac{4}{12} + 1 \cdot \frac{3}{12} + 2 \cdot \frac{3}{12} + 3 \cdot \frac{1}{12} + 4 \cdot 0 + 6 \cdot \frac{1}{12} = \frac{18}{12}.$$

We can easily verify that for this X and this Y, $E(X + Y) = E(X) + E(Y)$.

For any two random variables X and Y on the same sample space Ω, we can define, as above, the random variables:

$$X + Y = \{(\omega, x_i + y_j) | X(\omega) = x_i, Y(\omega) = y_j\}$$
and
$$XY = \{(\omega, x_i \, y_j) | X(\omega) = x_i, Y(\omega) = y_j\}.$$

The range of $X + Y$ is a set of numbers z_k where each z_k is the sum of one or more pairs x_i, y_j and the range of XY is a set of numbers ζ_k where each ζ_k is the product of one or more pairs x_i, y_j. Then

$$P_{X+Y} = \{(z_k, P(X + Y = z_k)) | z_k \in R_{X+Y}\}$$
and
$$P_{XY} = \{(\zeta_k, P(XY = \zeta_k)) | \zeta_k \in R_{XY}\}.$$

Furthermore,

$$P_{X+Y}(z_k) = P(X + Y = z_k) = P(\{\omega | X(\omega) + Y(\omega) = z_k\})$$

$$= P\left(\sum_{x_i + y_j = z_k} \{\omega | X(\omega) = x_i, Y(\omega) = y_j\} \right)$$

$$= \sum_{x_i + y_j = z_k} P(X = x_i, \, Y = y_j)$$

$$= \sum_{x_i + y_j = z_k} p_{ij}.$$

Similarly,

$$P_{XY}(\zeta_k) = P(XY = \zeta_k) = \sum_{x_i y_j = \zeta_k} p_{ij}.$$

The meaning of the notation becomes clear if we look at the joint distribution table of X and Y,

R_X $\backslash$ R_Y	x_1	x_2	x_3	$\cdots$
y_1	p_{11}	p_{12}	p_{13}	$\cdots$
y_2	p_{21}	p_{22}	p_{23}	$\cdots$
y_3	p_{31}	p_{32}	p_{33}	$\cdots$
$\vdots$	$\vdots$	$\vdots$	$\vdots$	

If $x_2 + y_3 = x_1 + y_5 = x_7 + y_2 = z_3$, then

$$P_{X+Y}(z_3) = p_{32} + p_{51} + p_{27};$$

and if $x_2 y_3 = x_4 y_2 = x_5 y_6 = \zeta_8$, then

$$P_{XY}(\zeta_8) = p_{32} + p_{24} + p_{65}.$$

Using Definition 6-3a_1,

$$E(X + Y) = \sum_{z_k \in R_{X+Y}} z_k P_{X+Y}(z_k)$$

$$= \sum_{z_k \in R_{X+Y}} z_k \left(\sum_{x_i + y_j = z_k} p_{ij} \right)$$

$$= \sum_{z_k \in R_{X+Y}} \left(\sum_{x_i + y_j = z_k} (x_i + y_j)p_{ij} \right).$$

Since every pair $(x,y) \in R_X \times R_Y$ sums to one and only one z this double sum can be replaced by

6-3c_4 $$E(X + Y) = \sum_{x_i \in R_X} \sum_{y_j \in R_Y} (x_i + y_j)p_{ij}.$$

Similarly from

$$E(XY) = \sum_{\zeta_k \in R_{XY}} \zeta_k P_{XY}(\zeta_k)$$

we can derive

6-3c_5 $$E(XY) = \sum_{x_i \in R_X} \sum_{y_j \in R_Y} x_i y_j p_{ij}.$$

In our numerical example we showed that $E(X + Y) = E(X) + E(Y)$, and we now prove that this is true for any X and Y.

6-3c_6 **Theorem.** $E(X + Y) = E(X) + E(Y)$.

Proof: $$E(X + Y) = \sum_{x_i \in R_X} \sum_{y_j \in R_Y} (x_i + y_j)p_{ij}$$

$$= \sum_{x_i \in R_X} \sum_{y_j \in R_Y} (x_i p_{ij} + y_j p_{ij})$$

$$= \sum_{x_i \in R_X} x_i \sum_{y_j \in R_Y} p_{ij} + \sum_{y_j \in R_Y} y_j \sum_{x_i \in R_X} p_{ij}$$

$$= \sum_{x_i \in R_1} x_i P_X(x_i) + \sum_{y_j \in R_Y} y_j P_Y(y_j)$$

$$= E(X) + E(Y)$$

where again the rearrangement of the terms is permissible since we assume $E(X)$ and $E(Y)$ to be finite. ∎

6-3c₇ Corollary. $E(X_1 + X_2 + \cdots + X_n) = E(X_1) + E(X_2) + \cdots + E(X_n).$

A die is rolled twice, what is the expectation of the sum of the faces? The sample space Ω is the cross-product of $D = \{1,2,3,4,5,6\}$ with itself and let X and Y be two random variables defined on Ω such that if $(a,b) \in \Omega$, $X((a,b)) = a$ and $Y((a,b)) = b$. $R_X = R_Y = D$ and for $i \in D$, $P_X(i) = P_Y(i) = \dfrac{1}{6}$.   Further

$$E(X) = E(Y) = \frac{1}{6}(1 + 2 + 3 + 4 + 5 + 6) = \frac{7}{2}$$

and

$$E(X) + E(Y) = \frac{7}{2} + \frac{7}{2} = 7 = E(X + Y)$$

which says that the expectation of the sum the faces is seven.

A less trivial example is the following.   Repeat a Bernoulli experiment with $T = \{S, F\}$ n times so that $\Omega = (X\ T)_n$ and define n random variables X_i so that $X_i = 1$ if the ith trial results in S and $X_i = 0$ if the ith trial results in F.   Each X_i satisfies the Bernoulli distribution with parameter p and for each i, $E(X_i) = p$

(6-3a₂). Hence the $\displaystyle\sum_{i=1}^{n} E(X_i) = np$. Moreover $\displaystyle\sum_{i=1}^{n} X_i$ is also defined on Ω and it counts the number of S's in each ω.   Hence the random variable $\displaystyle\sum_{i=1}^{n} X_i$ has the binomial distribution with parameters n and p and from 6-3a₃ we know $E\left(\displaystyle\sum_{i=1}^{n} X_i\right) = np = \displaystyle\sum_{i=1}^{n} E(X_i).$

If we try to prove an analogous theorem for the expectation of the product of two random variables,—that is, if we try to prove $E(XY) = E(X)E(Y)$, we immediately run into difficulty.   We find

$$E(XY) = \sum_{x_i \in R_X}\ \sum_{y_j \in R_Y} x_i y_j p_{ij} = \sum_{x_i \in R_X} x_i \left(\sum_{y_j \in R_Y} y_j p_{ij}\right)$$

and what can we do with $\displaystyle\sum_{y_j \in R_Y} y_j p_{ij}$?   At this point, nothing.

EXERCISES 6-3c

Return to Exercises 6-3b.   The problem numbers here are the same as those in that set and, for example, the random variable X referred to in the first exercise here is described in problem one of Exercises 6-3b.

1.   Compute $E(X^2)$ and $E(X^3)$
2.   Compute $E(Y + Z)$ and $E(Y^2)$.
3.   Compute $E(X^2)$.
4.   Compute $E(2Y - 3)$.
5.   Compute $E(Z^2 - E(Z))$ and $E[(X - E(X)^2)]$.

6-4   VARIANCE

a.   Definition

6-4a$_1$ Definition.   *The second moment of a random variable X about its mean is called the **variance** of X.   In symbols,*

$$\sigma^2 = V(X) = E[(X - E(X))^2]$$

*σ, the nonnegative square root of the variance, is called the **standard deviation** of X.*

As we did with the expectation, we shall assume that all sums involved exist without explicitly stating so.

Since we know $E(aX + b) = aE(X) + b$ (6-3c$_2$),

$$E((X - E(X)^2)) = E[X^2 - 2XE(X) + (E(X))^2]$$
$$= E(X^2) - 2E[X(E(X))] + E[(E(X))^2]$$
$$= E(X^2) - 2E(X)E(X) + (E(X))^2$$
$$= E(X^2) - (E(X))^2,$$

we have

6-4a$_2$ $V(X) = E(X^2) - (E(X))^2.$

A random variable β distributed according to the Bernoulli law has

$$E(\beta) = 0(1 - p) + 1p = p$$
$$E(\beta^2) = 0(1 - p) + 1^2p = p$$
$$E(\beta^2) - (E(\beta))^2 = p - p^2$$

and so

6-4a₃ $$V(\beta) = p(1 - p).$$

If X is a random variable with $R_X = \{1,2,3,4,5,6\}$ and $P_X(i) = \dfrac{1}{6}$ for $i \in R_X$,

then $E(X) = \dfrac{7}{2}$, $E(X^2) = \dfrac{91}{6}$ and $V(X) = E(X^2) - (E(X))^2 = \dfrac{91}{6} - \left(\dfrac{7}{2}\right)^2 = \dfrac{35}{12}$.

b. Properties of the Variance

Since by definition $V(X) = E[(X - E(X))^2] = \sum_i (x_i - E(X))^2 P_X(x_i) \geqslant 0$, the variance serves as a measure of dispersion. If $V(X)$ is small, each term is small and so large fluctuations from the expectation occur with small probability; while if it is large, the random variable takes values which differ appreciably from the mean with large probability.

The difference in spread of two random variables having the same expectation is illustrated by X_1 and X_2 whose ranges and probability distributions are given in the following tables.

X_1:

R_{X_1}	0	1	2	3	4
P_{X_1}	$\dfrac{1}{9}$	$\dfrac{2}{9}$	$\dfrac{1}{9}$	$\dfrac{3}{9}$	$\dfrac{2}{9}$

$E(X_1) = \dfrac{7}{3}$; $V(X_1) = \dfrac{16}{9} = 1.78$

X_2:

R_{X_2}	-3	-1	2	4	6
P_{X_2}	$\dfrac{2}{9}$	$\dfrac{1}{9}$	$\dfrac{1}{9}$	$\dfrac{2}{9}$	$\dfrac{3}{9}$

$E(X_2) = \dfrac{7}{3}$; $V(X_2) = \dfrac{114}{9} = 12.67$

In X_2 the range values vary from -3 to 6 while in X_1 they go only from 0 to 4, and although this is not evidenced in their expectations, it is clearly reflected in their variances.

We know, by 6-3c₂, that $E(aX + b) = aE(X) + b$, and this along with the definition of $V(X)$ yields

6-4b₁ $$V(aX + b) = a^2 V(X).$$

To find a counterpart of 6-3c₆, $E(X + Y) = E(X) + E(Y)$ for the variance, let us start with the definition. In what follows we use repeatedly the two properties of the expectation just mentioned,

$$E(X + Y) = E(X) + E(Y) \text{ and } E(aX + b) = aE(X) + b:$$

$$V(X + Y) = E((X + Y)^2) - (E(X + Y))^2$$
$$= E(X^2 + 2XY + Y^2) - (E(X) + E(Y))^2$$
$$= E(X^2) + 2E(XY) + E(Y^2) - [(E(X))^2 + 2E(X)E(Y) + (E(Y))^2]$$
$$= [E(X^2) - (E(X))^2] + [E(Y^2) - (E(Y))^2] + 2[E(XY) - E(X)E(Y)]$$
$$= V(X) + V(Y) + 2[E(XY) - E(X)E(Y)].$$

We call the quantity $E(XY) - E(X)E(Y)$ which appears here for the first time the covariance of X and Y; formally,

6-4b_2 Definition.   *The* **covariance** *of X and Y is the quantity*

$$E(XY) - E(X)E(Y) = \text{cov}(X,Y).$$

It exists whenever $E(X)$ and $E(Y)$ do. With this notion of covariance and the calculations above, we have proved:

6-4b_3 Theorem. $V(X + Y) = V(X) + V(Y) + 2\,\text{cov}(X,Y).$

If X and Y have identical Bernoulli distributions then $P_X(1) = P_Y(1) = p$, $0 < p < 1$, $P_X(0) = P_Y(0) = 1 - p$ and $E(X) = E(Y) = p$ (see 6-3a_2). Their joint distribution is given in Figure 6-4b_4.

6-4b_4 *Joint distribution table of X and Y*

$(q = 1 - p)$

R_Y \ R_X	0	1	P_Y
0	q^2	pq	q
1	pq	p^2	p
P_X	q	p	1

Then

$$E(XY) = 0 \cdot (q^2 + pq + pq) + 1 \cdot p^2 = p^2$$

and

$$\text{cov}(X,Y) = E(XY) - E(X)E(Y) = p^2 - p^2 = 0.$$

This last calculation together with the fact that Bernoulli distributed random variables have variance $p(1 - p)$ (see 6-4a_3) yields

$$V(X + Y) = 2p(1 - p).$$

To take this last example out of the abstract consider rolling a die twice and construct the usual probability space. Let X equal one if the first roll yields a four and equal zero otherwise and let Y equal one if the second roll yields a five and equal zero otherwise. X and Y are identically distributed random variables obeying a Bernoulli law with parameter $\frac{1}{6}$. The joint distribution of X and Y is given in Figure 6-4b_4 with $p = \frac{1}{6}$, $q = \frac{5}{6}$. We have

$$E(X) = E(Y) = \frac{1}{6}$$

$$V(X) = V(Y) = \frac{1}{6} \cdot \frac{5}{6} = \frac{5}{36}$$

$$V(X + Y) = 2\left(\frac{5}{36}\right) = \frac{5}{18}.$$

6-4b_5 Corollary.

$$V(X_1 + X_2 + \cdots + X_n) = \sum_{i=1}^{n} V(X_i) + 2 \sum_{i \neq j} \text{cov}\,(X_i, X_j).$$

Let an experiment consist of a sequence of Bernoulli trials and let X_k be a random variable on the sample space Ω, where for $\omega \in \Omega$, $X_k(\omega) = 1$ or $X_k(\omega) = 0$ depending upon whether the kth trial results in success or failure. For every k let $P_{X_k}(1) = p$, $P_{X_k}(0) = 1 - p$ so that, again for all k,

$$E(X_k) = p, \ V(X_k) = p(1 - p);$$

and for all possible pairs i,j let the joint distribution of X_i and X_j be that in Figure 6-4b_4. Let S be another random variable on Ω which counts the number of successes in n trials; $S = \sum_{i=1}^{n} X_i$. Since for these trials we have

$\text{cov}\,(X_k X_j) = 0$, $V(S) = \sum_{i=1}^{n} V(X_i) = np(1 - p)$. Furthermore S, as defined, is a binomial distribution with parameters n and p; hence the last result

$$V(S) = np(1 - p)$$

gives the variance of a binomially distributed random variable with those parameters. This same result can be computed directly from the definition.

If S is a random variable obeying the binomial law with parameters n and p,

$$V(S) = E(S^2) - (E(S))^2$$

$$= \sum_{k=0}^{n} k^2 \binom{n}{k} p^k (1-p)^{n-k} - (np)^2$$

$$= np[(n-1)p + 1] - (np)^2.$$

and in both cases we have for a binomially distributed random variable S with parameters n and p

6-4b_6 $$V(S) = np(1-p)$$

A certain student population* consists of b boys and g girls and from it a sample of size n is taken where once a student is selected for the sample he is not returned to the population (a sample without replacement). What is the expected number of boys in the sample and what is its variance? To answer these questions we can either consider a random variable Z which counts the number of boys in the sample of size n or we can define n random variables, X_k, where X_k equals zero or one depending upon whether the kth selection is a girl or a boy. X_k is then a Bernoulli trial and $E(X_k) = P_{X_k}(1)$, and, furthermore, we have

$$E\left(\sum_{k=1}^{n} X_k \right) = \sum_{k=1}^{n} (E(X_k)) = E(Z).$$ The number of points in a sample of size n is $(b + g)_n$ and assuming no prejudice in their selection

$$P_Z(k) = \binom{n}{k} \frac{(b)_k \, (g)_{n-k}}{(b+g)_n}$$

and

$$E(Z) = \sum_{k=0}^{n} k \binom{n}{k} \frac{(b)_k \, (g)_{n-k}}{(b+g)_n} = \frac{nb}{b+g} \sum_{k=1}^{n} \binom{n-1}{k-1} \frac{(b-1)_{k-1} \, (g)_{n-k}}{(b+g-1)_{n-1}}$$

(see Prob. 2a, Exercises 4-2a). Our problem is reduced to evaluating the last sum. To find $P_{X_k}(1)$ we observe that k students can be selected in $(b + g)_k$ ways and for $j < k$, the probability of j girls and $k - j - 1$ boys on the first $k - 1$ choices is $\binom{k-1}{j} \frac{(b)_{k-1-j}(g)_j}{(b+g)_{k-1}}$ and so

$$P_{X_k}(1) = \sum_{j=0}^{k-1} \binom{k-1}{j} \frac{(b)_{k-j-1}(g)_j}{(b+g)_{k-1}} \frac{b-k+j+1}{(b+g-k+1)}$$

$$= \frac{b}{b+g} \sum_{j=0}^{k-1} \binom{k-1}{j} \frac{(b-1)_{k-j-1}(g)_j}{(b+g-1)_{k-1}}.$$

*This paragraph contains an example involving some elaborate calculations. It may be omitted without loss of continuity.

The sum in $P_{X_k}(1)$ is exactly the same as the one appearing in $E(Z)$ and *if* that sum

is one, $P_{X_k}(1) = \dfrac{b}{b + g}$ for every k and

$$E(X_k) = P_{X_k}(1) = \frac{b}{b + g} \quad \text{and } E\left(\sum_{k=1}^{n} X_k\right) = \sum_{k=1}^{n} E(X_k) = \frac{nb}{b + g} = E(Z).$$

That this sum is one is proved in a note at the end of this chapter. Since

$$P_{X_k}(1) = \frac{b}{b + g}$$

for every k, we know too that $V(X_k) = \dfrac{bg}{(b + g)^2}$ and that

$$V\left(\sum_{k=1}^{n} X_k\right) = \sum_{k=1}^{n} V(X_k) + 2 \sum_{i \neq j} \operatorname{cov}(X_i, X_j)$$

$$= \frac{nbg}{(b + g)^2} + 2 \sum_{i \neq j} \operatorname{cov}(X_i, X_j).$$

Cov $(X_i, X_j) = E(X_i X_j) - E(X_i)E(X_j)$. To find $E(X_i X_j)$ we construct the table of
their joint distributions and observe that we really only need to know that
$P(X_i X_j = 1) = P(X_i = 1, X_j = 1)$. Set $(b + g)(b + g - 1) = q$, then

R_{X_j} \ R_{X_i}	0	1	P_{X_j}
0			$\dfrac{g}{b + g}$
1		$\dfrac{b(b - 1)}{q}$	$\dfrac{b}{b + g}$
P_{X_i}	$\dfrac{g}{b + g}$	$\dfrac{b}{b + g}$	

We know $P(X_i = 1) = \dfrac{b}{b + g}$ and having eliminated the ith trial from considera-

tion for all $j \neq i$, if $X_i = 1$, the space reduces to $b + g - 1$ elements of which $b - 1$

are boys and so, knowing $X_i = 1$, the probability X_j also equals 1 is $\dfrac{b - 1}{b + g - 1}$, and

$$P(X_i = 1, X_j = 1) = \frac{b(b - 1)}{(b + g)(b + g - 1)}.$$

(The simplest method of computing the other entries in the joint distribution table is to use the marginal distributions; for example, $P(X_i = 1, X_j = 0) = P_{X_i}(1) - P(X_i = 1, X_j = 1))$. Hence

$$E(X_iX_j) = \frac{b(b-1)}{(b+g)(b+g-1)}$$

and

$$\operatorname{cov}(X_i, X_j) = E(X_iX_j) - E(X_i)E(X_j)$$

$$= \frac{b(b-1)}{(b+g)(b+g-1)} - \frac{b^2}{(b+g)^2} = \frac{-bg}{(b+g)^2(b+g-1)}.$$

Hence

$$V\left(\sum_{k=1}^{n} X_k\right) = \frac{nbg}{(b+g)^2} - 2\binom{n}{2}\frac{bg}{(b+g)^2(b+g-1)}$$

$$= \frac{nbg}{(b+g)^2}\left\{1 - \frac{n-1}{b+g-1}\right\}.$$

c. Chebyschev's[*] Inequality

A relation of theoretical interest between the mean and variance of a random variable is given by *Chebyshev's Inequality*. Its practicality is limited; its importance lies in its application to all random variables.

6-4c_1   **Theorem.** (*Chebyshev's Inequality*). *Let X be a random variable. Then for any real number r,*

$$P(|X - E(X)| > r) \leqslant \frac{V(X)}{r^2}$$

Proof:

$$V(X) = \sum_{i=1}(x_i - E(X))^2 P_X(x_i)$$

$$\geqslant \sum_{|x_i-E(X)|>r}(x_i - E(X))^2 P_X(x_i)$$

$$\geqslant r^2 \sum_{|x_i-E(X)|>r} P(x_i)$$

$$= r^2 P\{|X - E(X)| > r\}.$$

To get the first inequality we dropped from the sum those terms in which $|x_i - E(X)| \leqslant r$ and to get the second we replaced every $|x_i - E(X)|$ by r itself; noting also that $\sum\limits_{|x_i-E(X)|>r} P(x_i) = P\{|X - E(X)| > r\}$ our proof is complete. ∎

*P. L. Chebyshev (1821–1894).

Variance is by definition the second moment of the random variable about its mean and as such it is a measure of spread. Chebyschev's Inequality gives us a guide to the likelihood that the random variable wanders far from its expectation. If for any random variable X we let r equal two standard deviations $(r = 2\sigma = 2\sqrt{V(X)})$, the inequality gives us

$$P\{|X - E(X)| > 2\sigma\} \leqslant \frac{V(X)}{4\sigma^2} = \frac{1}{4}.$$

That is, the probability of the set of range values of a random variable that differ from its mean by more than two standard deviations is less than one fourth.

6-5 NOTE: Proof that

$$\sum_{j=0}^{n} \binom{n}{j} \frac{(b)_j (g)_{n-j}}{(b + g)_n} = 1.$$

Our proof uses the Principle of Induction. If $n = 1$,

$$\sum_{j=0}^{1} \binom{1}{j} \frac{(b)_j (g)_{1-j}}{(b + g)} = \binom{1}{0} \frac{g}{b + g} + \binom{1}{1} \frac{b}{b + g} = \frac{b + g}{b + g} = 1.$$

Hence the statement is true for $n = 1$.

Let k be an integer such that

$$\sum_{j=0}^{k} \binom{k}{j} \frac{(b)_j (g)_{k-j}}{(b + g)_k} = 1.$$

We remark here that the sum is symmetrical with respect to b and g. Let us consider

$$\sum_{j=0}^{k+1} \binom{k+1}{j} \frac{(b)_j (g)_{k+1-j}}{(b + g)_{k+1}} = \sum_{j=0}^{k+1} \left\{ \binom{k}{j} + \binom{k}{j-1} \right\} \frac{(b)_j (g)_{k+1-j}}{(b + g)_{k+1}} \qquad \text{(see 4-2}a_1\text{).}$$

Looking at the two sums independently, we have

$$\sum_{j=0}^{k+1} \binom{k}{j} \frac{(b)_j (g)_{k+1-j}}{(b + g)_{k+1}} = \frac{g}{b + g} \sum_{j=0}^{k} \binom{k}{j} \frac{(b)_j (g-1)_{k-j}}{(b + g - 1)_k} = \frac{g}{b + g}$$

since the statement is true for k. Also

$$\sum_{j=0}^{k+1} \binom{k}{j-1} \frac{(b)_j (g)_{k+1-j}}{(b + g)_{k+1}} = \frac{b}{b + g} \sum_{j=1}^{k+1} \binom{k}{j-1} \frac{(b-1)_{j-1} (g)_{k+1-j}}{(b + g - 1)_k} = \frac{b}{b + g}.$$

Hence

$$\sum_{j=0}^{k+1} \binom{k+1}{j} \frac{(b)_j(g)_{k+1-j}}{(b+g)_{k+1}} = \frac{g}{b+g} + \frac{b}{b+g} = 1$$

and by the Principle of Induction our identity is proved.   ■

EXERCISES 6-5

The problems referred to in Probs. 1,2, and 3 are those Exercises 6-3b.

1.   Find the variance of the random variables in Probs. 1,2 and 9.

2.   Compute the cov (Y,Z) of Prob. 5.

3.   What are $V(Y)$, $V(Z)$ and $V(Y + Z)$ for the random variables in Prob. 5.

4.   A deck of 5 cards numbered 1, 2, 3, 4, 5 is shuffled and then turned over one by one.   If the card number i is the ith one turned over, we have a match.   What is the probability of a match in the ith place?   Of a least one match?   (This problem could be expressed in terms of 5 men having the hats they checked returned at random—what is the probability a man receives his own hat—or of a scatterbrain teacher haphazardly returning papers to his students.)   If X_i is a random variable which equals one if the ith card is in the ith place and zero if it is not and if $S_5 = X_1 + X_2 + X_3 + X_4 + X_5$, find $E(X_i)$ and $V(X_i)$ for $i = 1, 2, 3, 4, 5$.   Find also $E(S_5)$, cov (X_i,X_j), and $V(S_5)$.

5.   If in Prob. 4, there are n numbered cards and X_i is the same random variable and $S_n = \sum_{i=1}^{n} X_i$, find $E(X_i)$, $V(X_i)$, $E(S_n)$ and $V(S_n)$.   To what extend do $E(S_n)$ and $V(S_n)$ depend on n?

6.   (a) What is the probability of at least one match among the 5 cards? the n cards? (b) What is the probability of no matches?

7.   Prove: At least the fraction $1 - \left(\dfrac{1}{h^2}\right)$ of the total probability of a random variable lies within h standard deviations of the mean.

8.   Prove: $\sum_{j=0}^{n} k^2 \binom{n}{k} p^k(1 - p)^{n-k} = np\,[(n - 1)p + 1]$.   (This fills in the details of the calculation leading to 6-4b_6.)

9.   Show that cov $(X,Y) = E[(X - E(X))(Y - E(Y))]$.

chapter 7

Stochastic Independence

7-1. CONDITIONAL PROBABILITY

a. Definition

At the annual school picnic every member of the school community consisting of 752 students, 53 teachers, 6 administrators, and 10 staff personnel has the same chance of winning the door prize. We construct a probability space $\{\Omega, A_\Omega, P\}$ where the elements of the sample space Ω are the individual members of community and we define P on A_Ω by setting, for $A \in A_\Omega$, $P(A) = \dfrac{n(A)}{821}$. Of these 821 people 400 are women and 421 are men, and of the teachers 28 are women. If T and W are the events "winner is a teacher" and "winner is a woman," $P(T) = \dfrac{53}{821}$ and $P(W) = \dfrac{400}{821}$. If we know the winner is a woman, what is the probability she is a teacher? We would answer immediately $\dfrac{28}{400}$, for we limit our considerations to the four hundred women and assume they all have the same chance of being chosen. Then, since among the women the subset of teachers has 28 elements, the probability that the winner is a teacher is $\dfrac{28}{400}$. Thus, we have developed a new probability where for

$$T \in A_\Omega, \ P_W(T) = \frac{n(TW)}{n(W)}.$$

Hence $P_W(T) = \dfrac{28}{400}$. ($P_W(T)$ is read: the probability the winner is a teacher, given she is a woman.) In terms of our original P we have

$$P_W(T) = \frac{n(TW)}{n(W)} = \frac{\dfrac{n(TW)}{n(\Omega)}}{\dfrac{n(W)}{n(\Omega)}} = \frac{P(TW)}{P(W)}.$$

137

In a certain bridge game North has no king; let us determine the probability that South also has no king. The probability space for this experiment is $\{\Omega, A_\Omega, P\}$ where Ω has $\binom{52}{13}\binom{39}{13}$ elements and we choose P such that each has the same probability. If N is the event North has no king, $P(N) = \dfrac{\binom{48}{13}}{\binom{52}{13}}$;

and if S is the event South has no king, $P(S) = P(N)$ and $P(SN) = \dfrac{\binom{48}{13}\binom{35}{13}}{\binom{52}{13}\binom{39}{13}}$.

Knowing that N has no king, let us focus our attention on South's hand. It can come from 39 cards of which 35 are not kings. Assuming all subsets of 13 cards are equally likely to be dealt, it seems natural to set $P_N(S) = \dfrac{\binom{35}{13}}{\binom{39}{13}}$. Comparing this with our earlier calculations of $P(N)$ and $P(SN)$, we see that in terms of P we have chosen $P_N(S) = \dfrac{P(SN)}{P(S)}$.

In each of the preceding examples the probability of the events "winner is a teacher" and "South has no kings" were altered because we already knew the winner was a woman and that North had no king. This is not always the case. Let a coin be tossed three times and let A and B be the events "the third toss is a head" and "the first two tosses are the same."

$$\Omega = \{HHH, HHT, HTH, THH, HTT, THT, TTH, TTT\}$$

$$A = \{HHH, HTH, THH, TTH\}$$

$$AB = \{HHH, TTH\}$$

$$B = \{HHH, HHT, TTH, TTT\}$$

If we know the first two tosses were the same, then the probability the third toss is a head is $\dfrac{1}{2}$, for of the four points in B, only HHH and TTH are also in A.

Again it is convenient to write $P_B(A) = \dfrac{1}{2}$ and again we find $P_B(A) = \dfrac{P(AB)}{P(B)}$.

But in this case,

$$P_B(A) = \frac{1}{2} = P(A),$$

that is, knowledge of the occurrence of B does not change the probability of A. These several examples lead us to:

7-1a₁ **Definition.** *Given $\{\Omega, A_\Omega, P\}$ and $A \in A_\Omega$, $M \in A_\Omega$ with $P(M) > 0$. The* **conditional probability** *of A given M is*

$$P_M(A) = \frac{P(AM)}{P(M)}.$$

The phrases "on the hypothesis M" and "knowing that M has already occurred" are other ways of expressing "given M." Another common notation for $P_M(A)$ is $P(A|M)$.

It is implicit in the definition that P_M is a set function defined on A_Ω; let us show that this is indeed a probability. For $A \in A_\Omega$, $P_M(A) = \dfrac{P(AM)}{P(M)} \geqslant 0$ since $P(AM) \geqslant 0$ (P is a probability) and $P(M) > 0$ by definition. Let $\Sigma_i A_i$ be the sum of mutually disjoint sets A_i, where each $A_i \in A_\Omega$, then

$$P_M(\Sigma_i A_i) = \frac{P(M \Sigma_i A_i)}{P(M)} = \frac{P(\Sigma_i M A_i)}{P(M)} = \frac{\Sigma_i P(M A_i)}{P(M)} = \sum_i \frac{P(M A_i)}{P(M)} = \Sigma_i P_M(A_i).$$

Finally,

$$P_M(\Omega) = \frac{P(M\Omega)}{P(M)} = \frac{P(M)}{P(M)} = 1.$$

Since P_M is a probability, it satisfies all the properties of a probability (see Sec. 5-3b). We have, for example,

$$P_M(A \cup B) = P_M(A) + P_M(B) - P_M(AB)$$

and

$$P_M(A') = 1 - P_M(A).$$

The equation defining conditional probability can be rewritten to give an expression for the probability of the intersection of two events. From Def. 7-1a_1 we have immediately

7-1a₂ $$P(AM) = P_M(A)P(M).$$

We remark that 7-1a_2 holds even if $P(M) = 0$.

From

$$P(AB) = P_B(A)P(B)$$

we can move on to

7-1a₃ $$P(ABC) = P_{BC}(A)P(BC) = P_{BC}(A)P_C(B)P(C).$$

Since ABC is symmetric in all three letters, we can also write

$$P(ABC) = P(A)P_A(B)P_{AB}(C) = P(A)P_A(C)P_{AC}(B).$$

$$P(ABC) = P(B)P_B(A)P_{AB}(C) = P(B)P_B(C)P_{BC}(A).$$

The probability of the intersection of four or more events is an uncomplicated generalization of this.

EXERCISES 7-1a

1. Given the face of a die is even, what is the probability it is a four?

2. A card is drawn from a standard deck. It is black. What is the probability it is a club? a four? a face card?

3. Four balls are placed successively into four cells so that all 4^4 arrangements are equally likely. Given that the first two balls fall into two cells, what is the probability one cell contains exactly three balls?

4. Three dice are rolled. If no two are alike, what is the probability one is a 6?

5. A throw of five distinguishable dice produces at least one ace. What is the probability of two or more aces?

6. A die is rolled until a 6 appears. (a) If the first roll did not produce a 6, what is the probability it will require more than three rolls to get a 6? (b) If the number of rolls is even, what is the probability the number will be 2?

7. One man in 20 and one woman in 500 are color-blind. A color-blind person is chosen at random, what is the probability it is a woman? (Assume the population is half men and half women.)

8. In a toothbrush plant three machines X, Y, Z manufacture 20, 35, and 45 percent of the packing boxes. Of their output 4, 4, and 3 percent respectively are defective. A box is chosen at random and is found to be defective. What is the probability it was made by machine X? by Y? by Z?

9. In a bridge game North has no king. What is the probability his partner (South) has at least two kings?

10. In another bridge game North and South have 10 spades between them. (a) What is the probability all three remaining spades are in the same hand (either East's or West's)? (b) If it is known the queen is among the three outstanding spades, what is the probability she is unguarded (i.e., one opponent has the queen, the other the remaining two cards)?

11. An urn contains b black and w white balls. One is drawn and its color noted. The ball is returned to the urn along with n additional balls of the same color and m balls of the opposite color. (There are now $b + w + m + n$ balls in

the urn.) This is repeated two more times. What is the probability of getting three black balls?

12. In the previous problem, if the second ball was black, what is the probability the first was black?

13. If $n > 0$ and $m = 0$ we create a rough model (due to Polya) of phenomena such as contagious diseases where the occurrence of an event increases the probability of future occurrences. Show that a sequence of μ drawings resulting in μ_1 black and μ_2 white balls $(\mu_1 + \mu_2 = \mu)$ has the same probability as the event drawing first μ_1 black and then μ_2 white balls.

14. Balls are placed randomly into 6 cells. One ball is placed. What is the probability the second ball will fall in the same cell? In a different cell? If two balls have been placed, that after the placement of the third ball, one cell is occupied? two cells? three cells?

b. Bayes' Rule

In a certain city n different makes of automobiles are registered, let us designate them as $M_1, M_2, \ldots, M_n$. The probability that a car chosen at random will be an M_i is p_i $(p_1 + p_2 + \cdots + p_n = 1)$ and the probability that an M_i car is red is r_i. If a car is chosen at random, what is the probability it is red? If R is the event "the car is red,"

$$R = RM_1 + RM_2 + \cdots + RM_n$$

(we use $+$ since $(RM_i) \cap (RM_j) = \varnothing$, $i \neq j$) and hence

$$P(R) = \sum_{i=1}^{n} P(RM_i) = \sum_{i=1}^{n} P_{M_i}(R)P(M_i) = \sum_{i=1}^{n} r_i p_i.$$

With the same data we can also answer the question: if the car chosen is red, what is the probability its make is M_j?

$$P_R(M_j) = \frac{P(RM_j)}{P(R)} = \frac{P_{M_j}(R)P(M_j)}{P(R)} = \frac{r_j p_j}{\sum\limits_{i=1}^{n} r_i p_i}.$$

These results can be stated in a general way. A sample space Ω is *partitioned* by the collection of sets $S_1, S_2, \ldots, S_n$ if (i) $S_i \subset \Omega$ for all i, (ii) $S_i S_j = \varnothing$, $i \neq j$, and (iii) $S_1 + S_2 + \cdots + S_n = \Omega$. Then for any set $A \in A_\Omega$, we have

$$A = AS_1 + AS_2 + \cdots + AS_n$$

(since $S_i S_j = \varnothing$, $(AS_i)(AS_j) = (AS_i S_j) = A\varnothing = \varnothing$). Then

7-1b_1
$$P(A) = \sum_{i=1}^{n} P(AS_i) = \sum_{i=1}^{n} P_{S_i}(A)P(S_i).$$

We observe that the collection $AS_1, AS_2,\dots,AS_n$, where it is possible that for some i $AS_i = \emptyset$, is a partition of A. We illustrate the partitioning of a sample space in Figure 7-1b_2.

7-1b_2 Figure

Partition of a rectangular sample space into nine disjoint subsets.

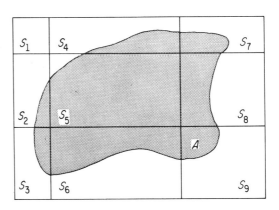

$$\Omega = S_1 + S_2 + \cdots + S_9$$

$$A = AS_2 + AS_3 + \cdots + AS_9 \qquad (AS_1 = \emptyset)$$

The probability that a litter of boxer puppies has exactly j pups is p_j ($p_1 + p_2 + p_3 + \cdots + p_m = 1$) and in each litter the sexes are distributed binomially with $p = \dfrac{1}{2}$. Thus, if there are n pups in a litter, the probability of k males and $n-k$ females is $\binom{n}{k}2^{-n}$. This is an example of a partitioned sample space where the partitions S_j are the litters with exactly j pups. If A is the event "the litter has three males," then

$$P(A) = p_3 2^{-3} + \binom{4}{3}p_4 2^{-4} + \binom{5}{3}p_5 2^{-5} + \cdots + \binom{m}{3}p_m 2^{-m}.$$

If S_5 is the event "the litter has five pups," then

$$P_A(S_5) = \frac{P(AS_5)}{P(A)} = \frac{P_{S_5}(A)P(S_5)}{P(A)} = \frac{\binom{5}{3}2^{-5}p_5}{p_3 2^{-3} + \binom{4}{3}p_4 2^{-4} + \cdots + \binom{m}{3}p_m 2^{-m}}$$

The last equation is an example of the classical

7-1b_3 Bayes' Rule. *Let Ω be partitioned by the collection $S_1, S_2, \ldots, S_n$ and let $A \in A_\Omega$ with $P(A) > 0$. Then*

$$P_A(S_j) = \frac{P(AS_j)}{P(A)} = \frac{P_{S_j}(A)P(S_j)}{\sum\limits_{i=1}^{n} P_{S_i}(A)P(S_i)}$$

Proof: Since the collection $S_j, j = 1, 2, \ldots, n$, is a partition of Ω, we can write for $A \in A_\Omega$

$$A = AS_1 + AS_2 + \cdots + AS_n.$$

For each j, from 7-1a_2,

$$P(AS_j) = P_{S_j}(A)P(S_j) = P_A(S_j)P(A).$$

Then 7-1b_1 and 7-1a_2,

$$P(A) = \sum_{j=1}^{n} P(AS_j) = \sum_{j=1}^{n} P_{S_j}(A)P(S_j)$$

$$P_A(S_j) = \frac{P(AS_j)}{P(A)} = \frac{P_{S_j}(A)P(S_j)}{\sum\limits_{i=1}^{n} P_{S_i}(A)P(S_i)}. \qquad \blacksquare$$

In many applied problems a sample space can be partitioned into subsets $S_1, S_2, \ldots, S_n$ with probabilities $P(S_1), P(S_2), \ldots, P(S_n)$ and for an event $A \in A_\Omega$ the values $P_{S_i}(A)$, $i = 1, 2, \ldots, n$ are readily computed. Bayes' rule enables us to use these available figures to find $P_A(S_i)$. A common type problem is illustrated in the following. A certain manufacturer uses three hauling firms A, B, and C with probabilities p_A, p_B, p_C ($p_A + p_B + p_C = 1$). The manufacturer knows that if he uses firm A the shipment will arrive late at its destination with probability $p_A(L)$; similarly, for B and C he knows $p_B(L)$ and $p_C(L)$. With Bayes' rule he can then determine the probability, if a shipment arrived late, whether the hauler was A, B, C. For our purposes memorizing the rule will gain us little; it will be both easier and safer to treat each problem independently.

EXERCISES 7-1b

1. In a certain country club 40 percent of the members are physicians, 30 percent lawyers, 15 percent teachers, 10 percent military and 5 percent none of these. 20 percent of the doctors are under 40 years of age, 30 percent of the lawyers, 40 percent of the teachers, 50 percent of the military and 10 percent of the others. (a) What is the probability a member is under forty? (b) If a member is under forty, what is the probability he is a lawyer?

2. In a nameless country 6 political parties have succeeded in electing members to the governing senate. The distribution by party is given in the accompanying table. (a) What is the probability a senator is a woman? (b) If a senator is a woman what is the probability she belongs to party C?

Party	A	B	C	D	E	F
Total number of senators	105	100	70	45	40	15
Number of women senators	15	20	10	5	3	2

3. In a large city it is found that its registered physicians have come from N medical schools: $S_1, S_2, \ldots, S_N$ and the probability that a physician chosen at random will have been educated at S_j is p_j. Furthermore, the probability that a S_j trained man will be a specialists is σ_j. What is the probability a physician chosen at random will be a specialist? If the physician chosen at random is a specialist, what is the probability he was educated at S_N?

4. The probability p_n that a tree survives n years is p^n where $n \geqslant 1$ and $p_0 = 1 - (p_1 + p_2 + p_3 + \cdots)$. Each year the tree blossoms with probability $\frac{1}{3}$. For $k \geqslant 1$, what is the probability the tree blossoms k times? [It is useful to know:

$$1 + \binom{k+1}{1}\left(\frac{2}{3}p\right) + \binom{k+2}{2}\left(\frac{2}{3}p\right)^2 + \binom{k+3}{3}\left(\frac{2}{3}p\right)^3 + \cdots$$

$$= \left(1 - \frac{2}{3}p\right)^{-(k+1)}].$$

5. If a tree is known to have blossomed at least once, what is the probability it blossoms two or more times?

7-2. STOCHASTIC INDEPENDENCE

a. Independent Events

If we draw a card from an ordinary bridge deck (and apply the usual probability space associated with this experiment) the probability the card is a king is $\frac{1}{13}$ and the probability the card is a heart is $\frac{1}{4}$. If we know the card drawn is a king, the probability it is a heart is still $\frac{1}{4}$; our knowledge that the card is a king does not alter the probability that it is a heart. The probability it is the king of hearts is, of course, $\frac{1}{52}$. Letting H and K represent respectively the events the card is a heart and the card is a king, the situation can be described by

$$P(HK) = \frac{1}{52} = P(H)P(K)$$

$$P_K(H) = \frac{P(HK)}{P(K)} = \frac{P(H)P(K)}{P(K)} = P(H)$$

A coin is tossed twice. In the framework of the usual probability space associated with this experiment, let H_1 and H_2 be respectively the events head on first toss and head on second toss. $P(H_1) = P(H_2) = \frac{1}{2}$ and $P(H_1H_2) = \frac{1}{4}$. and the events H_1 and H_2 are related (or unrelated) as are H and K in the preceding experiment. Again we have

$$P(H_1H_2) = P(H_1)P(H_2)$$

$$P_{H_2}(H_1) = P(H_1); \; P_{H_1}(H_2) = P(H_2).$$

The events H and K and H_1 and H_2 are called independent events.

7-2a_1 **Definition.** *Two events A and B are said to be* **stochastically independent** *if, and only if,*

$$P(AB) = P(A)P(B).$$

Since we shall speak of no other type of independence we will drop the use of the adjective "stochastically." Statistical independence is a synonym for stochastic independence. The definition is symmetric in A and in B and is valid even if $P(A) = 0$ or $P(B) = 0$. An equivalent definition of independence when $P(B) \neq 0$ is given by

7-2a_2 $$P_B(A) = P(A).$$

For then

$$P_B(A) = P(A) = \frac{P(A)P(B)}{P(B)}$$

and

$$P_B(A) = \frac{P(AB)}{P(B)},$$

and these together imply

$$P(AB) = P(A)P(B).$$

In the two examples already given it is intuitively clear that the knowledge that a card is a king tells nothing about the suit of the card drawn; nor if, in tossing a coin, we get a head on the first toss, are we more or less likely to get a head on the second. However, independence, or the lack of it, is not always so evident. Let a coin be tossed n times and let A be the event there is at most one head and B the event both heads and tails appear, so that AB is the event one head and $n - 1$ tails. $A = A_1 + A_2$ where A_1 and A_2 are the events no head and exactly one head and $B = \Omega - B'$ where B' is the event all heads or all tails. Since $A_1 A_2 = \varnothing$, $P(A_1 + A_2) = P(A_1) + P(A_2)$, and also $P(B) = 1 - P(B')$. For $n = 3$, $P(A) = P(A_1) + P(A_2) = \dfrac{1}{8} + \dfrac{3}{8} = \dfrac{1}{2}$, $P(B) = 1 - P(B') = 1 - \dfrac{2}{8} = \dfrac{3}{4}$ and $P(AB) = \dfrac{3}{8} = P(A)P(B)$. For $n = 4$, $P(A) = \dfrac{1}{16} + \dfrac{4}{16} = \dfrac{5}{16}$, $P(B) = 1 - \dfrac{2}{16} = \dfrac{7}{8}$ and $P(AB) = \dfrac{4}{16} \neq P(A)P(B)$. Thus, if $n = 3$ the events A and B are independent, while if $n = 4$, they are not.

b. Independence Of Three Or More Events

Definition of the independence of three or more events is a more difficult problem. In the experiment of tossing a coin twice let A be the event that head turns up on the first toss, B the event of head on second toss, and C the event that both tosses are the same. We know

$$P(A) = P(B) = P(C) = \frac{1}{2}$$

$$P(AB) = P(AC) = P(BC) = \frac{1}{4}$$

so that the events, A, B, and C are pairwise independent. However, information that both A and B have occurred guarantees that C has occurred; we have

$$P_{AB}(C) = \frac{P(ABC)}{P(AB)} = \frac{P(HH)}{P(HH)} = 1.$$

Hence C is not independent of A and B. Thus, in order that three events be independent we need not only their pairwise independence but also that

$$P(ABC) = P(A)P(B)P(C)$$

since this equation also insures that A is independent of BC, B is independent of AC, and C is independent of AB. If a coin is tossed three times and L,M, and N are respectively the events head on the first, second and third toss, then L, M, and N are independent events.

If n events, $A_1, A_2, \ldots, A_n$ are to be independent, they must be independent when taken two at a time, three at a time, four at a time and so forth. Actually, we define the independence of n events in this way.

7-2b_1 **Definition.** *The events $A_1, A_2, \ldots, A_n$ are said to be **mutually independent** if, for all combinations $1 \leqslant i < j < k < \cdots \leqslant n$, the following equations are satisfied:*

$$P(A_i A_j) = P(A_i)P(A_j)$$

$$P(A_i A_j A_k) = P(A_i)P(A_j)P(A_k)$$

$$\cdot$$
$$\cdot$$
$$\cdot$$

$$P(A_1 A_2 \cdots A_n) = P(A_1)P(A_2) \cdots P(A_n)$$

This is a formidable looking array of equations but in practice it is usually not necessary to check them all out.

EXERCISES 7-2b

1. Two distinguishable dice are rolled. A is the event ace on first die and B is the event even on second. Are A and B independent?

2. In the set of permutations of the four digits (1, 2, 3, 4) are the events "1 precedes 2" and "3 precedes 4" independent?

3. Consider families of n children and assume all the sex distributions equally probable. Are the events "family has children of both sexes" and "at most one boy" independent for $n = 3$? for $n = 4$?

4. If A and B are independent events prove that A and B', A' and B and A' and B' are also pairs of independent events.

5. An urn contains 7 balls of which 4 are white. A sample of size 4 is drawn with replacement (i.e., after the color of the ball drawn is noted, the ball is put back into the urn). Let H be the event the first ball drawn is white; K the event the fourth ball drawn is white; L the event exactly one of the first two balls are white; and M the event exactly two of the four balls are white. (a) Are H and K independent? (b) Are K and L independent? (c) Are $K, L,$ and M independent?

6. If in Prob. 6 the sampling is done without replacement, (a) are H and K independent? (b) are K and L independent?

7. Let $A_1, A_2, \ldots, A_n$ be independent events. Prove

$$P(A_1 \cup A_2 \cup \cdots \cup A_n) = 1 - P(A_1')P(A_2') \cdots P(A_n')$$

8. What is the probability that in eight tosses of a fair die the ace will appear at least once?

9. Cards are drawn from the standard deck with replacement. This is done four times. What is the probability at least one card will be a heart?

10. Again let $A_1, A_2, \ldots, A_n$ be independent events and let

$$P(A_i) = p_i, \quad i = 1, 2, \ldots, n.$$

Let p_0 be the probability none of the events occur. Prove

$$p_0 = (1 - p_1)(1 - p_2) \cdots (1 - p_n).$$

11. What is the probability that in eight tosses of a die none will produce an ace?

12. If four cards are drawn from a standard deck with replacement, what is the probability none is a heart?

c. Independent Trials

Consider n experiments, not necessarily different and each with a finite number of possible outcomes, which generate the probability spaces $\{E_1, A_{E_1}, P_1\}$, $\{E_2, A_{E_2}, P_2\}, \ldots, \{E_n, A_{E_n}, P_n\}$. Let the elementary events in E_i be designated by $e_{j_i}^i$, where $j_i = 1, 2, \ldots, v_i = n(E_i)$ and write

$$P_i(e_{j_i}^i) = p_{j_i}^i.$$

We remark

7-2c_1

$$\sum_{j_i=1}^{v_i} p_{j_i}^i = \sum_{j_i=1}^{v_i} P_i(e_{j_i}^i) = P_i(E_i) = 1.$$

We now look at

7-2c_2 $$\Omega = E_1 \times E_2 \times \cdots \times E_n$$

whose elementary events are the n-tuples $(e_{j_1}^1, e_{j_2}^2, \ldots, e_{j_n}^n)$, and where

$$n(\Omega) = (v_1)(v_2)\cdots(v_n) \qquad \text{(see 4-1a_3).}$$

In this context the ith experiment is referred to as the ith trial and $e_{j_i}^i$ as the outcome of the ith trial. For $\omega \in \Omega$, define

7-2c_3 $$P(\omega) = P((e_{j_1}^1, e_{j_2}^2, \ldots, e_{j_n}^n)) = p_{j_1}^1 p_{j_2}^2 \cdots p_{j_n}^n,$$

and for $A \in A_\Omega$, set

7-2c_4 $$P(A) = \sum_{\omega \in A} P(\omega).$$

We wish to prove that P is a probability.

In Sec. 5-4 we constructed analogous Ω and P for the case where the trials are identical and have only two possible outcomes S and F with probabilities p and $q = 1 - p$. In such cases the n-tuple, $\omega \in \Omega$, is an array of k S's and $n - k$ F's and $P(\omega) = p^k q^{n-k}$ and in 5-4a_2 we proved that P defined as in 7-2c_4 is a probability. The generalization here permits n different experiments each of which may have a different number of elementary events. For example, we may roll a die ($E_1 = \{1, 2, 3, 4, 5, 6\}$, $p_{j_1}^1 = \frac{1}{6}$), toss a coin ($E_2 = \{H, T\}$ $p_{j_2}^2 = \frac{1}{2}$), and draw a card for its suit ($E_3 = \{H, C, D, S\}$, $p_{j_3}^3 = \frac{1}{4}$).

$$\Omega = E_1 \times E_2 \times E_3 = \{(e_{j_1}^1, e_{j_2}^2, e_{j_3}^3) | e_i^i \in E_i, \, i = 1, 2, 3\}$$

and for $\omega \in \Omega$ we set

$$P(\omega) = P((e_{j_1}^1, e_{j_2}^2, e_{j_3}^3)) = p_{j_1}^1 p_{j_2}^2 p_{j_3}^3 = \frac{1}{48}.$$

For $A \in A_\Omega$,

$$P(A) = \sum_{\omega \in A} P(\omega) = \frac{n(A)}{48}$$

(since by its construction every $\omega \in \Omega$ has the same probability). Then by 5-2b_3 P is a probability and $\{\Omega, A_\Omega, P\}$ a probability space.

In what follows Ω and P are the sample space and probability defined by the sequence 7-2c_2, 7-2c_3, and 7-2c_4. Since for all i, $i = 1, 2, \ldots, n$, P_i is a probability, $P(\omega) \geqslant 0$, $\omega \in \Omega$, and hence for $A \in A_\Omega$,

7-2c_5
$$P(A) = \sum_{\omega \in A} P(\omega) \geqslant 0$$

since sums of nonnegative terms are nonnegative. If $A_1 A_2 = \emptyset$ where $A_1 \in A_\Omega$ and $A_2 \in A_\Omega$,

$$P(A_1 + A_2) = \sum_{\omega \in (A_1 + A_2)} P(\omega) = \sum_{\omega \in A_1} P(\omega) + \sum_{\omega \in A_2} P(\omega) = P(A_1) + P(A_2)$$

where the terms in the first sum can be rearranged and separated into $\omega \in A_1$ and $\omega \in A_2$ since any $\omega \in (A_1 + A_2)$ belongs to only one of the sets. Because A_Ω is finite ($n(A_\Omega) = 2^{n(\Omega)} = 2^{(v_1)(v_2)\cdots(v_n)}$), we can use the Principle of Induction to prove that if $A_i \in A_\Omega$, $i = 1, 2, \ldots, N \leqslant n(A_\Omega)$ and if $A_i A_j = \emptyset$ whenever $i \neq j$,

7-2c_6
$$P\left(\sum_{i=1}^{N} A_i\right) = \sum_{i=1}^{N} P(A_i).$$

It remains to show $P(\Omega) = 1$. We first introduce some notation.   The multiple sum

$$\sum_{j_1=1}^{v_1} \sum_{j_2=1}^{v_2} \ldots \sum_{j_n=1}^{v_n} p_{j_1}^1 p_{j_2}^2 \cdots p_{j_n}^n$$

is the sum over all possible arrangements of the subscripts $j_1, j_2, \ldots, j_n$.   There will be $(v_1)(v_2)\ldots(v_n) = n(\Omega)$ addends.   For example,

$$\sum_{i=1}^{2} \sum_{j=1}^{3} a_i b_j = a_1 b_1 + a_1 b_2 + a_2 b_1 + a_2 b_2 + a_3 b_1 + a_3 b_2$$

$$= a_1 \sum_{j=1}^{2} b_j + a_2 \sum_{j=1}^{2} b_j + a_3 \sum_{j=1}^{2} b_j$$

$$= b_1 \sum_{i=1}^{3} a_i + b_2 \sum_{i=1}^{3} a_i.$$

Here we can count the $3 \cdot 2 = 6$ terms.   As an aid in proving $P(\Omega) = 1$ we first establish

7-2c₇   Lemma

$$\sum_{j_1=1}^{v_1} \sum_{j_2=1}^{v_2} \cdots \sum_{j_n=1}^{v_n} p_{j_1}^1 p_{j_2}^2 \cdots p_{j_n}^n = \left(\sum_{j_1=1}^{v_1} p_{j_1}^1 \right) \left(\sum_{j_2=1}^{v_2} p_{j_2}^2 \right) \cdots \left(\sum_{j_n=1}^{v_n} p_{j_n}^n \right).$$

Proof: Our proof is by induction.   Let $n = 2$.

$$\sum_{j_1=1}^{v_1} \sum_{j=2}^{v_2} p_{j_1}^1 p_{j_2}^2 = p_1^1 \left(\sum_{j_2=1}^{v_2} p_{j_2}^2 \right) + p_2^1 \left(\sum_{j_2=1}^{v_2} p_{j_2}^2 \right) + \cdots + p_{v_1}^1 \left(\sum_{j_2=1}^{v_2} p_{j_2}^2 \right)$$

where from the $v_1 v_2$ terms in the left-hand sum we grouped together those which contain p_1, then those with p_2 and so forth.   Now taking out the common factor on the right,

$$\sum_{j_1=1}^{v_1} \sum_{j_2=1}^{v_2} p_{j_1}^1 p_{j_2}^2 = \left(\sum_{j_1=1}^{v_1} p_{j_1}^1 \right) \left(\sum_{j_2=1}^{v_2} p_{j_2}^2 \right).$$

So our lemma is true for $n = 2$.   Suppose it is also true for some $k > 2$.   Let us set

$$A = \sum_{j_1=1}^{v_1} \sum_{j_2=1}^{v_2} \cdots \sum_{j_k=1}^{v_k} p_{j_1}^1 p_{j_2}^2 \cdots p_{j_k}^k = \left(\sum_{j_1=1}^{v_1} p_{j_1}^1 \right) \left(\sum_{j_1=1}^{v_2} p_{j_2}^2 \right) \cdots \left(\sum_{j_k=1}^{v_k} p_{j_k}^k \right).$$

Then for $n = k + 1$, if we sum first over j_{k+1}, we have

$$\sum_{j_1=1}^{v_1} \sum_{j_2=1}^{v_2} \cdots \sum_{j_k=1}^{v_k} \sum_{j_{k+1}=1}^{v_{k+1}} p_{j_1}^1 p_{j_2}^2 \cdots p_{j_k}^k \, p_{j_{k+1}}^{k+1}$$

$$= p_1^{k+1} \cdot A + p_2^{k+2} \cdot A + \cdots + p_{v_{k+1}}^{k+1} \cdot A$$

$$= A \left(\sum_{j_{k+1}=1}^{v_{k+1}} p_{j_{k+1}}^{k+1} \right)$$

$$= \left(\sum_{j_1=1}^{v_1} p_{j_1}^1 \right) \left(\sum_{j_2=1}^{v_2} p_{j_2}^2 \right) \cdots \left(\sum_{j_k=1}^{v_k} p_{j_k}^k \right) \left(\sum_{j_{k+1}=1}^{v_{k+1}} p_{j_{k+1}}^{k+1} \right)$$

where the last step is possible since we assumed the theorem to be true for $n = k$. Thus whenever the lemma is true for $n = k$, it is true for $n = k + 1$, so by the Principle of Induction the lemma is true for all $n > 2$. ∎

We are now ready to prove $P(\Omega) = 1$. From 7-2c_3 and 7-2c_4,

$$P(\Omega) = \sum_{\omega \in \Omega} P(\omega) = \sum_{\omega \in \Omega} P((e_{j_1}^1, e_{j_2}^2, \ldots, e_{j_n}^n))$$

$$= \sum_{j_1=1}^{v_1} \sum_{j_2=1}^{v_2} \cdots \sum_{j_n=1}^{v_n} p_{j_1}^1 p_{j_2}^2 \cdots p_{j_n}^n$$

$$= \left(\sum_{j_1=1}^{v_1} p_{j_1}^1 \right) \left(\sum_{j_2=1}^{v_2} p_{j_2}^2 \right) \cdots \left(\sum_{j_n=1}^{v_n} p_{j_n}^n \right) \qquad \text{(from 7-2}c_7)$$

$$= (1)(1) \cdots (1) \qquad\qquad\qquad\qquad \text{(from 7-2}c_1).$$

Hence

7-2c_8 $\qquad\qquad\qquad\qquad\qquad P(\Omega) = 1$

and 7-2c_5, 7-2c_6 and 7-2c_8 together prove that P is a probability and hence that $\{\Omega, A_\Omega, P\}$ is a probability space.

Let us return to the experiment discussed earlier in this section. In it a die is rolled, a coin tossed, and a card drawn for its suit. $\Omega = E_1 \times E_2 \times E_3$ where $\{E_i, A_{E_i}, P_i\}$ $i = 1, 2, 3$, are the obvious probability spaces. If $\omega \in \Omega, P(\omega) = \frac{1}{48}$ and if $A \in A_\Omega$, $P(A) = \sum_{\omega \in A} P(\omega) = \frac{n(A)}{48}$. Let $A_1 \in A_\Omega$ be the event a three appears when the die is rolled, $A_1 = \{(3, e_2, e_3) | e_2 \in E_2, \, e_3 \in E_3\}$. To count the number of elements in A_1 we observe that the first element of every triple is fixed but there are two choices for the second and four for the third; hence $n(A_1) = 2 \cdot 4 = 8$. Thus

$$P(A_1) = \frac{n(A_1)}{48} = \frac{8}{48} = \frac{1}{6} = P_1(3),$$

and what happened on the die was unaffected by what occurred when the coin was tossed and the card drawn. We say that A_1 was determined by the first trial.

For the general case let $\{\Omega, A_\Omega, P\}$ be defined by 7-2c_2, 7-2c_3, and 7-2c_4, and let $A_k \in A_\Omega$ be such that for some $B \in A_{E_k}$

$$A_k = E_1 \times E_2 \times \cdots \times B \times \cdots \times E_n$$

$$= \{(e^1_{j_1}, e^2_{j_2}, \ldots, e^k_{j_k}, \ldots, e^n_{j_n}) \,|\, e^i_{j_i} \in E_i,\, i \neq k, e^k_{j_k} \in B\}.$$

In effect we express interest only in the outcome of the kth trial, and we say that the event A_k is *determined by* the kth trial. We prove

7-2c_9 $$P(A_k) = P_k(B).$$

$$P(A_k) = \sum_{\omega \in A_k} P(\omega) = \sum_{j_1=1}^{v_1} \cdots \sum_{e^k_{j_k} \in B} \cdots \sum_{j_n=1}^{v_n} p^1_{j_1} p^2_{j_2} \cdots p^k_{j_k} \cdots p^n_{j_n}$$

$$= \left(\sum_{j_1=1}^{v_1} p^1_{j_1} \right) \left(\sum_{j_2=1}^{v_2} p^2_{j_2} \right) \cdots \left(\sum_{e^k_{j_k} \in B} p^k_{j_k} \right) \cdots \left(\sum_{j_n=1}^{v_n} p^n_{j_n} \right) \qquad (7\text{-}2c_7)$$

$$= \sum_{e^k_{j_k} \in B} p^k_{j_k} = P_k(B).$$

We have shown that if the event A_k is determined by the kth trial, $P(A_k)$ depends only on P_k. For this reason the experiments $E_1, E_2, \ldots, E_n$ are referred to as *independent trials*, and $\Omega = E_1 \times E_2 \times \cdots \times E_n$ is a sample space composed of n independent trials. If $A_k \in A_\Omega$ is determined by the kth trial, $P(A_k)$ is independent of P_i for $i \neq k$.

A card is drawn from an ordinary bridge deck, its suit noted, and then returned to the deck and the deck thoroughly shuffled. This is done ten times, and we have ten independent identical trials. The probability that the third card is a heart is, by 7-2c_9, one fourth.

Furthermore, let A_k be determined by the kth trial and A_h by the hth $(k \neq h)$. Thus there exist $B \in A_{E_k}$ and $C \in A_{E_h}$ such that

$$A_k = \{(e^1_{j_1}, e^2_{j_2}, \ldots, e^n_{j_n}) \,|\, e^i_{j_i} \in E_i,\, i \neq k,\, e^k_{j_k} \in B\}$$

$$A_h = \{(e^1_{j_1}, e^2_{j_2}, \ldots, e^n_{j_n}) \,|\, e^i_{j_i} \in E_i,\, i \neq h,\, e^h_{j_h} \in C\}.$$

By the argument above $P(A_k) = P_k(B)$ and $P(A_h) = P_h(C)$. The event

$$A_k A_h = \{(e^1_{j_1}, e^2_{j_2}, \ldots, e^n_{j_n}) \,|\, e^i_{j_i} \in E_i,\, i \neq k, h,\, e^k_{j_k} \in B,\, e^h_{j_h} \in C\}$$

and we prove

7-2c$_{10}$ $\quad\quad\quad P(A_kA_h) = P_k(B)P_h(C) = P(A_k)P(A_h).$

$$P(A_kA_h) = \sum_{\omega \in A_kA_h} P(\omega)$$

$$= \sum_{j_1=1}^{v_1} \cdots \sum_{e_{j_k}^k \in B} \cdots \sum_{e_{j_h}^h \in C} \cdots \sum_{j_n=1}^{v_n} p_{j_1}^1 p_{j_2}^2 \cdots p_{j_n}^n$$

$$= \left(\sum_{e_{j_k}^k \in B} p_{j_k}^k \right) \left(\sum_{e_{j_h}^h \in C} p_{j_h}^h \right) \left(\sum_{j_1=1}^{v_1} p_{j_1}^1 \right) \cdots \left(\sum_{j_n=1}^{v_n} p_{j_n}^n \right)$$

$$= P_k(B)P_h(C)$$

$$= P(A_k)P(A_h)$$

Thus events determined by two independent trials are independent events and the probability of their intersection is the product of their probabilities.

If a die is rolled a hundred times the probability of a deuce on the second and eighteenth rolls is the probability of a deuce on the second times the probability of a deuce on the eighteenth. Again going back to the experiment with the die, the coin and the cards, if A_1 is the event the die turns up 3 and A_2 the event the coin is a head, $P(A_1) = \dfrac{1}{6}$, $P(A_2) = \dfrac{1}{2}$ and $P(A_1A_2) = P(A_1)P(A_2) = \dfrac{1}{12}$.

Not all experiments whose sample spaces are ordened n-tuples result from independent trials. From an urn containing three black and four white balls one is drawn out and a probability space $\{U,A_U,P_U\}$ is created $U = \{B,W\}$ and $P_U(B) = \dfrac{3}{7}$, $P_U(W) = \dfrac{4}{7}$. If a second ball is drawn before the first is replaced, a probability space $\{\Omega,A_\Omega,P\}$ is formed with $\Omega = \{(B,B),(B,W),(W,B),(W,W)\}$ and $P((B,B)) = \dfrac{6}{42}$ and $P = (B,W) = P(W,B) = P(W,W) = \dfrac{12}{42}$. $\Omega = U \times U$ but if $(a,b) \in \Omega$, $P((a,b)) \neq P_U(a)P_U(b)$ and so Ω is not the result of independent trials. On the other hand, if the first ball is replaced before the second ball is drawn and if we set $P_1(B) = \dfrac{9}{49}$, $P_1(W,B) = P_1(B,W) = \dfrac{24}{49}$, and $P_1(W,W) = \dfrac{16}{49}$, $\{\Omega,A_\Omega,P_1\}$ is the result of two independent trials. Another example of nonindependent trials appears in Sec. 6-4b where a committee of n-members was being chosen from b boys and g girls. At any trial the probability of selecting a boy was the same, $\dfrac{b}{b+g}$, but the probability of selecting a boy on two different trials was not $\left(\dfrac{b}{b+g}\right)^2$. In these two cases and in others we have and shall meet, the independence of the trials is usually evident.

EXERCISES 7-2c

1. A die is rolled one-hundred times. What is the probability of a two or a four on the first trial? on the fifth trial? on both the third and ninty-seventh trials? of no two, or no four appearing?

2.

Urn	Red	Blue
I	4	5
II	1	4
III	5	2

One ball is drawn from each urn. What is the probability that there are (a) two blue and one red balls? (b) at least two red balls? (c) more blue than red balls?

3. Die I has 2 black and 4 white faces and die II has 2 white and 4 black faces. A coin is tossed once. If it is heads, die I is used for the rest of the game; if it is tails, die II is used. (a) What is the probability of a black face on the first roll? on any roll? (b) If the first two rolls are black, what is the probability the third will also be black? (c) If the first n rolls are black, what is the probability die II is being used?

4. Given $E = \{a,b,c\}$ and define P_E on E so that $P_E(a) = p_a, P_E(b) = p_b$ and $P_E(c) = p_c$ with $p_a + p_b + p_c = 1$. If for $A \in A_E, P(A) = \sum_{\omega \in A} P_E(\omega)$, $\{E, A_E, P_E\}$ is a probability space. Let $\Omega = (X\, E)_n$ and define P on Ω as in 7-2c_3 and 7-2c_4. (a) What is the probability that in Ω a occurs i times, b, j times, and c, k times where $i + j + k = n$? (b) What is the probability of a on the vth trial, $v \leqslant n$? (c) What is the probability of a on the vth trial and b on the μth trial, $1 \leqslant v < \mu \leqslant n$?

5. A certain missile hits a target with probability $\dfrac{1}{2}$ and misses with probability $\dfrac{1}{2}$. If separate firings are considered to be independent, how many shots must be fired in order that the probability of a hit is greater than .95?

6. A baseball player's records show that each time he goes to bat, his probability of a hit is .3, of getting a walk is .2 and of being struck out is .5. Each time the player is at bat is an independent trial. In a certain game, he came to bat five times. What is the probability (a) of 2 hits and 3 outs? (b) of 2 hits, a walk, and 2 outs?

7. Prove 7-2c_6.

7-3. Independent Random Variables

Let X and Y be random variables defined on the same sample space with probability distributions P_X and P_Y and let their joint distribution be given by $p_{ij} = P(X = x_i, Y = y_j)$.

7-3a,   **Definition.**   *Two random variables on the same sample space are said to be* **independent** *if*

$$p_{ij} = P_X(x_i)P_Y(y_j).$$

In this case their joint distribution table becomes a sort of multiplication table. If X_k and X_h are random variables operating on the kth and hth trials of n independent trials then by 7-2c_{10}, X_k and X_h are independent. A die is rolled 10 times. Let $X_3 = 1$ if the third roll produces a two and $X_3 = 0$ otherwise. Let $X_7 = 1$ if the seventh roll is even and $X_7 = 0$ if it is odd. Then

$$P(X_3 = 1, X_7 = 1) = P_{X_3}(1)P_{X_7}(1) = \frac{1}{6} \cdot \frac{1}{2} = \frac{1}{12}.$$

The notion of mutual independence of three or more random variables is a straightforward extension of that for two. The random variables $X_1, X_2, \ldots, X_n$ defined on the same sample are said to be mutually independent if

$$p_{j_1 j_2 \cdots j_n} = P(X_1 = x_{j_1}, X_2 = x_{j_2}, \ldots, X_n = x_{j_n}) = P_{X_1}(x_{j_1})P_{X_2}(x_{j_2}) \cdots P_{X_n}(x_{j_n}).$$

The independence of random variables has many important consequences. We have already proved that for two random variables X and Y, $E(X + Y) = E(X) + E(Y)$; but we could say nothing about $E(XY)$. We now can show:

7-3a$_2$ **Theorem:**   *If X and Y are independent random variables, then*

$$E(XY) = E(X)E(Y).$$

Proof:

$$E(XY) = \sum_{x_i \in R_X} \sum_{y_j \in R_Y} x_i y_j \, p_{ij} = \sum_{x_i \in R_X} \sum_{y_j \in R_Y} x_i y_j P_X(x_i)P_Y(y_j)$$

$$= \sum_{x_i \in R_X} x_i P_X(x_i) \sum_{y_j \in R_Y} y_j P_Y(y_i) = E(X)E(Y). \qquad \blacksquare$$

We remark that if X and Y are not independent, we have

$$E(XY) = \sum_{x_i \in R_X} \sum_{y_j \in R_Y} x_i y_j \, p_{ij} = \sum_{x_i \in R_X} \sum_{y_j \in R_Y} x_i y_j P_{\{Y=y_j\}}(x_i)P_Y(y_i)$$

$$= \sum_{y_j \in R_Y} y_j P(y_j) \sum_{x_i \in R_X} x_i P_{\{Y=y_j\}}(x_i)$$

and we can simplify no further.

7-3a₃   **Corollary.** *If $X_1, X_2, \ldots, X_n$ are mutually independent random variables then*

$$E(X_1 X_2 \cdots X_n) = E(X_1)E(X_2)\cdots E(X_n).$$

Two other important properties of independent random variables are:

7-3a₄   **Theorem.** *The covariance of two independent random variables is zero,*

and

7-3a₅   **Corollary.** *If $X_1, X_2, \ldots, X_n$ are mutually independent variables with finite variances $V(X_1), V(X_2), \ldots, V(X_n)$, then the variance of the sum is the sum of the variances.*

Proof: (7-3a₄) By definition $\text{cov}(X, Y) = E(XY) - E(X)E(Y)$. Since X and Y are independent, by Sec. 7-3a₂, $E(XY) = E(X)E(Y)$ and so $\text{cov}(X, Y) = 0$.   ∎

Proof: (7-3a₅) By Theorem 6-4b₅,

$$V(X_1 + X_2 + \cdots + X_n) = \sum_{i=1}^{n} V(X_i) + 2 \sum_{i \neq j} \text{cov}(X_i X_j).$$

But since the random variables are mutually independent $\text{cov}(X_i X_j) = 0$ for all

i and j. Therefore $\sum_{i \neq j} \text{cov}(X_i X_j) = 0$ and

$$V\left(\sum_{i=1}^{n} X_i\right) = \sum_{i=1}^{n} V(X_i).$$ ∎

7-4   Conditional Expectation

Let X and Y be random variables on the same sample space Ω and let p_{ij} be their joint distribution. The conditional probability that $X = x_i$ when $Y = y_j$ has meaning, and by 7-1a₁

$$P_{\{Y = y_j\}}(X = x_i) = \frac{P(X = x_i, Y = y_j)}{P(Y = y_j)} = \frac{p_{ij}}{P_Y(y_j)}$$

7-4a₁   **Definition.** *The function*

$$P_{X|y_j} = \{(x_i, P_{\{Y=y_j\}}(X = x_i)| x_i \in R_X\}$$

is the **conditional distribution of X given Y** $= y_j$.

That $P_{X|y_j}$ is itself a probability on R_X is easily verified.

7-4a₂   **Definition.** *The* **conditional expectation of X given Y** $= y_j$ *is the sum*

$$E(X|y_j) = \sum_{x_i \in R_X} x_i P_{X|y_j}(x_i).$$

From this definition we can derive

$$E(X|y_j) = \sum_{x_i \in R_X} x_i P_{\{Y=y_j\}}(X = x_i)$$

$$= \sum_{x_i \in R_X} x_i \frac{p_{ij}}{P_Y(y_j)}.$$

Now if X and Y are independent random variables we use 7-3a_1 and obtain

7-4a_3 $$E(X|y_j) = \sum_{x_i \in R_X} x_i \frac{P_X(x_i)P_Y(y_j)}{P_Y(y_j)} = E(X),$$

a result that is not at all surprising.

Let us return to an example used in Sec. 6-1c. From an urn containing four balls numbered 1, 2, 3, 4 two are drawn. Let X and Y be random variables where X is the sum of the numbers drawn and Y is the larger of the numbers. The joint distribution of X and Y, P_X, and P_Y are given in Table 6-1c_2. What are $P_{X|4}(5)$, $P_{X|4}$, and $E(X|4)$?

$$P_{X|4}(5) = \frac{P(X = 5, Y = 4)}{P(Y = 4)} = \frac{\frac{1}{6}}{\frac{1}{2}} = \frac{1}{3}$$

$$P_{X|4} = \left\{ (3,0),(4,0), \left(5,\frac{1}{3}\right), \left(6,\frac{1}{3}\right), \left(7,\frac{1}{3}\right) \right\}$$

$$E(X|4) = \sum_{x_i \in R_X} x_i\, P_{X|4}(x_i) = \sum_{i=3}^{7} i P_{X|4}(i)$$

$$= 3.0 + 4.0 + 5 \cdot \frac{1}{3} + 6 \cdot \frac{1}{3} + 7 \cdot \frac{1}{3}$$

$$= \frac{1}{3}(18) = 6.$$

We remark

$$E(X) = \sum_{x_i \in R_X} x_i P_X(x_i) = \sum_{i=3}^{7} i P_X(i) = 5 \neq E(X|4)$$

In this case, where X and Y are not independent, knowledge that $Y = 4$ changed the expectation of X, for since $Y = 4$, we knew that certain values of X could not occur.

EXERCISES 7-4a

1. A coin is tossed n times. X_i, $i = 1,2,\ldots,n$, is the random variable which equals 1 if the ith toss is a head and 0 if it is a tail, and Y is the random variable which counts the number of heads.

  (a) Construct the joint distribution table for X_1 and X_3.
  (b) What is cov (X_1, X_3)?
  (c) Construct the joint distribution table for X_i and X_j, $i \neq j$.
  (d) What is cov (X_i, X_j)?
  (e) What are $E(X_i)$ and $V(X_i)$ for each i?

  (f) What are $\displaystyle\sum_{i=1}^{n} E(X_i)$, $E\left(\displaystyle\sum_{i=1}^{n} X_i\right)$, $E(Y)$?

  (g) What are $\displaystyle\sum_{i=1}^{n} V(x_i)$, $V\left(\displaystyle\sum_{i=1}^{n} X_i\right)$, $V(Y)$?

2. Three indistinguishable die are rolled. $(\Omega = \{(a,b,c) | a \leqslant b \leqslant c\}$ where $a,b,c \in \{1,2,\ldots,6\})$. Let X, Y and Z be random variables on Ω such that for $\omega = (a,b,c) \in \Omega, X(\omega) = a + b + c$, $Y(\omega) = \max(a,b,c)$, $Z(\omega) = \min(a,b,c)$. Are X and Y, X, and Z, Y, and Z independent variables?

3. A die with one white, two red, and three blue faces is rolled 10 times. X counts the number of white faces, Y counts the number of red faces, Z counts the number of blue faces, and W counts the number of nonblue faces. What are

$E(X | Y = 3)$, $E(Y | Z = 3)$, $P_{\{Z=7\}}(W = 3)$, $P_{\{Z=6\}}(W = 3)$, $P_{\{X=2\}}(W = 4)$, and $E(W | X = 2)$?

4. Balls bearing the integers 1, 2, 3, 4, 5 are put in a box. One is drawn out; then after removing those balls (if any) with numbers less than the one drawn, the first is put again in the box and a second number drawn. Let X and Y record the numbers on the first and second draws respectively. (a) Construct the joint distribution table of X and Y. (b) Find $P(X + Y > 7)$ and $P(Y - X > 0)$. (c) Construct the conditional distribution $P_{Y|3}$ and find $E(Y|3)$.

PART **III**

APPLICATIONS

chapter 8

Markov Chains

8-1. DEFINITION

a. An Example

Two friends A and B repeatedly play a game which cannot end in a tie and is such that A wins the game with probability p $(0 \leqslant p \leqslant 1)$ and B wins with probability $q = 1 - p \geqslant 0$. (If the game is tossing a coin once, and if A wins when a head comes up and B wins when a tail comes up, $p = q = \frac{1}{2}$.) At the end of each game the loser gives the winner a dollar, and they agree to stop playing whenever one of them has won three dollars. If the play has been in progress for some time when we begin our observation, at the end of the game then being played the contest might be in anyone of seven states:

E_1 A ahead by three games (contest over)
E_2 A ahead by two games
E_3 A ahead by one game
E_4 Contest tied, neither player ahead
E_5 B ahead by one game
E_6 B ahead by two games
E_7 B ahead by three games (contest over)

If at the end of the kth game the contest is in E_i, $i = 2,3,4,5,6$, at the end of the $(k + 1)$ st game it will be in E_{i-1} with probability p (A wins the $(k + 1)$ st game) or in E_{i+1} with probability q (B wins the $(k + 1)$st game). No other possibilities exist, and no matter how long the contest continues, the probability it will be in E_i at the end of a given game depends only upon the state it is in at the end of the previous game. If, for example, the contest is in E_4 and the subsequent games are won successively by

$$A \quad B \quad A \quad A \quad B \quad A \quad B \quad B$$

the contest goes through the states

$$E_3 \quad E_4 \quad E_3 \quad E_2 \quad E_3 \quad E_2 \quad E_3 \quad E_4 .$$

Let E_{j_k} be the state of the contest at the end of the kth game ($E_{j_k} \in \{E_1, E_2, E_3, E_4, E_5, E_6, E_7\}$) so that the chain of events $E_{j_1}, E_{j_2}, E_{j_3}, \ldots, E_{j_k}$ records the progress of the play. Then the probability that after the $(k + 1)$st game the contest is in E_i depends not on the entire history of the play, but only on E_{j_k}; notationally,

$$P(E_i | E_{j_1} E_{j_2} \cdots E_{j_k}) = P(E_i | E_{j_k}).$$

If at the end of the kth game the contest is in the state E_i where $i \neq 1,7$, at the end of the $(k + 1)$st game it will be in E_j, $j = 1,2,3,4,5,6,7$, with probabilities

8-1a_1
$$P(E_j | E_i) = \begin{cases} p & \text{if } j = i - 1 \quad (A \text{ wins}) \\ q & \text{if } j = i + 1 \quad (B \text{ wins}) \\ 0 & \text{otherwise.} \end{cases}$$

If $E_i = 1$ or $E_i = 7$,

8-1a_2
$$P(E_j | E_i) = \begin{cases} 1 & j = i \\ 0 & \text{otherwise} \end{cases}$$

where this last equation reflects the fact that the contest ends when it arrives in E_1 or E_7, and it will never more into another state. We further simplify notation by setting

8-1a_3
$$P(E_j | E_i) = p_{ij};$$

p_{ij} is called a *transition probability*, it is the probability the contest moves from E_i to E_j in one play of the game. Putting this notation into Eqs. 8-1a_1 and 8-1a_2, we have

8-1a_1
$$p_{ij} = \begin{cases} p & j = i - 1 \\ q & j = i + 1 \\ 0 & \text{otherwise} \end{cases} \quad i \neq 1,7$$

and

8-1a_2
$$p_{ij} = \begin{cases} 1 & i = j \\ 0 & \text{otherwise} \end{cases} \quad i = 1, i = 7$$

For example,

$$p_{34} = p_{56} = q, \qquad p_{32} = p_{54} = p$$

and

$$p_{31} = p_{24} = p_{33} = 0.$$

The probabilities can be arranged in a *matrix* [6] of seven rows and seven columns where the entry in the ith row and jth column ($i, j = 1,2,3,4,5,6,7$) is the transition probability p_{ij}.   In accord with this, the matrix is called a *transition matrix* and is generally designated by T.   For our example,

8-1a₄

$$T = \begin{pmatrix} 1 & 0 & 0 & 0 & 0 & 0 & 0 \\ p & 0 & q & 0 & 0 & 0 & 0 \\ 0 & p & 0 & q & 0 & 0 & 0 \\ 0 & 0 & p & 0 & q & 0 & 0 \\ 0 & 0 & 0 & p & 0 & q & 0 \\ 0 & 0 & 0 & 0 & p & 0 & q \\ 0 & 0 & 0 & 0 & 0 & 0 & 1 \end{pmatrix}$$

In T the sum of the entries in each row is one, reflecting the fact that no matter what state the contest is in at the end of the kth game it must be in one of the seven states at the end of the $k + 1$st game.   The fourth row of T:

$$0 \quad 0 \quad p \quad 0 \quad q \quad 0 \quad 0$$

says that if at the end of any game the contest is in the state E_4. it will with probability 0 be in E_1, E_2, E_4, E_6, E_7, with probability p be in E_3 with probability q be in E_5 at the end of the next game.

Throughout the remainder of this chapter we shall return from time to time to this example in order to illustrate certain points and calculations.   For this reason we shall henceforth refer to it as Contest I.

b.   Markov Chains

Contest I is an example of a Markov* Chain and we proceed to define some terms and to investigate the phenomenon more fully.

Consider a system made up of a finite set of states $E = \{E_1, E_2, \ldots, E_N\}$, an experiment with its associated probability space $\{\Omega, A_\Omega, P\}$, and a rule by which sequences or *chains* of the states are generated as a result of repeatedly performing the experiment.   In Contest I the system consisted of the set $E = \{E_1, E_2, \ldots, E_7\}$, a two-person game which could not end in a tie and a rule which said that the chain would move from $E_i \in E$ to E_{i+1} or E_{i-1} depending on the outcome of the game.

Since the orderly development of a chain results from events embedded in a sample space, it is reasonable to consider the probability that a chain unfolds

---

*Named for the Russian mathematician, A. A. Markov (1856–1922).

in a certain way. Since each repetition of the experiment adds a *link* to the chain, if the first i links are

$$E_{j_1} \, E_{j_2} \, E_{j_3} \cdots E_{j_i},$$

we certainly can ask what is the probability the next link (the $i + 1$st) is E_k? We want the conditional probability

$$P(E_k | E_{j_1} \, E_{j_2} \ldots E_{j_i}).$$

In Contest I, the first eight links of a chain we considered are

$$E_3 \, E_4 \, E_3 \, E_2 \, E_3 \, E_2 \, E_3 \, E_4.$$

What is the probability the ninth link is E_5? The chain will move into E_5 if B wins the next game, so

$$P(E_5 | E_3 \, E_4 \, E_3 \, E_2 \, E_3 \, E_2 \, E_3 \, E_4) = q.$$

We remark also that the succession of steps that brought the chain to E_4 has nothing to do with the probability that the next link is E_5, so in Contest I

$$P(E_5 | E_3 \, E_4 \, E_3 \, E_2 \, E_3 \, E_2 \, E_3 \, E_4) = P(E_5 | E_4) = q.$$

Had we asked what is the probability the ninth link is E_2, we would have

$$P(E_2 | E_3 \, E_4 \, E_3 \, E_2 \, E_3 \, E_2 \, E_3 \, E_4) = P(E_2 | E_4) = 0$$

since in one play of the game we can only move from one state to an adjacent one.

Let a system be made up of the set of states $E = \{E_1, E_2, E_3, E_4, E_5, E_6\}$; the experiment of rolling a die and the rule that the ith link is E_v if the ith trial produces a v in such a system. In this system, if the first $i - 1$ links are

$$E_{j_1} \, E_{j_2} \, E_{j_3} \cdots E_{j_{i-1}}$$

$$P(E_v | E_{j_1} \, E_{j_2} \cdots E_{j_{i-1}}) = P(E_v).$$

Unlike the chain in Contest I, the ith link here is completely independent of earlier links.

We look at one more example. This one differs from the first two in that the experiment, and consequently the probability space, vary slightly with each repetition. The system has two states E_r and E_b and a chain of these states is generated in the following way: from an urn containing r red and b black balls one is drawn, its color noted, and the ball plus $c > 0$ more of the color drawn are placed in the urn. The ith link at the chain is E_r or E_b depending upon whether a red ball or a black ball is drawn at the ith trial. Here

$$P(E_r|E_bE_b) = \frac{r}{r+b+2c} \quad \text{and} \quad P(E_r|E_rE_b) = \frac{r+c}{r+b+2c}.$$

These two calculations serve to illustrate the property of this system that the probability that the ith link of the chain is E_r or E_b depends on the entire history of the chain.

Our interest will center on those chains for which

$$P(E_k|E_{j_1} E_{j_2}\cdots E_{j_i}) = P(E_k|E_{j_i}),$$

and we begin with

8-1b_1 **Definition.** *Given a set of N states $E = \{E_1,E_2,\ldots,E_n\}$ and a chain of these states*

$$E_{j_1}E_{j_2}E_{j_3}\cdots$$

The chain is a **Markov chain** *if*

$$P(E_k|E_{j_1}E_{j_2}\cdots E_{j_i}) = P(E_k|E_{j_i}).$$

According to this definition the chain in Contest I is a Markov chain. So is the chain associated with rolling a die, for the fact that

$$P(E_k|E_{j_1}E_{j_2}\cdots E_{j_i}) = P(E_k|F_{j_i}) = P(E_k)$$

is not in contradistinction to the definition. On the other hand, the chain generated by our last experiment is not Markovian since the $(i + 1)$st state depends on more than the ith state.

We note immediately that the definition does not explicitly take into account the experiment used to generate the chain; however it is there implicitly in the conditional probability. Following this, our concern will not focus on how the chains are generated but rather on how they move from one state to another.

One way to look at Markov chains is to say they are produced by processes that have no past or are nonhereditary. Only the present state of the process determines its future, or more specifically, its next state. In using the terms past, present and future we introduce the concept of a chain's being generated at regular intervals in time. Instead of the ith link being determined by the outcome of the ith trial, it can be determined by events occurring at regular intervals of time t_0, t_1, t_2,..., and we speak of the chain being in the state E_{j_k} at time t_k.

A classical example of a Markov chain is a *random walk*. In the one-dimensional case the walker takes a step to the left with probability p and to the right with probability $q = 1 - p$. The length of each step is constant, and if the axis of the walk has bounds, the walk ends whenever the walker reaches one or the other of these bounds. If the walker can move among N points, the walk is a Markov chain with $E_1, E_2, \ldots, E_N$ (see Figure 8-1b_2) where if $j \neq 1$ or $j \neq N$,

$$P(E_{j-1}|E_j) = p$$
$$P(E_{j+1}|E_j) = q$$
$$P(E_k|E_j) = 0 \quad k \neq j \pm 1$$

and if $j = 1$ or $j = N$

$$P(E_j|E_j) = 1$$
$$P(E_k|E_j) = 0 \quad k \neq j.$$

The states E_1 and E_N are called *absorbing states,* since once they are entered they cannot be left.

8-1b_2 Figure
   Axis of a one-dimensional random walk with seven states.

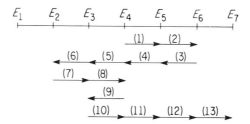

(The arrows trace the walk (chain)
   $E_4 \ E_5 \ E_6 \ E_5 \ E_4 \ E_3 \ E_2 \ E_3 \ E_4 \ E_3 \ E_4 \ E_5 \ E_6 \ E_7$)

If $N = 7$ as illustrated in Figure 8-1b_2 the Markov chain associated with the random walk is the same as that associated with Contest I. (A more colorful description of this one-dimensional random walk is given by the "inebriated gentleman" who takes a step toward his home with probability p and one toward his favorite pub with probability $q = 1 - p$. If he reaches his home or the pub he will stay for the night—he has reached one of the absorbing states and so the walk is over.)

c.   Transition Probabilities and Matrices

Along with the set of states $E_1, E_2, \ldots, E_N$, associated with every Markov chain, is the set of transition probabilities.

8-1c_1 Definition.   *The **transition probability,** p_{ij}, is the probability the system moves from E_i to E_j in one step (or at one trial, or in one time interval).*

In terms of our earlier notation $p_{ij} = P(E_j|E_i)$. Furthermore if the Markov chain has moved through the states $E_{k_1}, E_{k_2}, \ldots, E_{k_p}, E_i$, the probability it will be in E_j at the next step is, from 8-1b_1,

8-1c_2 $$P(E_j|E_{k_1}, E_{k_2}, \ldots, E_{k_p}, E_i) = P(E_j|E_i) = p_{ij}.$$

Since p_{ij} is a probability, for all i and j,

8-1c_3 $$p_{ij} \geqslant 0,$$

and for every i

8-1c_4 $$\sum_{j=1}^{N} p_{ij} = 1$$

since at every step the system moves from the state it is in to the same or to another state. If in a random walk with seven states $E_1\ E_2, \ldots, E_7$ (see Figure 8-1b_2), the walker moves to the right with probability $\frac{2}{3}$ and to the left with probability $\frac{1}{3}$, and if he is in E_4, then

$$p_{41} = p_{42} = p_{44} = p_{46} = p_{47} = 0$$

$$p_{43} = \frac{1}{3}, \ p_{45} = \frac{2}{3}$$

so that

$$\sum_{j=1}^{7} p_{4j} = p_{43} + p_{45} = \frac{1}{3} + \frac{2}{3} = 1$$

The set of all transition probabilities can be arranged in a transition matrix in which the entry in the ith row and jth column is p_{ij}.

8-1c_5 **Definition.** *The* **transition matrix** *of a system with* N *states,* $E_1, E_2, \ldots, E_N$ *and transition probabilities* p_{ij}, $i, j = 1, 2, \ldots, N$, *is*

$$T = \begin{pmatrix} p_{11} & p_{12} & p_{13} & \cdots & p_{1N} \\ p_{21} & p_{22} & p_{23} & \cdots & p_{2N} \\ p_{31} & p_{32} & p_{33} & \cdots & p_{3N} \\ \cdot & \cdot & \cdot & & \cdot \\ \cdot & \cdot & \cdot & & \cdot \\ \cdot & \cdot & \cdot & & \cdot \\ p_{N1} & p_{N2} & p_{N3} & \cdots & p_{NN} \end{pmatrix}$$

The ith row of the transition matrix is

$$p_{i1} \quad p_{i2} \quad p_{i3} \ \cdots \ p_{iN}$$

and the jth column is

$$
\begin{matrix}
p_{1j} \\
p_{2j} \\
p_{3j} \\
\cdot \\
\cdot \\
\cdot \\
p_{Nj}
\end{matrix}
$$

From 8-1c_3 all entries in a transition matrix are nonnegative, and from 8-1c_4 the row sum $\sum\limits_{j=1}^{N} p_{ij} = 1$ for all i. Furthermore, let us

8-1c_6 *Assume that* T *does not contain a column of zeros.*

(That is, there is no k such that $p_{ik} = 0$ for $i = 1,2,\ldots,N$) If the kth column had only zero entries, the probability of entering the state E_k would be zero and E_k would not appear as a link in any chain; hence we drop E_k from our set of states and future consideration. We will assume henceforth that there is a positive probability of entering all states.

For the random walk with seven states where the probability of moving to the right is $\dfrac{2}{3}$ and that of moving to the left is $\dfrac{1}{3}$, the transition matrix is

$$
T =
\begin{pmatrix}
1 & 0 & 0 & 0 & 0 & 0 & 0 \\
\dfrac{1}{3} & 0 & \dfrac{2}{3} & 0 & 0 & 0 & 0 \\
0 & \dfrac{1}{3} & 0 & \dfrac{2}{3} & 0 & 0 & 0 \\
0 & 0 & \dfrac{1}{3} & 0 & \dfrac{2}{3} & 0 & 0 \\
0 & 0 & 0 & \dfrac{1}{3} & 0 & \dfrac{2}{3} & 0 \\
0 & 0 & 0 & 0 & \dfrac{1}{3} & 0 & \dfrac{2}{3} \\
0 & 0 & 0 & 0 & 0 & 0 & 1
\end{pmatrix}
$$

The transition matrix for Contest I is exactly like this if we replace $\dfrac{1}{3}$ by p and $\dfrac{2}{3}$ by q (see 8-1a_4).

A series of Bernoulli trials will result in a sequence of successes and failures where a success occurs with probability p and a failure with probability $q = 1 - p$.

The series generates a chain of the states E_1, E_2, E_3, E_4, E_5 where the nth link is E_1 if the nth trial results in failure, it is E_k, $2 \leqslant k \leqslant 4$, if the last failure occurred at at trial $n - k + 1$, and it is E_5 if the last failure occurred at trial $n - k + 1 \geqslant 4$. The chain must start in E_1. For example, if the series of trials yields the sequence

$$F\ F\ S\ S\ F\ S\ S\ S\ S\ S\ F\ F\ F\ S\ S\ F\ S\ F\ S\ S$$

the resulting chain is

$$E_1\ E_1\ E_2\ E_3\ E_1\ E_2\ E_3\ E_4\ E_5\ E_5\ E_1\ E_1\ E_1\ E_2\ E_3\ E_1\ E_2\ E_1\ E_2\ E_3.$$

The transition probabilities for this chain are:

$$
\begin{aligned}
p_{k1} &= q, \ k = 1, 2, 3, 4, 5 \\
p_{k\,k+1} &= p, \ k = 1, 2, 3, 4 \\
p_{55} &\ -p \\
p_{kj} &= 0 \ \text{ in all other cases}
\end{aligned}
$$

The transition matrix is

8-1c₇

$$
T =
\begin{pmatrix}
q & p & 0 & 0 & 0 \\
q & 0 & p & 0 & 0 \\
q & 0 & 0 & p & 0 \\
q & 0 & 0 & 0 & p \\
q & 0 & 0 & 0 & p
\end{pmatrix}
$$

If the Bernoulli trial consists in rolling a pair of dice with succes being the appearance of a sum of seven, $p = \dfrac{1}{6}$ and $q = \dfrac{5}{6}$.

EXERCISES 8-1c

Four systems which generate Markov chains are now described; they will be used throughout this chapter to illustrate various properties of chains and as a basis for exercises. They will be referred to by the italicized title at the beginning of each statement.

A. *Tennis.* The sequence of points in the game of tennis is love, 15, 30, 40, game. To win, a player must lead by two points; hence if the score is 40–40, the game continues until one player wins two consecutive points. If the players are designated by A and B, the sequence of scores generates a chain of the five states: E_1, A wins the game; E_2, A leads by one stroke; E_3, A and B tied; E_4, B leads by one stroke and E_5, B wins the game.

B. *Urn I.* An urn contains 4 red, 3 white, and 6 blue balls. A ball is drawn, then a second; the first ball is replaced and a third ball drawn; the second is replaced then a fourth drawn, and so forth.   A chain of the states E_r, E_w, E_b is generated if, starting with the second ball drawn, the link is E_r if a red ball is drawn, E_w if the ball is white and E_b if it is blue.   If the sequence of draws is

$$w \quad r \quad r \quad b \quad b \quad w \quad w \quad r \quad w \quad b,$$
$$E_r \quad E_r \quad E_b \quad E_b \quad E_w \quad E_w \quad E_r \quad E_w \quad E_b$$

is the sequence of states.

C.   *Urn II.*   Add three green balls to Urn I and generate a chain in the same way.   The new chain has four states; E_r, E_w, E_b, and E_g.

D.   *Maze.*   A box contains nine compartments which open into each other so that a mouse, when stimulated, will move from the cell he occupies to an adjacent one.

If the mouse is in E_4 only the exits from E_4 open, so a stimulus will cause the mouse to leave E_4, but he will have to remain in the cell he first enters.   The sequence of moves taken by the mouse form a Markov chain with the states $E_1, E_2, \ldots, E_9$.

1.   A sequence of Bernoulli trials generate a chain whose links are either F or S.   If $p(S) = p$ and $p(F) = q = 1 - p$, construct the transition matrix.

2.   If in *Tennis* A wins a point with probability $\frac{3}{7}$ and B wins with probability $\frac{4}{7}$, construct the transition matrix. (a) What are $p_{12}, p_{24}, p_{54}, p_{43}$? (b) What is the probability A wins the game?

3.   If in *Tennis A* wins a point with probability p and B wins with probability $q = 1 - p$, construct the transition matrix.   What is the probability B wins the game?

4.   Construct the transition matrix for the chains in *Urn I*.   (Instead of p_{12}, p_{32}, etc., use p_{rw}, p_{wb}, etc.)

5.   Construct the transition matrix for the chain in *Urn II*.

6.   If in the *Maze* the mouse moves with equal probability in any available direction, construct the transition matrix.   (For example,

$$p_{12} = p_{14} = \frac{1}{2}, \quad p_{1j} = 0, j \neq 2, 4.)$$

7. The mouse's progress through the *Maze* is a two-dimensional random walk. (a) If the four corner cells, E_1, E_3, E_7, E_9 are absorbing, that is, once they are entered the walk is over, construct the transition matrix for this two-dimensional random walk. (b) If the center cell, E_5, is absorbing, construct the transition matrix.

8. Suppose the square used in the *Maze* is sectioned into 16 cells. Describe the Markov chain generated by a mouse treated as in the 9-cell case. Construct the transition matrix.

9. Do Prob. 7 for a 16-cell, two-dimensional walk. Note that in this case there are four interior cells, E_6, E_7, E_{10}, E_{11}.

8-2. PROBABILITY VECTORS

a. Initial Probability Vector

In studying a Markov chain it is frequently of interest to know the probability that the system moves through a prescribed sequence of states. Suppose we want the probability the system moves through E_2, E_4, E_3, E_7, and E_6 in that order. Using the general from for the probability of the intersection of events (see 7-1a_3),

$$P(E_2E_4E_3E_7E_6) = P(E_6|E_2E_4E_3E_7)P(E_7|E_2E_4E_3)P(E_3|E_2E_4)P(E_4|E_2)P(E_2),$$

and since we are working with a Markov chain, we have from 8-1c_2

$$P(E_2E_4E_3E_7E_6) = p_{76}p_{37}p_{43}p_{24}P(E_2) = P(E_2)p_{24}p_{43}p_{37}p_{76}.$$

$P(E_2)$ is the probability that the Markov chain being studied is in the state E_2 at the start of the observations. In general, a Markov chain will move through the sequence $E_{j_1}, E_{j_2}, E_{j_3}, \ldots, E_{j_k}$, where each E_{j_i} is a state of the system, with probability

8-2a_1
$$P(E_{j_1}E_{j_2}\cdots E_{j_k}) = P(E_{j_1})p_{j_1j_2}p_{j_2j_3}\cdots p_{j_{k-1}j_k}$$

$P(E_{j_1})$ is the probability that the system is in the state E_{j_1} when the observations are begun.

The appearance of $P(E_2)$ in the first example and $P(E_{j_1})$ in the second indicate that to describe a Markov chain we need to know more than the set of states $E_1, E_2, \ldots, E_N$ and the transition probabilities p_{ij}, $i, j = 1, 2, \ldots, N$. We need also to know at any given time the probability that the system is in a given state. Thus we need to know the nonnegative numbers $P^{(t)}(E_i)$, $i = 1, 2, \ldots, N$, which at time t (or step number t) are the probabilities that the system is in the state E_i. Since at time t the system is in some one of the states,

$$P^{(t)}(E_1) + P^{(t)}(E_2) + \cdots + P^{(t)}(E_N) = 1.$$

Notice that what we have here is a probability distribution over the states of the system; it is completely analogous to the concept of a probability distribution of a random variable. Since we are here concerned with the probability distribution at the beginning of a set of observations, we shall refer to the *initial probability distribution.* For convenience we write

$$P^0(E_i) = p_i,$$

so that for every i,

$$p_i \geqslant 0$$

and

8-2a$_2$
$$\sum_{i=1}^{N} p_i = 1.$$

(The process must begin in one of the states $E_1, E_2, \ldots, E_N$.)

These probabilities are generally displayed in a one-row matrix (a $1 \times N$ matrix)

$$(p_1 \quad p_2 \quad p_3 \quad \cdots \quad p_N)$$

One-row matrices are commonly called *vectors,* and where the entries are the elements of an initial probability distribution, it is called an *initial probability vector.* It is convenient to designate this vector by π_0, so that we have

8-2a$_3$
$$\pi_0 = (p_1 \quad p_2 \quad p_3 \quad \cdots \quad p_N)$$

Do we know, or can we compute, this initial probability vector? In many cases it is given along with the set of states and the transition probabilities. In Contest I, if the observations are begun at t_0, the system must be in E_4 (neither A nor B ahead) and so $p_4 = 1, p_k = 0, k \neq 4$. Hence

$$\pi_0 = (0 \quad 0 \quad 0 \quad 1 \quad 0 \quad 0 \quad 0).$$

Consequently, starting at t_0 it is impossible for the system to move through the sequence $E_5 \, E_6 \, E_5 \, E_4 \, E_3$, since

$$P(E_5 E_6 E_5 E_4 E_3) = p_5 p_{56} p_{65} p_{54} p_{43} = 0.$$

If we begin observations at t_1 (after the play of the first game), the system can be in E_3 with probability p and E_5 with probability q, so the corresponding initial distribution vector is

$$(0 \quad 0 \quad p \quad 0 \quad q \quad 0 \quad 0),$$

and if we start observations after two games have been played,

$$(0 \quad p^2 \quad 0 \quad 2pq \quad 0 \quad q^2 \quad 0)$$

is the initial probability vector. The tree diagram of Figure 8-2a_4 indicates one way the vectors can be computed if the observations are begun after the play of k games.

8-2a_4   **Figure**
      Tree diagram.

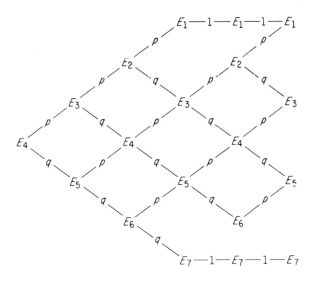

Games played	0	1	2	3	4	5
Time	t_0	t_1	t_2	t_3	t_4	t_5

Consider the Markov chain resulting from a series of Bernoulli trials discussed in Sec. 8-1c. If t_0 is taken to be the beginning of the process,

$$\pi_0 = (1 \quad 0 \quad 0 \quad 0 \quad 0)$$

since the process must start in E_1. If observations start after the third Bernoulli trial (the first must result in failure), the probability distribution will yield the vector

$$(q \quad qp \quad p^2 \quad 0 \quad 0).$$

Thus to describe a Markov chain we need a set of states, a set of transition probabilities—or, what is the same thing, a transition matrix—and an initial probability vector. As we have used the term, an initial probability distribution, or vector, is not necessarily the distribution at the time the process began, but rather the distribution at the time the observation began.

EXERCISES 8-2a

1. In *Tennis* what is π_0 if observations begin when the first 40–40 score is attained? What is the initial probability vector if observations begin three points later?

2. If a mouse is put into E_5 in the *Maze* and observations start before the first stimulus, $\pi_0 = (0\ 0\ 0\ 0\ 1\ 0\ 0\ 0\ 0)$. (a) What is π_0 if observations begin after the first stimulus? (b) After the second?

3. In a chain generated by a sequence of Bernoulli trials where S occurs with probability p and F with probability $q = 1 - p$ (see 8-1c_7), (a) what is the distribution vector at the start of the sequence? (b) after one trial? (c) after three trials?

4. If the behavior of the mouse in *Maze* is altered so that at each stimulus he does not move with probability $\dfrac{1}{2}$ and moves through any of the available openings with equal probability (for a total of $\dfrac{1}{2}$) and if the mouse is set in E_5, (a) what is the probability distribution after the first stimulus? (b) after the second?

b. Higher-Order Probability Vectors

Given the initial probability vector and the transition matrix of a Markov chain is it possible to determine the probability that the system will be in a given state at a given time? We need to answer this question positively if we are to be able to find the probability that A wins Contest I. Player A wins if the system is in the state E_1 and this is possible at $t_3, t_5, t_7 \ldots$ (A quick glance at Figure 8-2a_4 will convince one that A can win only at odd-numbered times.)

A Markov process can be in the state E_j at $t = 1$ if it was in E_i at $t = 0$ and then went from E_i to E_j. Furthermore, if $i \neq k$, the events the system moves from E_i to E_j and the system moves from E_k to E_j are disjoint, and hence the probability that a system with states $E_1, E_2, \ldots, E_N$ is in E_j at $t = 1$ is

$$\textbf{8-2}b_1 \qquad \sum_{i=1}^{N} P(E_i)P(E_j|E_i) = \sum_{i=1}^{N} p_i p_{ij} = p_j^{(1)}$$

where $p_j^{(1)}$ designates the probability that the system is in E_j at $t = 1$. Since

$p_j^{(1)}$ is a probability and since the system is in one of its possible states at $t = 1$, we have

$$p_j^{(1)} \geqslant 0 \quad \text{for all } j$$

and

$$\sum_{j=1}^{N} p_j^{(1)} = 1.$$

The set of $p_j^{(1)}$'s is the probability distribution of the system at $t = 1$ and analogous to the initial probability vector π_0, we display this distribution as a vector:

8-2b₂
$$\pi_1 = (p_1^{(1)} \ p_2^{(1)} \ p_3^{(1)} \ \cdots p_N^{(1)})$$

Again using Contest I for purposes of illustration, if

$$\pi_0 = (0 \quad 0 \quad 0 \quad 1 \quad 0 \quad 0 \quad 0)$$

and T is given in 8-1a_4, we calculate

$$p_1^{(1)} = \sum_{i=1}^{7} p_i p_{i1} = p_4 p_{41} = 0$$

$$p_2^{(1)} = \sum_{i=1}^{7} p_i p_{i2} = p_4 p_{42} = 0$$

$$p_3^{(1)} = \sum_{i=1}^{7} p_i p_{i3} = p_4 p_{43} = p$$

$$p_4^{(1)} = \sum_{i=1}^{7} p_i p_{i4} = p_4 p_{44} = 0$$

$$p_5^{(1)} = \sum_{i=1}^{7} p_i p_{i5} = p_4 p_{45} = q$$

$$p_6^{(1)} = \sum_{i=1}^{7} p_i p_{i6} = p_4 p_{46} = 0$$

$$p_7^{(1)} = \sum_{i=1}^{7} p_i p_{i7} = p_4 p_{47} = 0$$

$p_{4j} = 0$ unless $j = 4 \pm 1$. Hence only $p_3^{(1)}$ and $p_5^{(1)}$ do not vanish.) We see immediately that $p_j^{(1)} \geqslant 0$ for $j = 1, 2, \ldots, 7$, $\sum_{j=1}^{7} p_j^{(1)} = p + q = 1$ and

$$\pi_1 = (0 \quad 0 \quad p \quad 0 \quad q \quad 0 \quad 0).$$

Analogously, a process can be in E_j at $t = 2$ if it was in E_i at $t = 1$ and then moved from E_i to E_j. We can calculate the probabilities, $p_j^{(2)}$, that the system is in E_j at $t = 2$ from the $p_j^{(1)}$'s;

8-2b₃
$$p_j^{(2)} = \sum_{i=1}^{N} p_i^{(1)} p_{ij},$$

where, obviously,

$$p_j^{(2)} \geqslant 0 \quad \text{for all } j$$

and

$$\sum_{j=1}^{N} p_j^{(2)} = 1.$$

The set of $p_j^{(2)}$'s is the probability distribution over the states of the Markov chain at $t = 2$ and they can be displayed as the vector

8-2b₄
$$\pi_2 = (p_1^{(2)} \quad p_2^{(2)} \quad p_3^{(2)} \quad \cdots \quad p_N^{(2)}).$$

We note that we cannot find π_1 without knowledge of π_0 nor π_2 without knowledge of π_1. For Contest I,

8-2b₅
$$\pi_0 = (0 \ 0 \ 0 \ 1 \ 0 \ 0 \ 0)$$

$$\pi_1 = (0 \ 0 \ p \ 0 \ q \ 0 \ 0).$$

Let us compute π_2:

$$p_1^{(2)} = \sum_{i=1}^{7} p_i^{(1)} p_{i1} = p_3^{(1)} p_{31} + p_5^{(1)} p_{51} = 0$$

$$p_2^{(2)} = \sum_{i=1}^{7} p_i^{(1)} p_{i2} = p_3^{(1)} p_{32} + p_5^{(1)} p_{52} = p \cdot p = p^2$$

$$p_3^{(2)} = \sum_{i=1}^{7} p_i^{(1)} p_{i3} = p_3^{(1)} p_{33} + p_5^{(1)} p_{53} = 0$$

$$p_4^{(2)} = \sum_{i=1}^{7} p_i^{(1)} p_{i4} = p_3^{(1)} p_{34} + p_5^{(1)} p_{54} = p \cdot q + q \cdot p = 2pq$$

$$p_5^{(2)} = \sum_{i=1}^{7} p_i^{(1)} p_{i5} = p_3^{(1)} p_{35} + p_5^{(1)} p_{55} = 0$$

$$p_6^{(2)} = \sum_{i=1}^{7} p_i^{(1)} p_{i6} = p_3^{(1)} p_{36} + p_5^{(1)} p_{56} = q \cdot q = q^2$$

$$p_7^{(2)} = \sum_{i=1}^{7} p_i^{(1)} p_{i7} = p_3^{(1)} p_{37} + p_5^{(1)} p_{57} = 0$$

Hence

8-2b₆ $$\pi_2 = (0 \quad p^2 \quad 0 \quad 2pq \quad 0 \quad q^2 \quad 0)$$

Making a brief check of our calculations, we observe

$$p_j^{(2)} \geq 0 \quad \text{for every } j$$

and

$$\sum_{i=1}^{7} p_j^{(2)} = p^2 + 2pq + q^2 = (p + q)^2 = 1.$$

Once we know the distribution π_2, we can readily find π_3, since

$$p_j^{(3)} = \sum_{i=1}^{7} p_i^{(2)} p_{ij}$$

Continuing to build in this way we have that at $t - k$, the system is in E_j with probability

8-2b₇ $$p_j^{(k)} = \sum_{i=1}^{N} p_i^{(k-1)} p_{ij},$$

that is, the probability the chain is in E_j at the kth step (or trial or time interval) is the sum over all i of the probability the chain is in E_i at the $(k-1)$st step and in the next step moves from E_i to E_j. We have the distribution vector

$$\pi_k = (p_1^{(k)} \quad p_2^{(k)} \quad p_3^{(k)} \quad \cdots \quad p_N^{(k)})$$

For the Markov chain described by the transition matrix

$$T = \begin{pmatrix} \dfrac{1}{3} & \dfrac{1}{2} & \dfrac{1}{6} \\[2mm] \dfrac{1}{4} & \dfrac{3}{8} & \dfrac{3}{8} \\[2mm] \dfrac{1}{7} & \dfrac{2}{7} & \dfrac{4}{7} \end{pmatrix}$$

and $\pi_0 = \left(\dfrac{1}{3} \quad \dfrac{1}{4} \quad \dfrac{5}{12} \right)$, successive calculations give

$$p_1^{(1)} = \frac{1}{3} \cdot \frac{1}{3} + \frac{1}{4} \cdot \frac{1}{4} + \frac{5}{12} \cdot \frac{1}{7} = \frac{430}{2016} = .2331$$

$$p_2^{(1)} = \frac{1}{3} \cdot \frac{1}{2} + \frac{1}{4} \cdot \frac{3}{8} + \frac{5}{12} \cdot \frac{2}{7} = \frac{765}{2016} = .3795$$

$$p_3^{(1)} = \frac{1}{3} \cdot \frac{1}{6} + \frac{1}{4} \cdot \frac{3}{8} + \frac{5}{12} \cdot \frac{4}{7} = \frac{781}{2016} = .3874$$

$$\sum_{j=1}^{3} p_j^{(1)} = 1.0000$$

and

$$p_1^{(2)} = (.2331)\frac{1}{3} + (.3795)\frac{1}{4} + (.3874)\frac{1}{7} = .2279$$

$$p_2^{(2)} = (.2331)\frac{1}{2} + (.3795)\frac{3}{8} + (.3874)\frac{2}{7} = .3696$$

$$p_3^{(2)} = (.2331)\frac{1}{6} + (.3795)\frac{3}{8} + (.3874)\frac{4}{7} = \underline{.4025}$$

$$\sum_{j=1}^{3} p_j^{(2)} = 1.0000$$

c. A Second Approach to Computing the π_i

Let us look again at the Markov chains with

$$\pi_0 = \pi_0' = (1 \ 0 \ 0 \ 0 \ 0)$$

and transition matrices

$$T = \begin{pmatrix} q & p & 0 & 0 & 0 \\ q & 0 & p & 0 & 0 \\ q & 0 & 0 & p & 0 \\ q & 0 & 0 & 0 & p \\ q & 0 & 0 & 0 & p \end{pmatrix} \quad \text{and} \quad T' = \begin{pmatrix} \frac{1}{6} & \frac{5}{6} & 0 & 0 & 0 \\ \frac{1}{6} & 0 & \frac{5}{6} & 0 & 0 \\ \frac{1}{6} & 0 & 0 & \frac{5}{6} & 0 \\ \frac{1}{6} & 0 & 0 & 0 & \frac{5}{6} \\ \frac{1}{6} & 0 & 0 & 0 & \frac{5}{6} \end{pmatrix}$$

which are generated by sequences of Bernoulli trials (see the end of Sec. 8-1c). To find π_1 and π_1' we have from 8-2b_1.

$$p_1^{(1)} = \sum_{i=1}^{5} p_i p_{i1} = p_1 p_{11} = 1 \cdot q = q$$

$$p_2^{(1)} = \sum_{i=1}^{5} p_i p_{i2} = p_1 p_{12} = 1 \cdot p = p$$

since $p_1 + p_2 = 1, p_3 = p_4 = p_5 = 0$, and hence

$$\pi_1 = (q \ p \ 0 \ 0 \ 0).$$

Substituting $\frac{1}{6}$ for q and $\frac{5}{6}$ for p, we have

$$\pi_1' = \begin{pmatrix} \frac{1}{6} & \frac{5}{6} & 0 & 0 & 0 \end{pmatrix}.$$

To find π_2, and π_2' from 8-2b_3

$$p_1^{(2)} = \sum_{i=1}^{5} p_i^{(1)} p_{i1} = p_1^{(1)} p_{11} + p_2^{(1)} p_{21} = q \cdot q + p \cdot q = q(p + q) = q$$

$$p_2^{(2)} = \sum_{i=1}^{5} p_i^{(1)} p_{i2} = p_1^{(1)} p_{12} + p_2^{(1)} p_{22} = q \cdot p + p \cdot 0 = qp$$

$$p_3^{(2)} = \sum_{i-1}^{5} p_i^{(1)} p_{i3} = p_1^{(1)} p_{13} + p_2^{(1)} p_{23} = q \cdot 0 + p \cdot p = p^2$$

We note

$$p_1^{(2)} + p_2^{(2)} + p_3^{(2)} = q + qp + p^2 = q + p(q + p) = 1,$$

so that

$$p_4^{(2)} = p_5^{(2)} = 0.$$

Hence

$$\pi_2 = \begin{pmatrix} q & qp & p^2 & 0 & 0 \end{pmatrix} \quad \text{and} \quad \pi_2' = \begin{pmatrix} \frac{1}{6} & \frac{5}{36} & \frac{25}{36} & 0 & 0 \end{pmatrix}.$$

Going one step further,

$$p_1^{(3)} = \sum_{i=1}^{5} p_i^{(2)} p_{i1} = p_1^{(2)} p_{11} + p_2^{(2)} p_{21} + p_3^{(2)} p_{31} = q \cdot q + (qp) q + p^2 q = q$$

$$p_2^{(3)} = \sum_{i=1}^{5} p_i^{(2)} p_{i2} = p_1^{(2)} p_{12} + p_2^{(2)} p_{22} + p_3^{(2)} p_{32} = qp + qp \cdot 0 + p^2 \cdot 0 = qp$$

$$p_3^{(3)} = \sum_{i=1}^{5} p_i^{(2)} p_{i3} = p_1^{(2)} p_{13} + p_2^{(2)} p_{23} + p_3^{(2)} p_{33} = q \cdot 0 + (qp) p + p^2 0 = qp^2$$

$$p_4^{(3)} = \sum_{j=1}^{5} p_i^{(2)} p_{i4} = p_1^{(2)} p_{14} + p_2^{(2)} p_{24} + p_3^{(2)} p_{34} = q \cdot 0 + (qp) \cdot 0 + p^2 \cdot p = p^3.$$

Since

$$p_1^{(3)} + p_2^{(3)} + p_3^{(3)} + p_4^{(3)} = 1, \quad p_5^{(3)} = 0$$

and $\quad \pi_3 = (q \quad qp \quad qp^2 \quad p^3 \quad 0) \quad$ and $\quad \pi_3' = \left(\dfrac{1}{6} \quad \dfrac{5}{36} \quad \dfrac{25}{216} \quad \dfrac{125}{216} \quad 0 \right)$

Let us look more closely at the transition from π_0 to π_1, from π_1 to π_2 and so forth. Given a transition matrix T and an initial probability vector π_0, the elements of π_1 are computed by the formula

$$p_j^{(1)} = \sum_{i=1}^{5} p_i p_{ij} \tag{8-2b_1}$$

This says that to find the jth element of π_1 we multiply the ith element of π_0 by the ith element of the jth column of T and sum over i. We find it convenient to write this as

$$\pi_0 T = (1 \ 0 \ 0 \ 0 \ 0) \begin{pmatrix} q & p & 0 & 0 & 0 \\ q & 0 & p & 0 & 0 \\ q & 0 & 0 & p & 0 \\ q & 0 & 0 & 0 & p \\ q & 0 & 0 & 0 & p \end{pmatrix} = (q \ p \ 0 \ 0 \ 0) = \pi_1.$$

(This notation will be explained carefully in the next section.) Similarly, since

$$p_j^{(2)} = \sum_{i=1}^{5} p_i^{(1)} p_{ij}$$

the jth element of π_2 is calculated by multiplying the ith element of π_1 by the ith element of the jth column of T and summing over i. Using the same notation as above,

$$\pi_1 T = (q \ p \ 0 \ 0 \ 0) \begin{pmatrix} q & p & 0 & 0 & 0 \\ q & 0 & p & 0 & 0 \\ q & 0 & 0 & p & 0 \\ q & 0 & 0 & 0 & p \\ q & 0 & 0 & 0 & p \end{pmatrix}$$

$$= (q + qp \quad qp \quad p^2 \ 0 \ 0) = (q \quad qp \quad p^2 \ 0 \ 0) = \pi_2.$$

Let us try this method with Contest I, where $\pi_0 = (0 \ 0 \ 0 \ 1 \ 0 \ 0 \ 0)$ and

$$T = \begin{pmatrix} 1 & 0 & 0 & 0 & 0 & 0 & 0 \\ p & 0 & q & 0 & 0 & 0 & 0 \\ 0 & p & 0 & q & 0 & 0 & 0 \\ 0 & 0 & p & 0 & q & 0 & 0 \\ 0 & 0 & 0 & p & 0 & q & 0 \\ 0 & 0 & 0 & 0 & p & 0 & q \\ 0 & 0 & 0 & 0 & 0 & 0 & 1 \end{pmatrix}$$

Employing the new notation,

$$\pi_0 T = (0\ \ 0\ \ 0\ \ 1\ \ 0\ \ 0\ \ 0) \begin{pmatrix} 1 & 0 & 0 & 0 & 0 & 0 & 0 \\ p & 0 & q & 0 & 0 & 0 & 0 \\ 0 & p & 0 & q & 0 & 0 & 0 \\ 0 & 0 & p & 0 & q & 0 & 0 \\ 0 & 0 & 0 & p & 0 & q & 0 \\ 0 & 0 & 0 & 0 & p & 0 & q \\ 0 & 0 & 0 & 0 & 0 & 0 & 1 \end{pmatrix}$$

$$= (0\ \ 0\ \ p\ \ 0\ \ q\ \ 0\ \ 0) = \pi_1.$$

We write out explicitly only $p_1^{(1)}$ and $p_3^{(1)}$:

$$p_1^{(1)} = p_1 p_{11} + p_2 p_{21} + p_3 p_{31} + p_4 p_{41} + p_5 p_{51} + p_6 p_{61} + p_7 p_{71}$$
$$0\ \ = 0 \cdot 1 + 0 \cdot p + 0 \cdot 0 + 1 \cdot 0 + 0 \cdot 0 + 0 \cdot 0 + 0 \cdot 0$$

$$p_3^{(1)} = p_1 p_{13} + p_2 p_{23} + p_3 p_{33} + p_4 p_{43} + p_5 p_{53} + p_6 p_{63} + p_7 p_{13}$$
$$p\ \ = 0 \cdot 0 + 0 \cdot q + 0 \cdot 0 + 1 \cdot p + 0 \cdot 0 + 0 \cdot 0 + 0 \cdot 0$$

Also

$$\pi_1 T = (0\ \ 0\ \ p\ \ 0\ \ q\ \ 0\ \ 0) \begin{pmatrix} 1 & 0 & 0 & 0 & 0 & 0 & 0 \\ p & 0 & q & 0 & 0 & 0 & 0 \\ 0 & p & 0 & q & 0 & 0 & 0 \\ 0 & 0 & p & 0 & q & 0 & 0 \\ 0 & 0 & 0 & p & 0 & q & 0 \\ 0 & 0 & 0 & 0 & p & 0 & q \\ 0 & 0 & 0 & 0 & 0 & 0 & 1 \end{pmatrix}$$

$$= (0\ \ p^2\ \ 0\ \ 2pq\ \ 0\ \ q^2\ \ 0) = \pi_2.$$

As an example of the calculations,

$$p_4^{(2)} = p_1^{(1)}p_{14} + p_2^{(1)}p_{24} + p_3^{(1)}p_{34} + p_4^{(1)}p_{44} + p_5^{(1)}p_{54} + p_6^{(1)}p_{64} + p_7^{(1)}p_{74}$$

$$2pq = 0 \cdot 0 + 0 \cdot 0 + p \cdot q + 0 \cdot 0 + q \cdot p + 0 \cdot 0 + 0 \cdot 0$$

π_1 and π_2 agree with our previous results 8-2b_5 and 8-2b_6.

Let us now apply this to any Markov process with

$$\pi_0 = (p_1 \quad p_2 \quad p_3 \quad \cdots \quad p_N)$$

and

$$T = \begin{pmatrix} p_{11} & p_{12} & p_{13} & \cdots & p_{1N} \\ p_{21} & p_{22} & p_{23} & \cdots & p_{2N} \\ p_{31} & p_{32} & p_{33} & \cdots & p_{3N} \\ \cdot & \cdot & \cdot & & \cdot \\ \cdot & \cdot & \cdot & & \cdot \\ \cdot & \cdot & \cdot & & \cdot \\ p_{N1} & p_{N2} & p_{N3} & \cdots & p_{NN} \end{pmatrix}$$

Then this operation of multiplying the ith element of π_0 with the ith element of the jth column of T and summing over i would yield:

8-2c_1

$$\pi_0 = (p_1 \quad p_2 \quad p_3 \cdots p_N) \begin{pmatrix} p_{11} & p_{12} & p_{13} & \cdots & p_{1N} \\ p_{21} & p_{22} & p_{23} & \cdots & p_{2N} \\ p_{31} & p_{32} & p_{33} & \cdots & p_{3N} \\ \cdot & \cdot & \cdot & & \cdot \\ \cdot & \cdot & \cdot & & \cdot \\ \cdot & \cdot & \cdot & & \cdot \\ p_{N1} & p_{N2} & p_{N3} & \cdots & p_{NN} \end{pmatrix}$$

$$= \left(\sum_{i=1}^{N} p_i p_{i1} \quad \sum_{i=1}^{N} p_i p_{i2} \quad \sum_{i=1}^{N} p_i p_{i3} \cdots \sum_{i=1}^{N} p_i p_{iN} \right) = \pi_1$$

by 8-2b_1 and 8-2b_2.

For $\pi_0 = \left(\dfrac{1}{3} \quad \dfrac{1}{4} \quad \dfrac{5}{12} \right)$ and $T = \begin{pmatrix} \dfrac{1}{3} & \dfrac{1}{2} & \dfrac{1}{6} \\[2mm] \dfrac{1}{4} & \dfrac{3}{8} & \dfrac{3}{8} \\[2mm] \dfrac{1}{7} & \dfrac{2}{7} & \dfrac{4}{7} \end{pmatrix}$

the operation yields

$$\pi_0 T = \begin{pmatrix} \frac{1}{3} & \frac{1}{4} & \frac{5}{12} \end{pmatrix} \begin{pmatrix} \frac{1}{3} & \frac{1}{2} & \frac{1}{6} \\ \frac{1}{4} & \frac{3}{8} & \frac{3}{8} \\ \frac{1}{7} & \frac{2}{7} & \frac{4}{7} \end{pmatrix}$$

$$= \begin{pmatrix} \frac{1}{3} \cdot \frac{1}{3} + \frac{1}{4} \cdot \frac{1}{4} + \frac{5}{12} \cdot \frac{1}{7} & \frac{1}{3} \cdot \frac{1}{2} + \frac{1}{4} \cdot \frac{3}{8} + \frac{5}{12} \cdot \frac{2}{7} & \frac{1}{3} \cdot \frac{1}{6} + \frac{1}{4} \cdot \frac{3}{8} + \frac{5}{12} \cdot \frac{4}{7} \end{pmatrix}$$

$$= \begin{pmatrix} \frac{430}{2016} & \frac{765}{2016} & \frac{781}{2016} \end{pmatrix} = \pi_1$$

If we apply the operation to π_1 and T, we obtain

8-2c₂

$$\pi_1 T = \begin{pmatrix} p_1^{(1)} & p_2^{(1)} & p_3^{(1)} & \cdots p_N^{(1)} \end{pmatrix} \begin{pmatrix} p_{11} & p_{12} & p_{13} & \cdots & p_{1N} \\ p_{21} & p_{22} & p_{23} & \cdots & p_{2N} \\ p_{31} & p_{32} & p_{33} & \cdots & p_{3N} \\ \cdot & \cdot & \cdot & & \cdot \\ \cdot & \cdot & \cdot & & \cdot \\ p_{N1} & p_{N2} & p_{N3} & \cdots & p_{NN} \end{pmatrix}$$

$$= \begin{pmatrix} \sum_{i=1}^{N} p_i^{(1)} p_{i1} & \sum_{i=1}^{N} p_i^{(1)} p_{i2} & \sum_{i=1}^{N} p_i^{(1)} p_{i3} & \cdots & \sum_{i=1}^{N} p_i^{(1)} p_{iN} \end{pmatrix}$$

$$= \begin{pmatrix} p_1^{(2)} & p_2^{(2)} & p_3^{(2)} & \cdots p_N^{(2)} \end{pmatrix} = \pi_2$$

by 8-2b₃ and 8-2b₄. In the numerical example used above

$$\pi_1 T = \begin{pmatrix} \frac{470}{2016} & \frac{765}{2016} & \frac{781}{2016} \end{pmatrix} \begin{pmatrix} \frac{1}{3} & \frac{1}{2} & \frac{1}{6} \\ \frac{1}{4} & \frac{3}{8} & \frac{3}{8} \\ \frac{1}{7} & \frac{2}{7} & \frac{4}{7} \end{pmatrix}$$

$$\pi_1 T = \left(\frac{420}{2016} \cdot \frac{1}{3} + \frac{765}{2016} \cdot \frac{1}{4} + \frac{781}{2016} \cdot \frac{1}{7} \quad \frac{470}{2016} \cdot \frac{1}{2} + \frac{765}{2016} \cdot \frac{3}{8} + \frac{781}{2016} \cdot \frac{2}{7} \right.$$

$$\left. \frac{470}{2016} \cdot \frac{1}{6} + \frac{765}{2016} \cdot \frac{3}{3} + \frac{781}{2016} \cdot \frac{4}{7} \right)$$

$$= \left(\frac{77194}{338688} \quad \frac{125163}{338688} \quad \frac{136331}{338688} \right) = \pi_2$$

We can repeat this operation as often as we choose and for any $k > 0$,

8-2c₃

$$\pi_{k-1} T = \left(p_1^{(k-1)} p_2^{(k-1)} p_3^{(k-1)} \cdots p_N^{(k-1)} \right) \begin{pmatrix} p_{11} & p_{12} & p_{13} & \cdots & p_{1N} \\ p_{21} & p_{22} & p_{23} & \cdots & p_{2N} \\ p_{31} & p_{32} & p_{33} & \cdots & p_{3N} \\ \cdot & \cdot & \cdot & & \cdot \\ \cdot & \cdot & \cdot & & \cdot \\ \cdot & \cdot & \cdot & & \cdot \\ p_{N1} & p_{N2} & p_{N3} & \cdots & p_{NN} \end{pmatrix}$$

$$= \left(\sum_{i=1}^{N} p_i^{(k-1)} p_{i1} \quad \sum_{i=1}^{N} p_i^{(k-1)} p_{i2} \quad \sum_{i=1}^{N} p_i^{(k-1)} p_{i3} \quad \cdots \quad \sum_{i=1}^{N} p_i^{(k-1)} p_{ik} \right)$$

$$= \left(p_1^{(k)} \quad p_2^{(k)} \quad p_3^{(k)} \quad \cdots p_N^{(k)} \right) = \pi_k$$

where the last two equalities come from 8-2b₇ and 8-2b₈.

To illustrate the development of the π_k's we programmed the operation $\pi_{k-1} T = \pi_k$ for a large computer and starting with

$$T = \begin{pmatrix} \frac{1}{3} & \frac{1}{2} & \frac{1}{6} \\ \frac{1}{4} & \frac{3}{8} & \frac{3}{8} \\ \frac{1}{7} & \frac{2}{7} & \frac{4}{7} \end{pmatrix}$$

and

$$\pi_0 = \left(\frac{1}{3} \quad \frac{1}{4} \quad \frac{5}{12} \right)$$

generated the sequence of π_k's:

$$\pi_1 = (.2331349 \qquad .3794643 \qquad .3874008)$$
$$\pi_2 = (.2279207 \qquad .3695525 \qquad .4025268)$$
$$\pi_3 = (.2258655 \qquad .3675502 \qquad .4065843)$$
$$\pi_4 = (.2252595 \qquad .3669310 \qquad .4078095)$$
$$\pi_5 = (.2250778 \qquad .3667459 \qquad .4081764)$$
$$\pi_6 = (.2250233 \qquad .3666904 \qquad .4082863)$$
$$\pi_7 = (.2250070 \qquad .3666738 \qquad .4083193)$$
$$\pi_8 = (.2250021 \qquad .3666688 \qquad .4083291)$$
$$\pi_9 = (.2250006 \qquad .3666673 \qquad .4083321)$$
$$\pi_{10} = (.2250002 \qquad .3666669 \qquad .4083330)$$
$$\pi_{11} = (.2250001 \qquad .3666667 \qquad .4083332)$$
$$\pi_{12} = (.2250000 \qquad .3666667 \qquad .4083333)$$
$$\pi_{13} = (.2250000 \qquad .3666667 \qquad .4083333)$$

We see that $\pi_{12} = \pi_{13}$, and if we write them in fractional form we have

$$\pi_{12} = \pi_{13} = \left(\frac{9}{40} \quad \frac{11}{30} \quad \frac{49}{120} \right)$$

Furthermore, if we calculate π_{14} we have

$$\pi_{14} = \pi_{13}T = \left(\frac{9}{40} \quad \frac{11}{30} \quad \frac{49}{120} \right) \begin{pmatrix} \frac{1}{3} & \frac{1}{2} & \frac{1}{6} \\ \frac{1}{4} & \frac{3}{8} & \frac{3}{8} \\ \frac{1}{7} & \frac{2}{7} & \frac{4}{7} \end{pmatrix}$$

$$= \left(\frac{9}{120} + \frac{11}{120} + \frac{7}{120} \quad \frac{9}{80} + \frac{11}{80} + \frac{7}{60} \quad \frac{3}{80} + \frac{11}{80} + \frac{7}{30} \right)$$

$$= \left(\frac{9}{40} \quad \frac{11}{30} \quad \frac{49}{120} \right).$$

From this we can conclude that for this T, whenever $k \geqslant 12$,

$$\pi_k = \pi_k T = \left(\frac{9}{40} \quad \frac{11}{30} \quad \frac{49}{120} \right).$$

d. Product of a Vector and Matrix

Our probabilistic considerations have led us to a new kind of multiplication —an operation combining a row vector with a matrix. The product—the end result of this operation—is a second row vector.   In our examples from Markov chains the number of elements in the factor vector and the product vector were the same, but in the general definition this will not be required.

We begin with:

8-2d$_1$ Definition.   *The* **product of an n-element row vector with an n-element column vector** *is the sum of the products of the ith element of the row vector with the ith element of the column vector, i = 1,2,....,n; that is,*

$$(a_1 \ a_2 \ \cdots \ a_n) \begin{pmatrix} b_1 \\ b_2 \\ \cdot \\ \cdot \\ \cdot \\ b_n \end{pmatrix} = a_1 b_1 + a_2 b_2 + \cdots + a_n b_n = \sum_{i=1}^{n} a_i b_i.$$

This process is commonly referred to as *term-by-term* multiplication.   A simple example of the operation is

$$(4 \ 3 \ 2 \ 1) \begin{pmatrix} 0 \\ 8 \\ -7 \\ 3 \end{pmatrix} = 4 \cdot 0 + 3 \cdot 8 + 2(-7) + 1 \cdot 3 = 13.$$

Recalling that an *n*-element row vector is a $1 \times n$ matrix and an *n*-element column vector is a $n \times 1$ matrix, definition 8-2d$_1$ defines the multiplication of a $1 \times n$ matrix with an $n \times 1$ matrix as a number.   Were we to enclose this number in curved brackets, as $\left(\sum_{i=1}^{n} a_i b_i \right)$, we could consider it to be a 1×1 matrix.   In the example above we would write

$$(4 \ \ 3 \ \ 2 \ \ 1) \begin{pmatrix} 0 \\ 8 \\ -7 \\ 3 \end{pmatrix} = (13).$$

We shall now define the product of a *n*-element row vector with an $n \times m$ matrix as an extension of the row by column vector product.   What we do is to define the product as a *m*-element row vector each of whose elements is the product of the row vector and a column of the matrix treated as a column vector.

8-2d_2 **Definition.** *The* **product of a n-element row vector with an n $\times$ m matrix** *is a m-element row vector whose jth element is the product of the row vector with the jth column of the matrix treated as a column vector;* *we have*

$$(v_1 \ v_2 \ v_3 \ldots v_n) \begin{pmatrix} a_{11} & a_{12} & a_{13} & \cdots & a_{1m} \\ a_{21} & a_{22} & a_{23} & \cdots & a_{2m} \\ a_{31} & a_{32} & a_{33} & \cdots & a_{3m} \\ & \cdot & \cdot & \cdot & \cdot \\ & \cdot & \cdot & \cdot & \cdot \\ & \cdot & \cdot & \cdot & \cdot \\ a_{n1} & a_{n2} & a_{n3} & \cdots & a_{nm} \end{pmatrix}$$

$$= \left(\sum_{i=1}^{n} v_i a_{i1} \ \ \sum_{i=1}^{n} v_i a_{i2} \ \ \sum_{i=1}^{n} v_i a_{i3} \ \cdots \ \sum_{i=1}^{n} v_i a_{im} \right).$$

The definition is tantamount to defining the product of a $1 \times n$ matrix with a $n \times m$ matrix as a $1 \times m$ matrix.

In reviewing the probability that introduced us to this multiplicative operation, we find that we were looking for the probability distribution at $t = k$ when we knew the distribution at $t = k - 1$. Specifically, we came up with, for all integers $k > 0$,

$$p_j^{(k)} = \sum_{i=1}^{N} p_i^{(k-1)} p_{ij}. \tag{8-2b_7}$$

Combining 8-2b_7 and 8-2d_2, we have

$$(p_1^{(k)} \ p_2^{(k)} \ \cdots \ p_N^{(k)}) = (p_1^{(k-1)} \ p_2^{(k-1)} \ \cdots \ p_N^{(k-1)}) \begin{pmatrix} p_{11} & p_{12} & \cdots & p_{1N} \\ p_{21} & p_{22} & \cdots & p_{2N} \\ \cdot & \cdot & & \cdot \\ \cdot & \cdot & & \cdot \\ \cdot & \cdot & & \cdot \\ p_{N1} & p_{N2} & & p_{NN} \end{pmatrix}$$

or

8-2d_3 $$\pi_k = \pi_{k-1} T.$$

It was natural to write $p_i^{(k-1)} p_{ij}$ in that order, for this is the probability the system is in E_i at time $k - 1$ and goes to E_j at the next step. Following this pattern when we set the distribution vector beside the transition matrix, we set it to the left of the matrix. Definition 8-2d_2 reflects this, for it speaks of the product of a row vector *with* a matrix and not of a row vector and a matrix. It directs that the row vector be written to the left of the matrix and ignores completely any question of commutivity. And, temporarily at least, so will we. We shall employ only that operation which is defined—the product of a row vector with a matrix.

The definition contains no probabilistic requirements whatsoever. Hence we can apply it to any row vector and any matrix having compatible dimensions. For example, let us compute

$$(2\ \ 3\ \ -1\ \ 0\ \ 5)\begin{pmatrix} 7 & -2 & 4 \\ 3 & 6 & 8 \\ -2 & 8 & 0 \\ 4 & -1 & 1 \\ 0 & 3 & 1 \end{pmatrix} = (14+9+2 \quad -4+18-8+15 \quad 8+24+5)$$

$$= (25\ \ 21\ \ 37).$$

EXERCISES 8-2d

(*Tennis*, *Urn* I, and *Maze* are defined in Exercises 8-1c)

1. In *Tennis* A wins a point with probability $\frac{1}{3}$ and B wins one with probability $\frac{2}{3}$ and $\pi_0 = (0\ \ 0\ \ 1\ \ 0\ \ 0)$, compute the distribution vectors $\pi_1, \pi_2, \pi_3, \pi_4$, and π_5.

2. If in *Tennis* A wins a point with probability $\frac{3}{7}$ and B wins one with probability $\frac{4}{7}$ and $\pi_0 = (0\ \ 0\ \ 1\ \ 0\ \ 0)$, compute the distribution vectors $\pi_1, \pi_2, \pi_3, \pi_4$, and π_5.

3. (a) If the first ball drawn from *Urn* I is white, compute $\pi_0, \pi_1, \pi_2, \pi_3$, (b) If the first ball drawn is blue compute $\pi_0, \pi_1, \pi_2, \pi_3$.

4. If in *Maze* observations start when the mouse is in E_7, find π_1, π_2, π_3.

5. In 8-2c we introduced

$$T' = \begin{pmatrix} \frac{1}{6} & \frac{5}{6} & 0 & 0 & 0 \\ \frac{1}{6} & 0 & \frac{5}{6} & 0 & 0 \\ \frac{1}{6} & 0 & 0 & \frac{5}{6} & 0 \\ \frac{1}{6} & 0 & 0 & 0 & \frac{5}{6} \\ \frac{1}{6} & 0 & 0 & 0 & \frac{5}{6} \end{pmatrix}$$

and $\pi_0' = (1\ 0\ 0\ 0\ 0)$. Replace π_0' by $\pi_0'' = (0\ 0\ 1\ 0\ 0)$ and compute π_1'', π_2'', π_3''. How do these compare with π_1', π_2', π_3'?

6. Let $\pi_0 = \left(\dfrac{1}{10} \quad \dfrac{2}{10} \quad \dfrac{4}{10} \quad \dfrac{3}{10} \quad 0 \right)$ and compute π_1, π_2, π_3, where T' is the transition matrix in Prob. 5.

7. A chain is generated by rolling a fair die and defining the state of the chain to be E_n, $n = 1, 2, 3, 4, 5, 6$, if the largest face yet to have appeared is n. For example, if the rolls of the die yield

$$2,\ 3,\ 4,\ 1,\ 2,\ 5,\ 4,\ 3,\ 1,\ 6,\ 5,\ 4,$$

the successive links of the chain are

$$E_2\ E_3\ E_4\ E_4\ E_4\ E_5\ E_5\ E_5\ E_5\ E_6\ E_6\ E_6.$$

(Notice that E_6 is an absorbing state.) Construct the transition matrix. If we begin observations *after* the first throw, what are π_0, π_1, π_2? If the first throw is a six (i.e., $\pi_0 = (0\ 0\ 0\ 0\ 0\ 1)$) what are $\pi_1\ \pi_2$?

8-3. HIGHER-ORDER TRANSITION PROBABILITIES

a. Second-order transition probabilities

In computing $p_j^{(1)}$ and $p_j^{(2)}$, we found

$$p_j^{(1)} = \sum_{k=1}^{N} p_k\ p_{ki} \tag{8-2b_1}$$

and

$$p_j^{(2)} = \sum_{i-1}^{N} p_i^{(1)}\ p_{ij}. \tag{8-2b_3}$$

If we put 8-2b_1 into 8-2b_3, we have

$$p_j^{(2)} = \sum_{i=1}^{N} \left(\sum_{k=1}^{N} p_k p_{ki} \right) p_{ij}.$$

Since only finite sums are involved, we can change the order of summation and obtain

8-3a_1 $$p_j^{(2)} = \sum_{k=1}^{N} p_k \left(\sum_{i=1}^{N} p_{ki}\ p_{ij} \right).$$

What is $\sum_{i=1}^{N} p_{ki} p_{ij}$? The product $p_{ki} p_{ij}$ is the probability that the system goes from E_k to E_j in two steps via E_i. The sum of the $p_{ki}p_{ij}$ over all i is then the probability the system goes from E_k to E_j in two steps. We set

8-3a₂
$$p_{kj}^{(2)} = \sum_{i=1}^{N} p_{ki}p_{ij}$$

and this is the probability that the Markov process moves from E_k to E_j in two steps, and for all $k, j = 1, 2,...N$ the set $p_{kj}^{(2)}$ is called the set of *second-order transition probabilities* of the process. For example, in *Tennis* where $p = \frac{1}{3}, q = \frac{2}{3}$, and

$$T = \begin{pmatrix} 1 & 0 & 0 & 0 & 0 \\ \frac{1}{3} & 0 & \frac{2}{3} & 0 & 0 \\ 0 & \frac{1}{3} & 0 & \frac{2}{3} & 0 \\ 0 & 0 & \frac{1}{3} & 0 & \frac{2}{3} \\ 0 & 0 & 0 & 0 & 1 \end{pmatrix}$$

$$p_{33}^{(2)} = \sum_{i=1}^{5} p_{3i}p_{i3} = p_{31}p_{13} + p_{32}p_{23} + p_{33}p_{33} + p_{34}p_{43} + p_{35}p_{53}$$

$$= 0 \cdot 0 + \frac{1}{3} \cdot \frac{2}{3} + 0 \cdot 0 + \frac{2}{3} \cdot \frac{1}{3} + 0 \cdot 0$$

$$= \frac{4}{9}$$

and

$$p_{52}^{(2)} = \sum_{i=1}^{5} p_{5i}p_{i2} = p_{51}p_{12} + p_{52}p_{22} + p_{53}p_{32} + p_{54}p_{42} + p_{55}p_{52}$$

$$= 0 \cdot 0 + 0 \cdot 0 + 0 \cdot \frac{1}{3} + 0 \cdot 0 + 1 \cdot 0$$

$$= 0$$

For the process associated with Contest I we would expect that

$$p_{11}^{(2)} = p_{77}^{(2)} = 1$$

since once the system is in E_1 or E_7 it never leaves that state. But, also from $8\text{-}3a_2$ we have

$$p_{11}^{(2)} = \sum_{i=1}^{7} p_{1i}p_{i1} = p_{11}p_{11} + p_{12}p_{21} + p_{13}p_{31} + \cdots + p_{17}p_{71}$$

$$= 1 \cdot 1 + 0 \cdot p = 1$$

and

$$p_{77}^{(2)} = \sum_{i=1}^{7} p_{7i}p_{i7} = p_{71}p_{17} + p_{72}p_{27} + \cdots + p_{77}p_{77}$$

$$= 0 \cdot q + 1 \cdot 1 = 1.$$

For $j \neq 1$ and $k \neq 7$,

$$p_{1j}^{(2)} = \sum_{i=1}^{7} p_{1i}p_{ij} = 0 \quad \text{and} \quad p_{7k}^{(2)} = \sum_{i=1}^{7} p_{7i}p_{ik} = 0$$

since both $p_{1i} = 0$, $i \neq 1$, and $p_{1j} = 0$ and since both $p_{7i} = 0$, $i \neq 7$, and $p_{7k} = 0$. If $i \neq 1,7$, then $p_{i\,i+1} = q$, $p_{i\,i-1} = p$ and $p_{kj} = 0$, $j \neq k \pm 1$. Hence

$$p_{kj}^{(2)} = \sum_{k=1}^{7} p_{ki}p_{ij} = p_{kk-1}p_{k-1j} + p_{kk+1}p_{k+1j}$$

is different from zero if and only if $j = k$ or $j = k \pm 2$. A few examples are

$$p_{33}^{(2)} = \sum_{i=1}^{7} p_{3i}p_{i3} = p_{32}p_{23} + p_{34}p_{43} = pq + qp = 2pq$$

$$p_{57}^{(2)} = \sum_{i=1}^{7} p_{5i}p_{i7} = p_{56}p_{67} = q \cdot q = q^2$$

$$p_{42}^{(2)} = \sum_{i=1}^{7} p_{4i}p_{i2} = p_{43}p_{32} = p \cdot p = p^2$$

Looking at where the play might be after two steps when it starts in E_3, we have

$$p_{32}^{(2)} = p_{34}^{(2)} = p_{36}^{(2)} = p_{37}^{(2)} = 0$$

and

$$p_{31}^{(2)} = \sum_{i=1}^{7} p_{3i}p_{i1} = p_{32}p_{23} = p^2$$

$$p_{33}^{(2)} = \sum_{i=1}^{7} p_{3i}p_{i3} = p_{32}p_{23} + p_{34}p_{43} = 2pq$$

$$p_{35}^{(2)} = \sum_{i=1}^{7} p_{3i}p_{i5} = p_{34}p_{45} = q^2.$$

We note that

$$\sum_{j=1}^{7} p_{3j}^{(2)} = p^2 + 2pq + q^2 = (p + q)^2 = 1$$

reflecting the fact that if it is in E_3 to start, Contest I must be in some one of the states after two steps.

Putting 8-3a_2 into 8-3a_1 (and hence into 8-2b_3) we obtain

8-3a_3 $$p_j^{(2)} = \sum_{k=1}^{N} p_k p_{kj}^{(2)}$$

and this says that the probability the system is in the state E_j after two steps is the sum over all k of the probabilities that it starts in E_k then goes from E_k to E_j in two steps.

When in 8-2b and 8-2c we spoke about $p_j^{(2)}$, we showed that

$$\sum_{j=1}^{N} p_j^{(2)} = 1$$

(as is necessary, since after two steps the Markov process must be in one of the states). As a check on our work we show this remains true when we define $p_j^{(2)}$ by the sum 8-3a_3. Consider

8-3a_4 $$\sum_{j=1}^{N} p_j^{(2)} = \sum_{j=1}^{N} \left(\sum_{k=1}^{N} p_k p_{kj}^{(2)} \right).$$

Since the sums are finite, their order of summation can be reversed; hence

8-3a_4' $$\sum_{j=1}^{N} p_j^{(2)} = \sum_{k=1}^{N} p_k \sum_{j=1}^{N} p_{kj}^{(2)}$$

Into $\sum_{j=1}^{N} p_{kj}^{(2)}$ we substitute 8-3a_2,

$$\sum_{j=1}^{N} p_{kj}^{(2)} = \sum_{j=1}^{N} \sum_{i=1}^{N} p_{ki}\, p_{ij}$$

$$= \sum_{j=1}^{N} p_{ij} \sum_{i=1}^{N} p_{ki} = (1)(1) = 1. \qquad (8\text{-}1c_4)$$

(In both sums we use the fact that the sum of the elements of any row of a transition matrix equals one.)   Returning to 8-3a'_4, we have

8-3a_5

$$\sum_{j=1}^{N} p_j^{(2)} = \sum_{k=1}^{N} p_k \sum_{j=1}^{N} p_{kj}^{(2)} = \sum_{k=1}^{N} p_k = 1. \qquad (8\text{-}2a_4)$$

EXERCISES 8-3a

1. Find all nonzero second-order transition probabilities for the tennis game with

$$T = \begin{pmatrix} 1 & 0 & 0 & 0 & 0 \\ \dfrac{1}{3} & 0 & \dfrac{2}{3} & 0 & 0 \\ 0 & \dfrac{1}{3} & 0 & \dfrac{2}{3} & 0 \\ 0 & 0 & \dfrac{1}{3} & 0 & \dfrac{2}{3} \\ 0 & 0 & 0 & 0 & 1 \end{pmatrix}$$

2. Find all nonzero second-order transition probabilities for the transition matrix,

$$T = \begin{pmatrix} \dfrac{1}{6} & \dfrac{5}{6} & 0 & 0 & 0 \\ \dfrac{1}{6} & 0 & \dfrac{5}{6} & 0 & 0 \\ \dfrac{1}{6} & 0 & 0 & \dfrac{5}{6} & 0 \\ \dfrac{1}{6} & 0 & 0 & 0 & \dfrac{5}{6} \\ \dfrac{1}{6} & 0 & 0 & 0 & \dfrac{5}{6} \end{pmatrix}$$

3. Find the second-order transition probabilities for the Markov chain with

$$T = \begin{pmatrix} \frac{1}{3} & \frac{1}{2} & \frac{1}{6} \\ \frac{1}{4} & \frac{3}{8} & \frac{3}{8} \\ \frac{1}{7} & \frac{2}{7} & \frac{4}{7} \end{pmatrix}$$

Is $\sum_{j=1}^{3} p_{ij}^{(2)} = 1$ for $i = 1, 2, 3$?

b. nth Order Transition Probabilities

We recall that

$$p_j^{(3)} = \sum_{i=1}^{N} p_i^{(2)} p_{ij}$$

and putting 8-3a_3 into this and then changing the order of summation we have

$$p_j^{(3)} = \sum_{i=1}^{N} \left(\sum_{k=1}^{N} p_k p_{ki}^{(2)} \right) p_{ij}$$

8-3b_1
$$p_j^{(3)} = \sum_{k=1}^{N} p_k \sum_{i=1}^{N} p_{ki}^{(2)} p_{ij} .$$

Since each term in the second summation of 8-3b_1 is the probability the system goes from E_k to E_i in two steps and then from E_i to E_j in the third, we write

8-3b_2
$$p_{kj}^{(3)} = \sum_{i=1}^{N} p_{ki}^{(2)} p_{ij}$$

and for all $k, j = 1,2,\ldots,N$, the $p_{kj}^{(3)}$ are the *third-order transition probabilities*.

In 8-3b_2 let us express $p_{ki}^{(2)}$ as a sum of first-order transition probabilities (see 8-3a_2);

$$p_{kj}^{(3)} = \sum_{i=1}^{N} p_{ki}^{(2)} p_{ij} = \sum_{i=1}^{N} \left(\sum_{v=1}^{N} p_{kv} p_{vi} \right) p_{ij}$$

$$= \sum_{v=1}^{N} p_{kv} \left(\sum_{i=1}^{N} p_{vi} p_{ij} \right) = \sum_{v=1}^{N} p_{kv} p_{vj}^{(2)}$$

We have

8-3b_3
$$p_{kj}^{(3)} = \sum_{i=1}^{N} p_{ki}^{(2)} p_{ij} = \sum_{i=1}^{N} p_{ki} p_{ij}^{(2)}.$$

From these calculations we see that in computing the probability of going from E_k to E_j in three steps we can either consider going from E_k to E_i in two steps and from E_i to E_j in the third, or we can consider going first from E_k to E_i and then from E_i to E_j in two steps.

Again return for an example to the *Tennis* matrix,

$$
T = \begin{pmatrix}
1 & 0 & 0 & 0 & 0 \\
\dfrac{1}{3} & 0 & \dfrac{2}{3} & 0 & 0 \\
0 & \dfrac{1}{3} & 0 & \dfrac{2}{3} & 0 \\
0 & 0 & \dfrac{1}{3} & 0 & \dfrac{2}{3} \\
0 & 0 & 0 & 0 & 1
\end{pmatrix}
$$

We found in Prob. 1 of Exercises 8-3a the second–order transition probabilities. Arranged in a matrix these are

$$
\begin{pmatrix}
1 & 0 & 0 & 0 & 0 \\
\dfrac{3}{9} & \dfrac{2}{9} & 0 & \dfrac{4}{9} & 0 \\
\dfrac{1}{9} & 0 & \dfrac{4}{9} & 0 & \dfrac{4}{9} \\
0 & \dfrac{1}{9} & 0 & \dfrac{2}{9} & \dfrac{6}{9} \\
0 & 0 & 0 & 0 & 1
\end{pmatrix}
$$

where the entry in the ith row and jth column is $p_{ij}^{(2)}$. Using these $p_{ij}^{(2)}$'s, we can compute

$$
p_{11}^{(3)} = \sum_{i=1}^{5} p_{1i}^{(2)} p_{i1} = p_{11}^{(2)} p_{11} + p_{12}^{(2)} p_{21} + p_{13}^{(2)} p_{31} + p_{14}^{(2)} p_{41} + p_{15}^{(2)} p_{51}
$$

$$
= 1 \cdot 1 + 0 \cdot \frac{1}{3} + 0 \cdot 0 + 0 \cdot 0 + 0 \cdot 0 = 1.
$$

For a check we look at

$$
p_{11}^{(3)} = \sum_{i=1}^{5} p_{1i} p_{i1}^{(2)} = p_{11} p_{11}^{(2)} + p_{12} p_{21}^{(2)} + p_{13} p_{31}^{(2)} + p_{14} p_{41}^{(2)} + p_{15} p_{51}^{(2)}
$$

$$
= 1 \cdot 1 + 0 \cdot \frac{3}{9} + 0 \cdot \frac{1}{9} + 0 \cdot 0 + 0 \cdot 0 = 1.
$$

Also

$$p_{24}^{(3)} = \sum_{i=1}^{5} p_{2i}^{(2)} p_{i4} = p_{21}^{(2)} p_{14} + p_{22}^{(2)} p_{24} + p_{23}^{(2)} p_{34} + p_{24}^{(2)} p_{44} + p_{25}^{(2)} p_{52}$$

$$= \frac{3}{9} \cdot 0 + \frac{2}{9} \cdot 0 + 0 \cdot \frac{2}{3} + \frac{4}{9} \cdot 0 + 0 \cdot 0 = 0.$$

$$p_{24}^{(3)} = \sum_{i=1}^{5} p_{2i} p_{i4}^{(2)} = p_{21} p_{14}^{(2)} + p_{22} p_{24}^{(2)} + p_{23} p_{34}^{(2)} + p_{24} p_{44}^{(2)} + p_{25} p_{52}^{(2)}$$

$$= \frac{1}{3} \cdot 0 + 0 \cdot \frac{4}{9} + \frac{2}{3} \cdot 0 + 0 \cdot \frac{2}{9} + 0 \cdot 0 = 0.$$

As a final calculation,

and

$$p_{34}^{(2)} = \sum_{i=1}^{5} p_{3i}^{(2)} p_{i4} = \frac{1}{9} \cdot 0 + 0 \cdot 0 + \frac{4}{9} \cdot \frac{2}{3} + 0 \cdot 0 + \frac{4}{9} \cdot 0 = \frac{8}{27}$$

$$p_{34}^{(2)} = \sum_{i=1}^{5} p_{3i} p_{i4}^{(2)} = 0 \cdot 0 + \frac{1}{3} \cdot \frac{4}{9} + 0 \cdot 0 + \frac{2}{9} \cdot \frac{2}{9} + 0 \cdot 0 = \frac{8}{27}.$$

In practice we use one form or the other, not both; these were carried out only to illustrate 8-3b_3.

In 8-3b_1 we had

$$p_j^{(3)} = \sum_{k=1}^{N} p_k \sum_{i=1}^{N} p_{ki}^{(2)} p_{ij}$$

and combining this with 8-3b_2, we obtain

8-3b_4
$$p_j^{(3)} = \sum_{k=1}^{N} p_k p_{kj}^{(3)}$$

As we did with the second-order transition probabilities we can show

$$\sum_{j=1}^{N} p_{kj}^{(3)} = 1$$

and

$$\sum_{j=1}^{N} p_j^{(3)} = 1.$$

Using $p_{kj}^{(2)}$ and $p_{kj}^{(3)}$ as models, we let $p_{kj}^{(n)}$ denote the probability that the system goes from E_k to E_j in n steps. Using the Principle of Induction we can prove both

8-3b_5
$$p_{kj}^{(n)} = \sum_{i=1}^{N} p_{ki} p_{ij}^{(n-1)} = \sum_{i=1}^{N} p_{ki}^{(n-1)} p_{ij}$$

and

8-3b_6
$$p_j^{(n)} = \sum_{k=1}^{N} p_k p_{kj}^{(n)}$$

For all $k, j = 1,2,\ldots,N$, the numbers $p_{kj}^{(n)}$ are the *nth order transition probabilities* and the set of all *n*th order transition probabilities can be displayed in a matrix.

8-3b_7   **Definition.**   *The* **nth order transition matrix** *is*

$$T^n = \begin{vmatrix} p_{11}^{(n)} & p_{12}^{(n)} & p_{13}^{(n)} & \cdots & p_{1N}^{(n)} \\ p_{21}^{(n)} & p_{22}^{(n)} & p_{23}^{(n)} & \cdots & p_{2N}^{(n)} \\ p_{31}^{(n)} & p_{32}^{(n)} & p_{33}^{(n)} & \cdots & p_{3N}^{(n)} \\ \cdot & \cdot & \cdot & & \cdot \\ \cdot & \cdot & \cdot & & \cdot \\ \cdot & \cdot & \cdot & & \cdot \\ p_{N1}^{(n)} & p_{N2}^{(n)} & p_{N3}^{(n)} & \cdots & p_{NN}^{(n)} \end{vmatrix}$$

(When $n = 1$, $p_{ij}^{(1)} = p_{ij}$, the given transition probabilities, and in this case the superscript is suppressed.) We earlier (8-2b_8) defined the *n*th order probability distribution vector

$$\pi_n = (p_1^{(n)} \quad p_2^{(n)} \quad p_3^{(n)} \quad \cdots \quad p_N^{(n)}).$$

Combining 8-3b_6 and 8-3b_7 with the definition 8-2d_2 of the product of a vector with a matrix, we have

$$(p_1 \quad p_2 \quad p_3 \quad \cdots \quad p_N) \begin{vmatrix} p_{11}^{(n)} & p_{12}^{(n)} & p_{13}^{(n)} & \cdots & p_{1N}^{(n)} \\ p_{21}^{(n)} & p_{22}^{(n)} & p_{23}^{(n)} & \cdots & p_{2N}^{(n)} \\ p_{31}^{(n)} & p_{32}^{(n)} & p_{33}^{(n)} & \cdots & p_{3N}^{(n)} \\ \cdot & \cdot & \cdot & & \cdot \\ \cdot & \cdot & \cdot & & \cdot \\ \cdot & \cdot & \cdot & & \cdot \\ p_{N1}^{(n)} & p_{N2}^{(n)} & p_{N3}^{(n)} & \cdots & p_{NN}^{(n)} \end{vmatrix} = (p_1^{(n)} \quad p_2^{(n)} \quad p_3^{(n)} \quad \cdots \quad p_N^{(n)})$$

or, more concisely,

8-3b$_8$ $$\pi_0 T^n = \pi_n.$$

We can prove by induction on n that

8-3b$_9$ $$\sum_{j=1}^{N} p_{kj}^{(n)} = 1.$$

We again appeal to a large computer to help us illustrate some of these things we have been talking about. We start with the transition matrix

$$T = \begin{pmatrix} 1 & 0 & 0 & 0 & 0 \\ \dfrac{1}{3} & 0 & \dfrac{2}{3} & 0 & 0 \\ 0 & \dfrac{1}{3} & 0 & \dfrac{2}{3} & 0 \\ 0 & 0 & \dfrac{1}{3} & 0 & \dfrac{2}{3} \\ 0 & 0 & 0 & 0 & 1 \end{pmatrix}$$

and program the operation 8-3b$_5$ $(p_{kj}^{(n)} = \sum_{i=1}^{N} p_{ki} p_{ij}^{(n-1)})$ and from this calculate T^2, T^3, and T^4.

$$T = \begin{pmatrix} 1.000000 & 0.000000 & 0.000000 & 0.000000 & 0.000000 \\ .333333 & 0.000000 & .666667 & 0.000000 & 0.000000 \\ 0.000000 & .333333 & 0.000000 & .666667 & 0.000000 \\ 0.000000 & 0.000000 & .333333 & 0.000000 & .666667 \\ 0.000000 & 0.000000 & 0.000000 & 0.000000 & 1.000000 \end{pmatrix}$$

$$T^2 = \begin{pmatrix} 1.000000 & 0.000000 & 0.000000 & 0.000000 & 0.000000 \\ .333333 & .222222 & 0.000000 & .444444 & 0.000000 \\ .111111 & 0.000000 & .444444 & 0.000000 & .444444 \\ 0.000000 & .111110 & 0.000000 & .222222 & .666667 \\ 0.000000 & 0.000000 & 0.000000 & 0.000000 & 1.000000 \end{pmatrix}$$

$$T^3 = \begin{pmatrix} 1.000000 & 0.000000 & 0.000000 & 0.000000 & 0.000000 \\ .407407 & 0.000000 & .296296 & 0.000000 & .296296 \\ .111111 & .148148 & 0.000000 & .296296 & .444444 \\ .037037 & 0.000000 & .148148 & 0.000000 & .814815 \\ 0.000000 & 0.000000 & 0.000000 & 0.000000 & 1.000000 \end{pmatrix}$$

$$T^4 = \begin{pmatrix} 1.000000 & 0.000000 & 0.000000 & 0.000000 & 0.000000 \\ .407407 & 098765 & 0.000000 & .197531 & .296296 \\ .160494 & 0.000000 & .197531 & 0.000000 & .641975 \\ .037037 & .049383 & 0.000000 & .098765 & .814815 \\ 0.000000 & 0.000000 & 0.000000 & 0.000000 & 1.000000 \end{pmatrix}$$

We also put the matrix T in the program we used earlier (see material following 8-2c_3, $\pi_{k-1} T = \pi_k$) and found starting with $\pi_0 = (0\ \ 0\ \ 1\ \ 0\ \ 0)$,

$$\pi_0 T = (0.000000 \quad .333333 \quad 0.000000 \quad .666667 \quad 0.000000) = \pi_1$$

$$\pi_1 T = (.111111 \quad 0.000000 \quad .444444 \quad 0.000000 \quad .444444) = \pi_2$$

$$\pi_2 T = (.111111 \quad .148148 \quad 0.000000 \quad .296296 \quad .444444) = \pi_3$$

$$\pi_3 T = (.160494 \quad 0.000000 \quad .197531 \quad 0.000000 \quad .641975) = \pi_4$$

Instead of performing four successive multiplications to obtain π_4 we could have used 8-3b_8, $\pi_0 T^n = \pi_n$.   Then

$$\pi_0 T^4 = (0\ 0\ 1\ 0\ 0) \begin{pmatrix} 1.000000 & 0.000000 & 0.000000 & 0.000000 & 0.000000 \\ .407407 & .098765 & 0.000000 & .197531 & .296296 \\ .160494 & 0.000000 & .197531 & 0.000000 & .641975 \\ .037037 & .049383 & 0.000000 & .098765 & .814815 \\ 0.000000 & 0.000000 & 0.000000 & 0.000000 & 1.000000 \end{pmatrix}$$

$$= (.160494 \quad .000000 \quad .197531 \quad .000000 \quad .641975)$$

$$= \pi_4.$$

It is a simple matter to check that the row sums in the matrices T^2, T^3, and T^4 are all one, as are the sum of the row vectors π_1, π_2, π_3, and π_4.

In what has been done so far, the computation of $p_{kj}^{(n)} = \sum_{i=1}^{N} p_{ki} p_{ij}^{(n-1)}$ depends, for all $i = 1, 2, \ldots, N$, upon the given transition probabilities p_{ki} and the computed $p_{ij}^{(n-1)}$.   We now show that if σ and τ are any two nonnegative integers such that $\sigma + \tau = n$,

$$p_{kj}^{(n)} = \sum_{i=1}^{N} p_{ki}^{(\sigma)} p_{ij}^{(\tau)}.$$

Our attack is direct.   Since all sums involved are finite,

$$p_{kj}^{(n)} = \sum_{i=1}^{N} p_{ki} p_{ij}^{(n-1)} = \sum_{i=1}^{N} p_{ki} \left(\sum_{l=1}^{N} p_{il} p_{lj}^{(n-2)} \right)$$

$$p_{kj}^{n} = \sum_{l=1}^{N} \left(\sum_{i=1}^{N} p_{ki} p_{il} \right) p_{lj}^{(n-2)} = \sum_{l=1}^{N} p_{kl}^{(2)} p_{lj}^{(n-2)}$$

$$= \sum_{l=1}^{N} p_{kl}^{(2)} \left(\sum_{i=1}^{N} p_{li} p_{ij}^{(n-3)} \right) = \sum_{i=1}^{N} \left(\sum_{l=1}^{N} p_{kl}^{(2)} p_{li} \right) p_{ij}^{(n-3)}$$

$$= \sum_{l=1}^{N} p_{kl}^{(3)} p_{ij}^{(n-3)}$$

Suppose $\sigma \leqslant \tau < n$; by repeating the above argument $\sigma - 1$ times we obtain

$$p_{kj}^{(n)} = \sum_{i=1}^{N} p_{ki}^{(\sigma)} p_{ij}^{(n-\sigma)} = \sum_{i=1}^{N} p_{ki}^{(\sigma)} p_{ij}^{(\tau)}.$$

If $\sigma < \tau$ with $\sigma + \mu = \tau$, we continue for $\mu - 1$ additional repetitions of the argument to show that

8-3b_{10}   If $\sigma, \tau \in I^{+}$ and $\sigma + \tau = n$,

$$p_{kj}^{(n)} = \sum_{i=1}^{N} p_{ki}^{(\sigma)} p_{ij}^{(\tau)} = \sum_{i=1}^{N} p_{ki}^{(\tau)} p_{ij}^{(\sigma)} = p_{kj}^{(\sigma+\tau)}.$$

This result enables us to reduce some of the work required to find $p_{kj}^{(n)}$ for large values of n.   For example, to find the set of $p_{kj}^{(16)}$'s for a given chain we can proceed as follows:

$$p_{kj}^{(2)} = \sum_{i=1}^{N} p_{ki} p_{ij};$$

$$p_{kj}^{(4)} = \sum_{i=1}^{N} p_{ki}^{(2)} p_{ij}^{(2)};$$

$$p_{kj}^{(8)} = \sum_{i=1}^{N} p_{ki}^{(4)} p_{ij}^{(4)};$$

$$p_{kj}^{(16)} = \sum_{i=1}^{N} p_{ki}^{(8)} p_{ij}^{(8)}.$$

This is certainly simpler than finding all the sets $p_{kj}^{(v)}$, $v = 2, 3, \ldots, 16$.   We remark that we have not succeeded in doing what would be most satisfactory of all, that is, expressing for any n, $p_{kj}^{(n)}$ in terms of the initial probability distribution, the transition probabilities and n itself.

EXERCISES 8-3b

1. Starting with

$$T = \begin{pmatrix} 1 & 0 & 0 & 0 & 0 \\ \dfrac{1}{3} & 0 & \dfrac{2}{3} & 0 & 0 \\ 0 & \dfrac{1}{3} & 0 & \dfrac{2}{3} & 0 \\ 0 & 0 & \dfrac{1}{3} & 0 & \dfrac{2}{3} \\ 0 & 0 & 0 & 0 & 1 \end{pmatrix}$$

show that the sequence of calculation T^2, T^4, T^8 yields the same result as T^2, T^3, T^5, T^8.

2. Using the computer calculated values of T^2 and T^3 which are given following 8-3b_9 and $\pi_0 = (0\ 0\ 1\ 0\ 0)$ verify that $\pi_0 T^2 = \pi_2$ and $\pi_0 T^3 = \pi_3$. (In our example we computed π_2 and π_3 by the formula $\pi_{k-1} T = \pi_k$.)

3. If

$$T = \begin{pmatrix} \dfrac{1}{4} & \dfrac{1}{2} & \dfrac{1}{4} \\ \dfrac{1}{2} & 0 & \dfrac{1}{2} \\ 0 & \dfrac{1}{3} & \dfrac{2}{3} \end{pmatrix}$$

compute the transition matrices T^2, T^3 and T^5. If $\pi_0 = \left(\dfrac{1}{3}\ \dfrac{1}{6}\ \dfrac{1}{2} \right)$, calculate π_3 and π_5.

4. Prove by induction on n: $\displaystyle\sum_{j=1}^{N} p_{kj}^{(n)} = 1$ (8-3b_9).

5. (a) Using the data and results of Prob. 3, show that $\pi_3 T^5 = \pi_8$. (b) Prove that $\pi_v T^\mu = \pi_{v+\mu}$, $v, \mu \in I^+$.

6. A chain is generated by rolling a fair die and defining the state of the chain to be E_n. $n = 1, 2, \ldots, 6$, if the largest face yet to appear is n. In Prob. 7, Exercises 8-2d we set up the transition matrix T. Compute now T^2 and T^3. What is T^n for any n? Verify that if $\sigma + \tau = n$, $p_{ij}^{(n)} = \displaystyle\sum_{k=1}^{6} p_{ik}^{(\sigma)} p_{kj}^{(\tau)}$.

8-4. TRANSITION MATRICES

a. Matrix Multiplication

Let us look again at the second-order transition probabilities. We have from 8-3a_2

$$p_{kj}^{(2)} = \sum_{i=1}^{N} p_{ki}p_{ij},$$

and we observe that $p_{kj}^{(2)}$ is the term-by-term product of the kth row of the transition matrix T with the jth column of T. Schematically,

$$
\begin{pmatrix}
p_{11} & p_{12} & \cdots & p_{1j} & \cdots & p_{1N} \\
p_{21} & p_{22} & \cdots & p_{2j} & \cdots & p_{2N} \\
\cdot & \cdot & & \cdot & & \cdot \\
\cdot & \cdot & & \cdot & & \cdot \\
\cdot & \cdot & & \cdot & & \cdot \\
\boxed{p_{k1} \quad p_{k2} \quad \cdots \quad p_{kj} \quad \cdots \quad p_{kN}} \\
\cdot & \cdot & & \cdot & & \cdot \\
\cdot & \cdot & & \cdot & & \cdot \\
\cdot & \cdot & & \cdot & & \cdot \\
p_{N1} & p_{N2} & \cdots & p_{Nj} & \cdots & p_{NN}
\end{pmatrix}
\begin{pmatrix}
p_{11} & p_{12} & \cdots & \boxed{p_{1j}} & \cdots & p_{1N} \\
p_{21} & p_{22} & \cdots & \boxed{p_{2j}} & \cdots & p_{2N} \\
\cdot & \cdot & & \cdot & & \cdot \\
\cdot & \cdot & & \cdot & & \cdot \\
\cdot & \cdot & & \cdot & & \cdot \\
p_{k1} & p_{k2} & & \boxed{p_{kj}} & \cdots & p_{kN} \\
\cdot & \cdot & & \cdot & & \cdot \\
\cdot & \cdot & & \cdot & & \cdot \\
\cdot & \cdot & & \cdot & & \cdot \\
p_{N1} & p_{N2} & & \boxed{p_{Nj}} & & p_{NN}
\end{pmatrix}
$$

$$
=
\begin{pmatrix}
p_{11}^{(2)} & p_{12}^{(2)} & \cdots & p_{1j}^{(2)} & \cdots & p_{1N}^{(2)} \\
p_{21}^{(2)} & p_{22}^{(2)} & \cdots & p_{2j}^{(2)} & \cdots & p_{2N}^{(2)} \\
\cdot & \cdot & & \cdot & & \cdot \\
\cdot & \cdot & & \cdot & & \cdot \\
\cdot & \cdot & & \cdot & & \cdot \\
p_{k1}^{(2)} & p_{k2}^{(2)} & \cdots & \boxed{p_{kj}^{(2)}} & \cdots & p_{kN}^{(2)} \\
\cdot & \cdot & & \cdot & & \cdot \\
\cdot & \cdot & & \cdot & & \cdot \\
p_{N1}^{(2)} & p_{N2}^{(2)} & \cdots & p_{Nj}^{(2)} & \cdots & p_{NN}^{(2)}
\end{pmatrix}
$$

In fact, any element of the kth row of T^2 is the term-by-term product of the kth row of T with the appropriate column of T; $p_{k1}^{(2)}$ is the term-by-term product of the kth row of T with the first column of T; $p_{k2}^{(2)}$ is the term-by-term product of the kth row of T with the second column of T, and so on. This is an extension of the multiplication of row vectors and matrices which was defined in 8-2d_2.

We have also

$$p_{kj}^{(3)} = \sum_{i=1}^{N} p_{ki}^{(2)} p_{ij} \qquad (8\text{-}3b_2)$$

and from this we see that the element in the kth row and jth column of T^3 is the term-by-term product of the kth row of T^2 with the jth column of T. And since by 8-3b_3,

$$p_{kj}^{(3)} = \sum_{i=1}^{N} p_{ki} p_{ij}^{(2)}$$

the element in the kth row and jth column of T^3 is also the term-by-term product of the kth row of T with the jth column of T^2. We write

$$T^3 = T^2 T = T T^2.$$

Similarly, from 8-3b_5,

$$p_{kj}^{(n)} = \sum_{i=1}^{N} p_{ki}^{(n-1)} p_{ij},$$

and so the element in the kth row and jth column of T^n is the term-by-term product of the kth row of T^{n-1} with the jth column of T.

8-4a_1 $\quad T^n = T^{n-1} T$

$$=
\begin{pmatrix}
p_{11}^{(n-1)} & p_{12}^{(n-1)} & \cdots & p_{1j}^{(n-1)} & \cdots & p_{1N}^{(n-1)} \\
p_{21}^{(n-1)} & p_{22}^{(n-1)} & \cdots & p_{2j}^{(n-1)} & \cdots & p_{2N}^{(n-1)} \\
\cdot & \cdot & & \cdot & & \cdot \\
\cdot & \cdot & & \cdot & & \cdot \\
p_{k1}^{(n-1)} & p_{k2}^{(n-1)} & \cdots & p_{kj}^{(n-1)} & \cdots & p_{kN}^{(n-1)} \\
\cdot & \cdot & & \cdot & & \cdot \\
p_{N1}^{(n-1)} & p_{N2}^{(n-1)} & \cdots & p_{Nj}^{(n-1)} & \cdots & p_{NN}^{(n-1)}
\end{pmatrix}
\begin{pmatrix}
p_{11} & p_{12} & \cdots & p_{1j} & \cdots & p_{1N} \\
p_{21} & p_{22} & \cdots & p_{2j} & \cdots & p_{2N} \\
\cdot & \cdot & & \cdot & & \cdot \\
\cdot & \cdot & & \cdot & & \cdot \\
p_{k1} & p_{k2} & \cdots & p_{kj} & \cdots & p_{kN} \\
\cdot & \cdot & & \cdot & & \cdot \\
p_{N1} & p_{N2} & \cdots & p_{Nj} & \cdots & p_{NN}
\end{pmatrix}$$

$$= \begin{pmatrix} p_{11}^{(n)} & p_{12}^{(n)} & \cdots & p_{1j}^{(n)} & \cdots & p_{1N}^{(n)} \\ p_{21}^{(n)} & p_{22}^{(n)} & \cdots & p_{2j}^{(n)} & \cdots & p_{2N}^{(n)} \\ \cdot & \cdot & & \cdot & & \cdot \\ \cdot & \cdot & & \cdot & & \cdot \\ \cdot & \cdot & & \cdot & & \cdot \\ p_{k1}^{(n)} & p_{k2}^{(n)} & \cdots & \boxed{p_{kj}^{(n)}} & \cdots & p_{kN}^{(n)} \\ \cdot & \cdot & & \cdot & & \cdot \\ \cdot & \cdot & & \cdot & & \cdot \\ p_{N1}^{(n)} & p_{N2}^{(n)} & \cdots & p_{kN}^{(n)} & \cdots & p_{NN}^{(n)} \end{pmatrix}$$

As a numerical example, consider the matrix

$$T = \begin{pmatrix} \frac{1}{4} & \frac{1}{2} & \frac{1}{4} \\ \frac{1}{2} & 0 & \frac{1}{2} \\ 0 & \frac{3}{8} & \frac{5}{8} \end{pmatrix}$$

$$T^2 = T \times T$$

$$= \begin{pmatrix} \frac{1}{4} & \frac{1}{2} & \frac{1}{4} \\ \frac{1}{2} & 0 & \frac{1}{2} \\ 0 & \frac{3}{8} & \frac{5}{8} \end{pmatrix} \begin{pmatrix} \frac{1}{4} & \frac{1}{2} & \frac{1}{4} \\ \frac{1}{2} & 0 & \frac{1}{2} \\ 0 & \frac{3}{8} & \frac{5}{8} \end{pmatrix}$$

$$= \begin{pmatrix} \frac{1}{16} + \frac{1}{4} & \frac{1}{8} + \frac{3}{32} & \frac{1}{16} + \frac{1}{4} + \frac{5}{32} \\ \frac{1}{8} & \frac{1}{4} + \frac{3}{16} & \frac{1}{8} + \frac{5}{16} \\ \frac{3}{16} & \frac{15}{64} & \frac{3}{16} + \frac{25}{64} \end{pmatrix}$$

$$T^2 = \begin{pmatrix} \dfrac{10}{32} & \dfrac{7}{32} & \dfrac{15}{32} \\[2mm] \dfrac{2}{16} & \dfrac{7}{16} & \dfrac{7}{16} \\[2mm] \dfrac{12}{64} & \dfrac{15}{64} & \dfrac{37}{64} \end{pmatrix}$$

As a check we sum across the rows and find each row sum is one. Let us repeat the operation; then

$$T^3 = T^2 \times T = \begin{pmatrix} \dfrac{20}{64} & \dfrac{14}{64} & \dfrac{30}{64} \\[2mm] \dfrac{8}{64} & \dfrac{28}{64} & \dfrac{28}{64} \\[2mm] \dfrac{12}{64} & \dfrac{15}{64} & \dfrac{37}{64} \end{pmatrix} \begin{pmatrix} \dfrac{2}{8} & \dfrac{4}{8} & \dfrac{2}{8} \\[2mm] \dfrac{4}{8} & 0 & \dfrac{4}{8} \\[2mm] 0 & \dfrac{3}{8} & \dfrac{5}{8} \end{pmatrix}$$

As we perform the row by column multiplication each denominator will be $64 \cdot 8 = 512$. For simplicity in writing let us set $q = \dfrac{1}{512}$; we have

$$T^3 = T^2 \times T$$

$$= \begin{pmatrix} (40 + 56)q & (80 + 90)q & (40 + 56 + 150)q \\ (16 + 112)q & (32 + 84)q & (16 + 112 + 140)q \\ (24 + 60)q & (48 + 111)q & (24 + 60 + 185)q \end{pmatrix} = \begin{pmatrix} 96q & 170q & 246q \\ 128q & 116q & 268q \\ 84q & 159q & 269q \end{pmatrix}$$

Here again the row sums are one. From 8-3b_{10} we have

$$p_{kj}^{(\sigma+\tau)} = \sum_{i=1}^{N} p_{ki}^{(\sigma)} p_{ij}^{(\tau)} = \sum_{i=1}^{N} p_{ki}^{(\tau)} p_{ij}^{(\sigma)}$$

so that the element in the kth row and jth column of $T^{\sigma+\tau}$ is either the term-by-term product of the kth row of T^{σ} with the jth column of T^{τ}, or the term-by-term product of the kth row of T^{τ} with the jth column of T^{σ}

8-4a_2 $T^{\sigma+\tau} = T^\sigma T^\tau = T^\tau T^\sigma$

$$
=
\begin{pmatrix}
p_{11}^{(\sigma)} & p_{12}^{(\sigma)} & \cdots & p_{1j}^{(\sigma)} & \cdots & p_{1N}^{(\sigma)} \\
p_{21}^{(\sigma)} & p_{22}^{(\sigma)} & \cdots & p_{2j}^{(\sigma)} & \cdots & p_{2N}^{(\sigma)} \\
\cdot & \cdot & & \cdot & & \cdot \\
\cdot & \cdot & & \cdot & & \cdot \\
\cdot & \cdot & & \cdot & & \cdot \\
\boxed{p_{k1}^{(\sigma)} \quad p_{k2}^{(\sigma)} \quad \cdots \quad p_{kj}^{(\sigma)} \quad \cdots \quad p_{kN}^{(\sigma)}} \\
\cdot & \cdot & & \cdot & & \cdot \\
\cdot & \cdot & & \cdot & & \cdot \\
\cdot & \cdot & & \cdot & & \cdot \\
p_{N1}^{(\sigma)} & p_{N2}^{(\sigma)} & \cdots & p_{Nj}^{(\sigma)} & \cdots & p_{NN}^{(\sigma)}
\end{pmatrix}
\begin{pmatrix}
p_{11}^{(\tau)} & p_{12}^{(\tau)} & \cdots & \boxed{p_{1j}^{(\tau)}} & \cdots & p_{1N}^{(\tau)} \\
p_{21}^{(\tau)} & p_{22}^{(\tau)} & \cdots & \boxed{p_{2j}^{(\tau)}} & \cdots & p_{2N}^{(\tau)} \\
\cdot & \cdot & & \cdot & & \cdot \\
\cdot & \cdot & & \cdot & & \cdot \\
p_{k1}^{(\tau)} & p_{k2}^{(\tau)} & \cdots & \boxed{p_{kj}^{(\tau)}} & \cdots & p_{kN}^{(\tau)} \\
\cdot & \cdot & & \cdot & & \cdot \\
\cdot & \cdot & & \cdot & & \cdot \\
\cdot & \cdot & & \cdot & & \cdot \\
p_{N1}^{(\tau)} & p_{N2}^{(\tau)} & \cdots & \boxed{p_{Nj}^{(\tau)}} & \cdots & p_{NN}^{(\tau)}
\end{pmatrix}
$$

$$
=
\begin{pmatrix}
p_{11}^{(\tau)} & p_{12}^{(\tau)} & \cdots & p_{1j}^{(\tau)} & \cdots & p_{1N}^{(\tau)} \\
p_{21}^{(\tau)} & p_{22}^{(\tau)} & \cdots & p_{2j}^{(\tau)} & \cdots & p_{2N}^{(\tau)} \\
\cdot & \cdot & & \cdot & & \cdot \\
\cdot & \cdot & & \cdot & & \cdot \\
\cdot & \cdot & & \cdot & & \cdot \\
\boxed{p_{k1}^{(\tau)} \quad p_{k2}^{(\tau)} \quad \cdots \quad p_{kj}^{(\tau)} \quad \cdots \quad p_{kN}^{(\tau)}} \\
\cdot & \cdot & & \cdot & & \cdot \\
\cdot & \cdot & & \cdot & & \cdot \\
\cdot & \cdot & & \cdot & & \cdot \\
p_{N1}^{(\tau)} & p_{N2}^{(\tau)} & \cdots & p_{Nj}^{(\tau)} & \cdots & p_{NN}^{(\tau)}
\end{pmatrix}
\begin{pmatrix}
p_{11}^{(\sigma)} & p_{12}^{(\sigma)} & \cdots & \boxed{p_{1j}^{(\sigma)}} & \cdots & p_{1N}^{(\sigma)} \\
p_{21}^{(\sigma)} & p_{22}^{(\sigma)} & \cdots & \boxed{p_{2j}^{(\sigma)}} & & p_{2N}^{(\sigma)} \\
\cdot & \cdot & & \cdot & & \cdot \\
\cdot & \cdot & & \cdot & & \cdot \\
p_{k1}^{(\sigma)} & p_{k2}^{(\sigma)} & \cdots & \boxed{p_{kj}^{(\sigma)}} & \cdots & p_{kN}^{(\sigma)} \\
\cdot & \cdot & & \cdot & & \cdot \\
\cdot & \cdot & & \cdot & & \cdot \\
\cdot & \cdot & & \cdot & & \cdot \\
p_{N1}^{(\sigma)} & p_{N2}^{(\sigma)} & \cdots & \boxed{p_{Nj}^{(\sigma)}} & \cdots & p_{NN}^{(\sigma)}
\end{pmatrix}
$$

$$
=
\begin{pmatrix}
p_{11}^{(\sigma+\tau)} & p_{12}^{(\sigma+\tau)} & \cdots & p_{1j}^{(\sigma+\tau)} & p_{1N}^{(\sigma+\tau)} \\
p_{21}^{(\sigma+\tau)} & p_{22}^{(\sigma+\tau)} & \cdots & p_{2j}^{(\sigma+\tau)} & p_{2N}^{(\sigma+\tau)} \\
\cdot & \cdot & & \cdot & \cdot \\
\cdot & \cdot & & \cdot & \cdot \\
\cdot & \cdot & & \cdot & \cdot \\
p_{k1}^{(\sigma+\tau)} & p_{k2}^{(\sigma+\tau)} & \cdots & \boxed{p_{kj}^{(\sigma+\tau)}} & p_{kN}^{(\sigma+\tau)} \\
\cdot & \cdot & & \cdot & \cdot \\
\cdot & \cdot & & \cdot & \cdot \\
\cdot & \cdot & & \cdot & \cdot \\
p_{N1}^{(\sigma+\tau)} & p_{N2}^{(\sigma+\tau)} & \cdots & p_{Nj}^{(\sigma+\tau)} & p_{NN}^{(\sigma+\tau)}
\end{pmatrix}
$$

If we generalize this matrix multiplication which we have here developed through probabilistic considerations, to any pair of appropriate matrices—the precise meaning of the work "appropriate" will be explained shortly—we have the classical method of multiplying matrices. We start with

8-4a$_3$ **Definition.** *Let A and B be the matrices*

$$A = \begin{pmatrix} a_{11} & a_{12} & \cdots & a_{1n} \\ a_{21} & a_{22} & \cdots & a_{2n} \\ & & & \\ & & & \\ a_{m1} & a_{m2} & \cdots & a_{mn} \end{pmatrix} \quad and \ B = \begin{pmatrix} b_{11} & b_{12} & \cdots & b_{1p} \\ b_{21} & b_{22} & \cdots & b_{2p} \\ & & & \\ & & & \\ b_{n1} & b_{n2} & \cdots & b_{np} \end{pmatrix}$$

Then the **product of A and B** *is*

$$AB = \begin{pmatrix} c_{11} & c_{12} & \cdots & c_{1p} \\ c_{21} & c_{22} & \cdots & c_{2p} \\ & & & \\ & & & \\ c_{m1} & c_{m2} & \cdots & c_{mp} \end{pmatrix}$$

where for $k = 1, 2, \ldots, m$ *and* $j = 1, 2, \ldots, p$

$$c_{kj} = \sum_{i=1}^{n} a_{ki} b_{ij}.$$

It is implicit in the definition that the product exists only if the number of columns in A equals the number of rows in B—this is what is meant above by the phrase "pair of appropriate matrices"—otherwise the term-by-term multiplication which defines the elements of AB would not exist. Matrix multiplication is frequently referred to as row-by-column multiplication.

Let us apply definition to:

$$A = \begin{pmatrix} 4 & 8 & -1 \\ -2 & 0 & 3 \end{pmatrix} \quad and \ B = \begin{pmatrix} -1 & 0 & 1 & 2 \\ 1 & 0 & -2 & 1 \\ 4 & 3 & 0 & -4 \end{pmatrix}$$

$$AB = \begin{pmatrix} 4 & 8 & -1 \\ -2 & 0 & 3 \end{pmatrix} \begin{pmatrix} -1 & 0 & 1 & -2 \\ 1 & 0 & -2 & 1 \\ 4 & 3 & 0 & -4 \end{pmatrix}$$

$$= \begin{pmatrix} -4+8-4 & -3 & 4-16 & -8+8+4 \\ 2+12 & 9 & -2 & 4-12 \end{pmatrix}$$

$$= \begin{pmatrix} 0 & -3 & -12 & 4 \\ 14 & 9 & -2 & -8 \end{pmatrix}$$

When we spoke of the multiplication of a row vector with a matrix, we noted that the number of elements in the vector had to equal the number of rows in the matrix. In its pattern, at least, a matrix is a collection of row vectors each with the same number of elements. Thus our matrix product is a straightforward extension of our multiplication of a vector with a matrix.

If A is a $m \times n$ and B is an $n \times p$ matrix, their product AB is a $m \times p$ matrix. We observe at once that matrix multiplication is not commutative for if A is an $m \times n$ matrix and B is a $n \times p$ matrix, BA is not even defined for $m \neq p$. If $m = p$, so that A is $m \times n$ and B is $n \times m$, AB is $m \times m$ and BA is $n \times n$ and hence, if $m \neq n$, $AB \neq BA$. Furthermore, a simple example will show that if $m = n = p$, AB is not necessarily equal to BA. Consider:

$$\begin{pmatrix} 3 & 2 & 1 \\ 1 & 0 & 1 \\ 1 & 2 & 0 \end{pmatrix} \begin{pmatrix} 1 & 1 & 3 \\ 0 & 2 & 1 \\ 2 & 2 & 1 \end{pmatrix} = \begin{pmatrix} 5 & 9 & 12 \\ 3 & 3 & 4 \\ 1 & 5 & 5 \end{pmatrix}$$

and

$$\begin{pmatrix} 1 & 1 & 3 \\ 0 & 2 & 1 \\ 2 & 2 & 1 \end{pmatrix} \begin{pmatrix} 3 & 2 & 1 \\ 1 & 0 & 1 \\ 1 & 2 & 0 \end{pmatrix} = \begin{pmatrix} 7 & 8 & 2 \\ 3 & 2 & 2 \\ 9 & 6 & 4 \end{pmatrix}$$

b. Multiplicative Properties of Transition Matrices

We begin by pointing out that, in general, transition matrices do not commute. If

$$T_1 = \begin{pmatrix} \dfrac{1}{4} & \dfrac{1}{2} & \dfrac{1}{4} \\[2mm] \dfrac{1}{2} & 0 & \dfrac{1}{2} \\[2mm] 0 & \dfrac{1}{3} & \dfrac{2}{3} \end{pmatrix} \quad \text{and } T_2 = \begin{pmatrix} \dfrac{1}{2} & \dfrac{1}{2} & 0 \\[2mm] \dfrac{1}{2} & 0 & \dfrac{1}{2} \\[2mm] 0 & \dfrac{1}{2} & \dfrac{1}{2} \end{pmatrix}$$

$$T_1 T_2 \neq T_2 T_1.$$

However, for any transition matrix T we have already shown in 8-4a_2 that

$$T^\sigma T^\tau = T^\tau T^\sigma$$

and to this limited extent transition matrices do commute. Furthermore, in 8-4a_2 we also have that

$$T^{\sigma+\tau} = T^\sigma T^\tau$$

and as far as applications go this will be the more useful result.

EXERCISES 8-4b

1. In 8-4b above what are $T_1 T_2$ and $T_2 T_1$?
2. In the text we computed T^2 and $T^3 = T^2 T$ where

$$T = \begin{pmatrix} \dfrac{1}{4} & \dfrac{1}{2} & \dfrac{1}{4} \\[2mm] \dfrac{1}{2} & 0 & \dfrac{1}{2} \\[2mm] 0 & \dfrac{3}{8} & \dfrac{5}{8} \end{pmatrix}$$

Compute $T^3 = TT^2$ and compare it to the earlier result for T^3. Compute $T^4 = T^2 T^2$, $T^5 = TT^4$ and $T^5 = T^3 T^2$. How do the computations of T^5 compare? Is the result surprising?

3. In the text we proved that if $n = \sigma + \tau$, $T^n = T^\sigma T^\tau = T^\tau T^\sigma$ whenever T is a transition matrix. Prove that if σ, τ, μ, v, are positive integers such that $\sigma + \tau = \mu + v = n$, $T^n = T^\sigma T^\tau = T^\mu T^v$.

(HINT: Without loss of generality we can assume $\sigma \leqslant \mu \leqslant \nu \leqslant \tau$. Then since

$$p_{ik}^{(u)} = \sum_{\alpha=1}^{N} p_{i\alpha}^{(u-1)} p_{\alpha k}, \text{ we can show}$$

$$p_{ij}^{(n)} = \sum_{k=1}^{N} p_{ik}^{(u)} p_{kj}^{(v)} = \sum_{\alpha=1}^{N} p_{i\alpha}^{(u-1)} p_{\alpha j}^{(v+1)}.$$

4. Is matrix multiplication associative? That is, if A, B and C are appropriate matrices, does $A \times (B \times C) = (A \times B) \times C$? Could this result be used to solve Prob. 3?

5. If A is *not* a transition matrix and if $A \times A$ exists, do the powers of A commute? Why? (Note that for

$$A = \begin{pmatrix} 1 & 0 & 2 \\ 3 & -1 & 0 \\ 3 & 0 & 1 \end{pmatrix}, \quad A^2 A = A A^2).$$

c. Ergodic Chains

Let us look at progressively higher powers of the transition matrix:

$$T = \begin{pmatrix} 0 & \dfrac{1}{3} & \dfrac{2}{3} \\ 1 & 0 & 0 \\ 0 & 1 & 0 \end{pmatrix}$$

$$T^2 = \begin{pmatrix} \dfrac{1}{3} & \dfrac{2}{3} & 0 \\ 0 & \dfrac{1}{3} & \dfrac{2}{3} \\ 1 & 0 & 0 \end{pmatrix}$$

$$T^4 = \begin{pmatrix} \dfrac{1}{9} & \dfrac{4}{9} & \dfrac{4}{9} \\ \dfrac{6}{9} & \dfrac{1}{9} & \dfrac{2}{9} \\ \dfrac{3}{9} & \dfrac{6}{9} & 0 \end{pmatrix}$$

$$T^8 = \begin{pmatrix} \dfrac{37}{81} & \dfrac{32}{81} & \dfrac{12}{81} \\[2mm] \dfrac{18}{81} & \dfrac{37}{81} & \dfrac{26}{81} \\[2mm] \dfrac{39}{81} & \dfrac{18}{81} & \dfrac{24}{81} \end{pmatrix}$$

The transition matrix T^8 is called a positive matrix since it contains no zero elements, and in general we have

8-4c₁ **Definition.** *If T is a transition matrix all of whose entries are positive, T is called a **positive matrix**.*

(We shall have a lot more to say about positive matrices in Sec. 8-5.)

Consider any Markov process with transition matrix T such that for some integer $v \geqslant 1$, T^v is positive. That T^{v+1} is also positive is easily proved. The elements of $T^{v+1} = T^v T$ have the form

$$p_{kj}^{(v+1)} = \sum_{i=1}^{N} p_{ki}^{(v)} p_{ij}$$

where for every k and i, $p_{ki}^{(v)} > 0$ (T^v is positive) and for each j at least one $p_{ij} > 0$ (no column of T has only zero elements—see 8-1c_6). Hence the sum $\sum_{i=1}^{N} p_{ki}^{(v)} p_{ij}$ has at least one nonzero addend and so $p_{kj}^{(v+1)} > 0$ for all $k, j = 1, 2, \ldots, N$ and T^{v+1} is positive.

The significance of this result in terms of a Markov process is readily seen. Given an initial distribution π_0 and a transition matrix T with T^v positive for some integer $v \geqslant 1$, then for all $\mu > v$, $\pi_\mu = \pi_0 T^\mu$ (see 8-3b_8) has only positive elements. Thus after the process has continued for a certain length of time, for every $k = 1, 2, \ldots, N$ the probability that the system is in E_k is positive.

In the example at the beginning of the section T^8 was positive so that no matter what the initial distribution after eight steps there is a positive probability of the system's being in each of the states E_1, E_2, and E_3. It is easily calculated that in this example T^5 is also positive (hence T^6 and T^7 as well as T^8 and those which follow); however, in trying to determine whether a given matrix is positive we square at each step to get to as high an order as possible as soon as possible.

No power of the transition matrix

$$T = \begin{pmatrix} 1 & 0 & 0 & 0 & 0 & 0 & 0 \\ p & 0 & q & 0 & 0 & 0 & 0 \\ 0 & p & 0 & q & 0 & 0 & 0 \\ 0 & 0 & p & 0 & q & 0 & 0 \\ 0 & 0 & 0 & p & 0 & q & 0 \\ 0 & 0 & 0 & 0 & p & 0 & q \\ 0 & 0 & 0 & 0 & 0 & 0 & 1 \end{pmatrix}$$

associated with Contest I will be positive since for all integers $r \geqslant 1$ and all $j = 2, 3,\ldots,7$, $p_{1j}^{(r)} = 0$ and $p_{11}^{(r)} = 1$, that is, the first row of T^r will always be

$$\begin{array}{ccccccc} 1 & 0 & 0 & 0 & 0 & 0 & 0 \end{array}$$

That $p_{11}^{(r)} = 1$ for all integers $r \geqslant 1$ can be proved by induction. This is not surprising since once the contest goes into E_1 it is over, it is never possible to go from E_1 to any other state.

Consider now the matrix

$$T = \begin{pmatrix} 0 & p & 0 & 0 & 0 & q \\ q & 0 & p & 0 & 0 & 0 \\ 0 & q & 0 & p & 0 & 0 \\ 0 & 0 & q & 0 & p & 0 \\ 0 & 0 & 0 & q & 0 & p \\ p & 0 & 0 & 0 & q & 0 \end{pmatrix}$$

where $p + q = 1, p > 0, q > 0$.

$$T^2 = \begin{pmatrix} 2pq & 0 & p^2 & 0 & q^2 & 0 \\ 0 & 2pq & 0 & p^2 & 0 & q^2 \\ q^2 & 0 & 2pq & 0 & p^2 & 0 \\ 0 & q^2 & 0 & 2pq & 0 & p^2 \\ p^2 & 0 & q^2 & 0 & 2pq & 0 \\ 0 & p^2 & 0 & q^2 & 0 & 2pq \end{pmatrix}$$

and

$$T^4 = \begin{pmatrix} 6p^2q^2 & 0 & 4p^3q+q^4 & 0 & 4pq^3+p^4 & 0 \\ 0 & 6p^2q^2 & 0 & 4p^3q+q^4 & 0 & 4pq^3+p^4 \\ 4pq^3+p^4 & 0 & 6p^2q^2 & 0 & 4p^3q+q^4 & 0 \\ 0 & 4pq^3+p^4 & 0 & 6p^2q^2 & 0 & 4p^3q+q^4 \\ 4p^3q+q^4 & 0 & 4pq^3+p^4 & 0 & 6p^2q^2 & 0 \\ 0 & 4p^3q+q^4 & 0 & 4pq^3+p^4 & 0 & 6p^2q^2 \end{pmatrix}$$

We note that T^2 and T^4 have zero elements in the same rows and columns and examination of the pattern suggests that for all positive integers v, T^{2v} will have the format

$$\begin{pmatrix} + & 0 & + & 0 & + & 0 \\ 0 & + & 0 & + & 0 & + \\ + & 0 & + & 0 & + & 0 \\ 0 & + & 0 & + & 0 & + \\ + & 0 & + & 0 & + & 0 \\ 0 & + & 0 & + & 0 & + \end{pmatrix}$$

where $+$ indicates a·nonzero entry. This is easily proved by Induction. Thus no even power of T will ever be positive. However, it might be worth looking at the odd powers of T.

$$T^3 = T^2T = \begin{pmatrix} 0 & 3p^2q & 0 & p^3+q^3 & 0 & 3pq^2 \\ 3pq^2 & 0 & 3p^2q & 0 & p^3+q^3 & 0 \\ 0 & 3pq & 0 & 3p^2q & 0 & p^3+q^3 \\ p^3+q^3 & 0 & 3pq^2 & 0 & 3p^2q & 0 \\ 0 & p^3+q^3 & 0 & 3pq^2 & 0 & 3p^2q \end{pmatrix}$$

In T^3 the zero entries are in the location of the nonzero entries of T^2 and vice versa. Another direct proof by Induction will show that all the odd powers of T have the format

0	+	0	+	0	+
+	0	+	0	+	0
0	+	0	+	0	+
+	0	+	0	+	0
0	+	0	+	0	+
+	0	+	0	+	0

The fact that the even and odd powers of T "alternate" their zero and non-zero entries means that in a Markow chain with T as a transition matrix any state may be reached from any other state, but it may take at least two steps. For instance $p_{35} = 0$, but $p_{35}^{(2)} > 0$ and $p_{35}^{(4)} > 0$.

Assume a disk is divided into six equal sectors numbered in a clockwise sense 1, 2, 3, 4, 5, 6 and that a small ball will move at regular intervals of time one sector in clockwise direction with probability p and in a counterclockwise direction with probability $q = 1 - p$. The sectors form the states of a Markov process, a cyclic random walk, which has as its transition matrix the matrix just discussed.

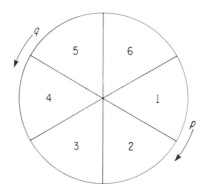

If the ball starts out in E_3 it can reach E_5 in two steps only if it goes $E_3 \rightarrow E_4 \rightarrow E_5$ so that $p_{35}^{(2)} = p^2$. In three steps a ball starting in E_3 can reach only E_2, E_4, or E_6 so that $p_{35}^{(3)} = 0$ and in four steps its can again be in E_5. If we list the possibilities, we have

$$P(E_3 E_2 E_1 E_6 E_5) = q^4$$
$$P(E_3 E_4 E_5 E_6 E_5) = p^3 q$$
$$P(E_3 E_4 E_3 E_4 E_5) = p^3 q$$
$$P(E_3 E_4 E_5 E_4 E_5) = p^3 q$$
$$P(E_3 E_2 E_3 E_4 E_5) = p^3 q$$

so that $p_{35}^{(4)} = q^4 + 4p^3q$ in agreement with our earlier calculation of T^4. A numerical example of such a chain is given in Table 8-4c_3.

Chains, like this last one, in which every state can be reached from every other state are given a special designation.

8-4c_2 Definition. *A Markov chain is called* **ergodic** *if it is possible to move from every state to every other state.*

Obviously any chain for which some integral power of the transition matrix is positive is ergodic, however, the example makes clear that an ergodic chain need not produce a positive matrix. The chain in Contest I is not ergodic since it is impossible to move out of E_1 or E_7. However, if we look at the second- and fourth-order transition matrices for that chain we find

$$T^2 = \begin{pmatrix} 1 & 0 & 0 & 0 & 0 & 0 & 0 \\ p & pq & 0 & q^2 & 0 & 0 & 0 \\ p^2 & 0 & 2pq & 0 & q^2 & 0 & 0 \\ 0 & p^3 & 0 & 2pq & 0 & q^2 & 0 \\ 0 & 0 & p^2 & 0 & 2pq & 0 & q^2 \\ 0 & 0 & 0 & p^2 & 0 & pq & q \\ 0 & 0 & 0 & 0 & 0 & 0 & 1 \end{pmatrix}$$

and

$$T^4 = \begin{pmatrix} 1 & 0 & 0 & 0 & 0 & 0 & 0 \\ p+p^2q & 2p^2q^2 & 0 & 4pq^3 & 0 & q^4 & 0 \\ p^2+2p^3q & 0 & 5p^2q^3 & 0 & 4pq^3 & 0 & q^4 \\ p^3 & 3p^3q & 0 & 6p^3q^2 & 0 & 3pq^3 & q^3 \\ p^4 & 0 & 4p^3q & 0 & 5p^2q^2 & 0 & 2pq^3+q^2 \\ 0 & p^4 & 0 & 4p^3q & 0 & 2p^2q^2 & pq^2+q \\ 0 & 0 & 0 & 0 & 0 & 0 & 1 \end{pmatrix}$$

If for the moment we limit our considerations to the states E_2, E_3, E_4, E_5, E_6 we see that the probability of movement among them produces at even-numbered steps the same pattern of alternating zeros and positive entries as in the transition matrix of the preceding example. Suppose we compute T^5.

$$T^5 = T^4T = \begin{pmatrix} 1 & 0 & 0 & 0 & 0 & 0 & 0 \\ p+p^2q+2p^3q^2 & 0 & 5p^2q^3 & 0 & 4pq^4 & 0 & q^5 \\ p^2+2p^3q & 5p^3q^2 & 0 & 9p^2q^3 & 0 & 4pq^4 & q^4 \\ p^3+3p^4q & 0 & 9p^3q^2 & 0 & 9p^2q^3 & 0 & 3pq^4+q^3 \\ p^4 & 4p^4q & 0 & 9p^3q^2 & 0 & 5p^2q^3 & 2pq^3+q^2 \\ p^5 & 0 & 4p^4q & 0 & 5p^3q^2 & 0 & 2p^2q^3+pq^2+q \\ 0 & 0 & 0 & 0 & 0 & 0 & 1 \end{pmatrix}$$

Comparing T^4 and T^5 we see that for $k, j = 2,3,4,5,6$, if $p_{kj}^{(4)} = 0$, $p_{kj}^{(5)} \neq 0$ and if $p_{kj}^{(4)} \neq 0$, $p_{kj}^{(5)} = 0$. Thus each of the states E_2, E_3, E_4, E_5, E_6 can be reached from any other one of them. For this reason they are called ergodic. In general, if E is a subset of the set of states of a Markov chain with the property that for every pair of states E_i, E_j in E there is a positive integer μ such that $p_{ij}^{(\mu)} > 0$, E is an *ergodic* subset of the set of states of the chain and the elements of E are called *ergodic states*. The chain we have just discussed is an example of a nonergodic chain which contains a subset of ergodic states.

8-4c$_3$ Table. *Successive powers of the transition matrix of a ergodic chain.*

$$T = \begin{pmatrix} 0 & .4 & 0 & 0 & 0 & .6 \\ .6 & 0 & .4 & 0 & 0 & 0 \\ 0 & .6 & 0 & .4 & 0 & 0 \\ 0 & 0 & .6 & 0 & .4 & 0 \\ 0 & 0 & 0 & .6 & 0 & .4 \\ .4 & 0 & 0 & 0 & .6 & 0 \end{pmatrix}$$

$$T^2 = \begin{pmatrix} .48 & 0 & .16 & 0 & .36 & 0 \\ 0 & .48 & 0 & .16 & 0 & .36 \\ .36 & 0 & .48 & 0 & .16 & 0 \\ 0 & .36 & 0 & .48 & 0 & .16 \\ .16 & 0 & .36 & 0 & .48 & 0 \\ 0 & .16 & 0 & .36 & 0 & .48 \end{pmatrix}$$

$$T^3 = \begin{pmatrix} 0 & .288 & 0 & .28 & 0 & .432 \\ .432 & 0 & .288 & 0 & .28 & 0 \\ 0 & .432 & 0 & .288 & 0 & .28 \\ .28 & 0 & .432 & 0 & .288 & 0 \\ 0 & .28 & 0 & .432 & 0 & .288 \\ .288 & 0 & .28 & 0 & .432 & 0 \end{pmatrix}$$

$$T^4 = \begin{pmatrix} .3456 & 0 & .2832 & 0 & .3712 & 0 \\ 0 & .3456 & 0 & .2832 & 0 & .3712 \\ .3712 & 0 & .3456 & 0 & .2832 & 0 \\ 0 & .3712 & 0 & .3456 & 0 & .2832 \\ .2832 & 0 & .3712 & 0 & .3456 & 0 \\ 0 & .2832 & 0 & .3712 & 0 & .3456 \end{pmatrix}$$

$$T^5 = \begin{pmatrix} 0 & .30816 & 0 & .336 & 0 & .35584 \\ .35584 & 0 & .30816 & 0 & .336 & 0 \\ 0 & .35584 & 0 & .30816 & 0 & .336 \\ .336 & 0 & .35584 & 0 & .30816 & 0 \\ 0 & .336 & 0 & .35584 & 0 & .30816 \\ .30816 & 0 & .336 & 0 & .35584 & 0 \end{pmatrix}$$

$$T^6 = \begin{pmatrix} .327232 & 0 & .324864 & 0 & .347904 & 0 \\ 0 & .327232 & 0 & .324864 & 0 & .347904 \\ .347904 & 0 & .327232 & 0 & .324864 & 0 \\ 0 & .347904 & 0 & .327232 & 0 & .324864 \\ .324864 & 0 & .347904 & 0 & .327232 & 0 \\ 0 & .324864 & 0 & .347904 & 0 & .327232 \end{pmatrix}$$

The two states E_1 and E_7 of Contest I have the property that once the system enters either of them it remains forevermore in that state. For this reason they are called *absorbing* states. Let E_j be a state of a Markov chain such that for every positive integer $\mu > 0$, $p_{jj}^{(\mu)} = 1$ (and hence $p_{jk}^{(\mu)} = 0$, $k \neq j$) then E_j is said to be *absorbing*. A chain may have none or one or more such states.

EXERCISES 8-4c

1. What are the absorbing states in the following matrices?

$$\begin{pmatrix} 0 & \frac{1}{3} & \frac{2}{3} \\ 1 & 0 & 0 \\ 0 & 1 & 0 \end{pmatrix} ; \quad \begin{pmatrix} \frac{1}{2} & 0 & \frac{1}{4} & \frac{1}{4} \\ 0 & 1 & 0 & 0 \\ \frac{1}{7} & \frac{1}{7} & \frac{3}{7} & \frac{2}{7} \\ 0 & 0 & 0 & 1 \end{pmatrix} ; \quad \begin{pmatrix} 0 & 0 & 0 & 1 \\ 0 & 1 & 0 & 0 \\ \frac{1}{2} & 0 & \frac{1}{4} & \frac{1}{4} \\ \frac{1}{7} & \frac{1}{7} & \frac{3}{7} & \frac{2}{7} \end{pmatrix}$$

2. Prove that if in the transition matrix T, $p_{jj} = 1$ (and hence $p_{jk} = 0$, $k \neq j$), $p_{jj}^{(n)} = 1$. This is the formal mathematical statement of an absorbing state.

3. Consider a one-dimensional random walk with seven states $E_1, E_2, \ldots, E_7$. If E_1 and E_7 are absorbing states and if for $i \neq 1, 7$, $p_{i\,i-1} = p$, $p_{i\,i+1} = q (p + q = 1)$ and $p_{ik} = 0$, $k \neq i \pm 1$, the transition matrix of this walk is the same as that in Contest I. Let us alter the design of the walk by removing the absorbing property of E_1 and E_7 and have rather that $p_{12} = p_{73} = 1$; that is as soon as the walk enters E_1 or E_7, it is reflected back into the interior. What is the transition matrix? Does this change the ergodic properties of the chain or of the states?

4. Consider a second one-dimensional random walk in which the probability of moving to the left or to the right is proportional to the distance from either end. Thus if the walk has five states E_1, E_2, E_3, E_4, E_5, $p_{32} = p_{34} = \frac{1}{2}$, $p_{21} = \frac{1}{4}$, $p_{23} = \frac{3}{4}$, and whenever the walk reaches E_1 or E_5, it is reflected back to E_3. What is the transition matrix? What are its ergodic properties?

5. In the text we discussed a cyclic random walk in which the walk moved in a clockwise direction with probability p and in a counter-clockwise direction with probability $q = 1 - p$. (The transition matrix is like that in Table 8-4c_3

with .4 replaced by p and .6 by q.) If the probability of the direction of the move depends on the current state of the chain, i.e., if $p_{i\,i-1} = \zeta_i$, $p_{i\,i+1} = \xi_i = 1 - \zeta_i$ (if $i = 1$, $i - 1 = 6$ and if $i = 6$, $i + 1 = 1$), are the ergodic properties of the chain altered?

8-5. POSITIVE MATRICES

a. Steady-State Matrices

Recall the chain generated by a series of Bernoulli trials in which the probability of success is $\dfrac{5}{6}$ and that of failure is $\dfrac{1}{6}$. The chain is described in Sec. 8-1c; it has the transition matrix

$$T = \begin{pmatrix} \dfrac{1}{6} & \dfrac{5}{6} & 0 & 0 & 0 \\[2mm] \dfrac{1}{6} & 0 & \dfrac{5}{6} & 0 & 0 \\[2mm] \dfrac{1}{6} & 0 & 0 & \dfrac{5}{6} & 0 \\[2mm] \dfrac{1}{6} & 0 & 0 & 0 & \dfrac{5}{6} \\[2mm] \dfrac{1}{6} & 0 & 0 & 0 & \dfrac{5}{6} \end{pmatrix}$$

If we compute T^2 and T^4 we find

$$T^2 = \begin{pmatrix} \dfrac{1}{6} & \dfrac{5}{6} & \dfrac{25}{36} & 0 & 0 \\[2mm] \dfrac{1}{6} & \dfrac{5}{6} & 0 & \dfrac{25}{36} & 0 \\[2mm] \dfrac{1}{6} & \dfrac{5}{6} & 0 & 0 & \dfrac{25}{36} \\[2mm] \dfrac{1}{6} & \dfrac{5}{6} & 0 & 0 & \dfrac{25}{36} \\[2mm] \dfrac{1}{6} & \dfrac{5}{6} & 0 & 0 & \dfrac{25}{36} \end{pmatrix} \quad \text{and} \quad T^4 = \begin{pmatrix} \dfrac{1}{6} & \dfrac{5}{6} & \dfrac{25}{216} & \dfrac{125}{1296} & \dfrac{625}{1296} \\[2mm] \dfrac{1}{6} & \dfrac{5}{6} & \dfrac{25}{216} & \dfrac{125}{1296} & \dfrac{625}{1296} \\[2mm] \dfrac{1}{6} & \dfrac{5}{6} & \dfrac{25}{216} & \dfrac{125}{1296} & \dfrac{625}{1296} \\[2mm] \dfrac{1}{6} & \dfrac{5}{6} & \dfrac{25}{216} & \dfrac{125}{1296} & \dfrac{625}{1296} \\[2mm] \dfrac{1}{6} & \dfrac{5}{6} & \dfrac{25}{216} & \dfrac{125}{1296} & \dfrac{625}{1296} \end{pmatrix}$$

We see immediately not only that T^4 is positive but also that all its rows are identical. What does this mean in terms of the Markov process associated with the chain? Since $p_{1j}^{(4)} = p_{2j}^{(4)} = p_{3j}^{(4)} = p_{4j}^{(4)} = p_{5j}^{(4)}$, $j = 1, 2, 3, 4, 5$, the probability of being in E_j after four steps is independent of the state in which the chain started. Thus, if at some step the distribution vector is $(\alpha_1 \ \alpha_2 \ \alpha_3 \ \alpha_4 \ \alpha_5)$:

$$\pi T^4 = (\alpha_1 \ \alpha_2 \ \alpha_3 \ \alpha_4 \ \alpha_5) \begin{pmatrix} \dfrac{1}{6} & \dfrac{5}{6} & \dfrac{25}{216} & \dfrac{125}{1296} & \dfrac{625}{1296} \\[2mm] \dfrac{1}{6} & \dfrac{5}{6} & \dfrac{25}{216} & \dfrac{125}{1296} & \dfrac{625}{1296} \\[2mm] \dfrac{1}{6} & \dfrac{5}{6} & \dfrac{25}{216} & \dfrac{125}{1296} & \dfrac{625}{1296} \\[2mm] \dfrac{1}{6} & \dfrac{5}{6} & \dfrac{25}{216} & \dfrac{125}{1296} & \dfrac{625}{1296} \\[2mm] \dfrac{1}{6} & \dfrac{5}{6} & \dfrac{25}{216} & \dfrac{125}{1296} & \dfrac{625}{1296} \end{pmatrix}$$

$$= \left(\frac{1}{6} \sum_{i=1}^{5} \alpha_i \quad \frac{5}{6} \sum_{i=1}^{5} \alpha_i \quad \frac{25}{216} \sum_{i=1}^{5} \alpha_i \quad \frac{125}{1296} \sum_{i=1}^{5} \alpha_i \quad \frac{625}{1296} \sum_{i=1}^{5} \alpha_i \right)$$

$$= \left(\frac{1}{6} \quad \frac{5}{6} \quad \frac{25}{216} \quad \frac{125}{1296} \quad \frac{625}{1296} \right).$$

In the same way we can show that if T^k is any power of T,

$$T^{k+4} = T^k T^4 = T^4,$$

and hence for all $\mu \geqslant 4$, $T^\mu = T^4$. As a consequence of this, for any π_0, if $\mu \geqslant 4$,

$$\pi_\mu = \pi_0 T^\mu = \pi_0 T^4 = \left(\frac{1}{6} \quad \frac{5}{6} \quad \frac{215}{216} \quad \frac{125}{1296} \quad \frac{625}{1296} \right),$$

that is, $p_j^{(\mu)} = p_j^{(4)}$, $j = 1, 2, 3, 4, 5$. This Markov process has reached a "steady state" in the sense that once the process has been going on sufficiently long, the probability of its being in the state E_j is independent of the number of the steps.

Is this property peculiar to this transition matrix? Or is it shared by others? It is certainly not a property of all matrices for in the preceding section we showed that for

$$T = \begin{pmatrix} 0 & p & 0 & 0 & 0 & 0 & q \\ q & 0 & p & 0 & 0 & 0 & 0 \\ 0 & q & 0 & p & 0 & 0 & 0 \\ 0 & 0 & q & 0 & p & 0 & 0 \\ 0 & 0 & 0 & q & 0 & p & 0 \\ 0 & 0 & 0 & 0 & q & 0 & p \\ p & 0 & 0 & 0 & 0 & q & 0 \end{pmatrix}$$

the zeros in T^{2v}, $v \geqslant 1$, alternated with the zeros in T^{2v+1}. Hence there can be no v such that for every positive integer k, $T^{v+k} = T^v$.

8-5a_1 Definition.   *Let S be an $N \times N$ transition matrix with N identical rows,—that is,*

$$S = \begin{pmatrix} p_1 & p_2 & \cdots & p_N \\ p_1 & p_2 & \cdots & p_N \\ \cdot & \cdot & & \cdot \\ \cdot & \cdot & & \cdot \\ p_1 & p_2 & \cdots & p_N \end{pmatrix}$$

*S is called a **steady state** matrix.*

In the paragraph above, T^4 is a steady-state matrix.

Several properties of steady-state matrices are easy to prove.

8-5a_2 Theorem.   *If S is an $N \times N$ steady-state matrix, then for all $N \times N$ transition matrices T,*

$$T \times S = S.$$

Proof:

$$T \times S = \begin{pmatrix} p_{11} & p_{12} & \cdots & p_{1N} \\ p_{21} & p_{22} & \cdots & p_{2N} \\ \cdot & \cdot & & \cdot \\ \cdot & \cdot & & \cdot \\ \cdot & \cdot & & \cdot \\ p_{N1} & p_{N2} & \cdots & p_{NN} \end{pmatrix} \begin{pmatrix} p_1 & p_2 & \cdots & p_N \\ p_1 & p_2 & \cdots & p_N \\ \cdot & \cdot & & \cdot \\ \cdot & \cdot & & \cdot \\ \cdot & \cdot & & \cdot \\ p_1 & p_2 & \cdots & p_N \end{pmatrix}$$

$$= \begin{vmatrix} P_1 \sum_{j=1}^{N} p_{1j} & P_2 \sum_{j=1}^{N} p_{1j} & \cdots & P_N \sum_{j=1}^{N} p_{1j} \\ P_1 \sum_{j=1}^{N} p_{2j} & P_2 \sum_{j=1}^{N} p_{2j} & \cdots & P_N \sum_{j=1}^{N} p_{2j} \\ \cdot & \cdot & & \cdot \\ \cdot & \cdot & & \cdot \\ P_1 \sum_{j=1}^{N} p_{Nj} & P_2 \sum_{j=1}^{N} p_{Nj} & \cdots & P_N \sum_{j=1}^{N} p_{Nj} \end{vmatrix}$$

Each sum in the product matrix is a row sum of T and so is equal to one. Hence

$$T \times S = \begin{pmatrix} p_1 & p_2 & \cdots & p_N \\ p_1 & p_2 & \cdots & p_N \\ \cdot & \cdot & & \cdot \\ \cdot & \cdot & & \cdot \\ \cdot & \cdot & & \cdot \\ p_1 & p_2 & \cdots & p_N \end{pmatrix} = S$$

and our proof is complete. ■

8-5a₃ Corollary. *If S is an $N \times N$ steady-state matrix with rows*

$$p_1 \ p_2 \ p_3 \ \cdots p_N$$

and π is a $1 \times N$ probability vector, then

$$\pi S = (p_1 \ \ p_2 \ \ p_3 \dots p_N).$$

EXERCISES 8-5a

1. Prove Corollary 8-5a₃.

2. Show by direct calculation that if T is the matrix at the beginning of Sec. 8-5a and $k \in I^+$, the statement in the text, $T^{k+4} = T^4$ is valid.

3. Compute T^2, T^4, T^8, and T^{16} for

$$T = \begin{pmatrix} p & q & 0 & 0 & 0 \\ p & 0 & q & 0 & 0 \\ p & 0 & 0 & q & 0 \\ p & 0 & 0 & 0 & q \\ p & 0 & 0 & 0 & q \end{pmatrix}$$

4. Take the matrix in Prob. 3 and extend its pattern to an $N \times N$ matrix (the first column will be p's, and $p_{12} = p_{23} = p_{34} = \cdots = p_{N-1 \, N} = p_{NN} = q$). Compute T^2, T^4, and T^8. Do you think that some power of this matrix will be a steady-state matrix? If so, what will it be?

b.   Behavior Of Positive Matrices

We study first a few numerical examples of positive matrices in hopes of discovering some of their characteristics.   We begin with

$$T = \begin{pmatrix} 3d & 3d & 6d & 3d \\ 4d & 2d & 6d & 3d \\ 4d & 3d & 5d & 3d \\ 4d & 3d & 6d & 2d \end{pmatrix}$$

where $d = \dfrac{1}{15}$.   Notice that the rows are "almost" alike and that the greatest difference between the largest and smallest element in any column is d.

$$T^2 = \begin{pmatrix} 57d^2 & 42d^2 & 84d^2 & 42d^2 \\ 56d^2 & 43d^2 & 84d^2 & 42d^2 \\ 56d^2 & 42d^2 & 85d^2 & 42d^2 \\ 56d^2 & 42d^2 & 84d^2 & 43d^2 \end{pmatrix}$$

and the difference between the maximum and minimum elements in any column is $d^2 = \dfrac{1}{225} < d$.

$$T^4 = \begin{pmatrix} 12657d^4 & 9492d^4 & 18984d^4 & 9492d^4 \\ 12656d^4 & 9493d^4 & 18984d^4 & 9492d^4 \\ 12656d^4 & 9492d^4 & 18985d^4 & 9492d^4 \\ 12656d^4 & 9492d^4 & 18984d^4 & 9493d^4 \end{pmatrix}$$

Here the rows are even more alike than in T and T^2, for in any column the difference between the greatest and smallest element is $d^4 < 2.0 \times 10^{-5} < d^2 < d$. If we compute T^8 we get

$$T^8 = \begin{pmatrix} 640{,}722{,}657d^8 & 480{,}541{,}922d^8 & 961{,}083{,}984d^8 & 480{,}541{,}922d^8 \\ 640{,}722{,}656d^8 & 480{,}541{,}923d^8 & 961{,}083{,}984d^8 & 480{,}541{,}922d^8 \\ 640{,}722{,}656d^6 & 480{,}541{,}922d^8 & 961{,}083{,}985d^8 & 480{,}541{,}922d^8 \\ 640{,}722{,}656d^8 & 480{,}541{,}922d^8 & 961{,}083{,}984d^8 & 480{,}541{,}923d^8 \end{pmatrix}$$

and we see that the difference between the maximum and minimum elements in any column is d^8 which is less than 1.0×10^{-9}. Thus while we have not attained identical rows the discrepancies between them have become very small. We can ask ourselves some questions. If we were to compute still higher powers of T, would the variations continue to decrease? Would some higher power of T be a steady-state matrix? On the basis of what we have already seen, I think we would anticipate that no power of T would be a steady-state matrix but that the difference between the maximum and minimum elements in any column would become increasingly less significant as the powers increase.

Let us start again with the same T, this time with its elements expressed in decimal form, and run it through a computer which has been programmed to compute $T \times T = T^2$ and give the entries in the product matrix to seven decimal places.

For convenience, set

8-5b₁ $$\Delta_j^{(k)} = \max_i p_{ij}^{(k)} - \min_i p_{ij}^{(k)}$$

(that is, $\Delta_j^{(k)}$ is the difference between the maximum and minimum elements in the jth column of T^k). Also set

8-5b₂ $$\Delta^{(k)} = \max_j \Delta_j^{(k)}$$

(that is, $\Delta^{(k)}$ is the greatest of the $\Delta_j^{(k)}$'s.)

$$T = \begin{pmatrix} .2000000 & .2000000 & .4000000 & .2000000 \\ .2666667 & .1333333 & .4000000 & .2000000 \\ .2666667 & .2000000 & .3333333 & .2000000 \\ .2666667 & .2000000 & .4000000 & .1333333 \end{pmatrix}$$

$$\Delta = .0666667 \left(= d = \frac{1}{15} \right)$$

$$T^2 = \begin{pmatrix} .2533333 & .1866667 & .3733333 & .1866667 \\ .2488889 & .1911111 & .3733333 & .1866667 \\ .2488889 & .1866667 & .3777777 & .1866667 \\ .2488889 & .1866667 & .3733333 & .1911111 \end{pmatrix}$$

$$\Delta^{(2)} = .0044444 \left(= d^2 = \frac{1}{225} \right)$$

$$T^4 = \begin{pmatrix} .2500148 & .1874963 & .3749926 & .1874963 \\ .2499951 & .1875160 & .3749926 & .1874963 \\ .2499951 & .1874963 & .3750123 & .1874963 \\ .2499951 & .1874963 & .3749926 & .1875160 \end{pmatrix}$$

$$\Delta^{(4)} = .0000197 \; (= d^4)$$

$$T^8 = \begin{pmatrix} .2500000 & .1875000 & .3750000 & .1875000 \\ .2500000 & .1875000 & .3750000 & .1875000 \\ .2500000 & .1875000 & .3750000 & .1875000 \\ .2500000 & .1875000 & .3750000 & .1875000 \end{pmatrix}$$

Now whatever $\Delta^{(8)}$ is, it is certainly less than 1.0×10^{-7}. Since in this form the rows are identical, further multiplication will yield the same product (see 8-5a_2). Hence all we can say about higher powers of T is that the difference between the maximum and minimum elements in any column is less than 1.0×10^{-7}.

Consider another example.

$$T = \begin{pmatrix} .300000 & .000000 & .700000 \\ .000000 & .600000 & .400000 \\ .200000 & .300000 & .500000 \end{pmatrix}$$

T, we note, is not a positive matrix; however

$$T^2 = \begin{pmatrix} .230000 & .210000 & .560000 \\ .080000 & .480000 & .440000 \\ .160000 & .330000 & .510000 \end{pmatrix}$$

is positive. We have $\Delta_1^{(2)} = .23 - .08 = .15$, $\Delta_2^{(2)} = .48 - .21 = .27$ and $\Delta_3^{(2)} = .56 - .44 = .12$, so that $\Delta^{(2)} = \Delta_2^{(2)} = .27$.

$$
T^4 = \begin{pmatrix} .159300 & .333900 & .506800 \\ .127200 & .392400 & .480400 \\ .144800 & .360300 & .494900 \end{pmatrix}
$$

$\Delta_1^{(4)} = .1593 - .1272 = .0321$, $\Delta_2^{(4)} = .3924 - .3339 = 0585$, $\Delta_3^{(4)} = .5068 - .4804 = .0264$, so that $\Delta^{(4)} = \Delta_2^{(4)} = .0585 < \Delta^{(2)}$.

$$
T^8 = \begin{pmatrix} .141233 & .366813 & .491954 \\ .139738 & .369538 & .490724 \\ .140558 & .368043 & .491399 \end{pmatrix}
$$

$\Delta^{(8)} = .002725 < \Delta^{(4)} < \Delta^{(2)}$

As in the first example we considered, the $\Delta^{(k)}$'s decrease with increasing k, but without following so discernable a pattern. If we compute T^{16}, limiting ourselves to six significant figures, we find

$$
T^{16} = \begin{pmatrix} .140353 & .368417 & .491230 \\ .140350 & .368423 & .491227 \\ .140351 & .368420 & .491229 \end{pmatrix}
$$

where $\Delta^{(16)} = .000006$. It seems reasonable to expect that if we were to compute T^{32} and T^{64} to six significant figures $\Delta^{(32)}$ and $\Delta^{(64)}$ would become so small that the rows would look alike, and we could only say that the maximum difference is less than 1.0×10^{-6}.

In addition to the steady decrease in $\Delta^{(k)}$ we observe also in both the numerical examples that in any column

$$
\min_i p_{ij}^{(2k)} < \min_i p_{ij}^{(2k+1)} < \max_i p_{ij}^{(2k+1)} < \max_i p_{ij}^{(2k)}.*
$$

Hence the entries in any given column seem to be boxing in some number. What is the significance of this for the Markov process associated with the original transition matrix? Let $\pi_0 = (p_1 \; p_2 \; p_3)$ be any distribution vector, and consider T^8 of our last example. Then

$$
\pi_0 T^8 = \pi_8 = (p_1^{(8)} \; p_2^{(8)} \; p_3^{(8)})
$$

$$
p_1^{(8)} = .141233 \, p_1 + .139738 \, p_2 + .140558 \, p_3
$$

*$\min_i p_{ij}^{(2k)}$ is the smallest element of the jth column of T^{2k}.

Since $p_1 + p_2 + p_3 = 1$, we have

$$p_1^{(8)} = .139738 + .001495\, p_1 + .000820\, p_3 > .139738$$

Also

$$p_1^{(8)} = .141233 - .001495\, p_2 - .000675\, p_3 < .141233$$

Similarly for $p_2^{(8)}$ and $p_3^{(8)}$ we have

$$.366813 < p_2^{(8)} < .369538$$

$$.490724 < p_3^{(8)} < .491954$$

Furthermore, in the same example,

$$\pi_0 T^{16} = (p_1^{(16)} \quad p_2^{(16)} \quad p_3^{(16)})$$

where

$$.140350 < p_1^{(16)} < .140353$$

$$.368417 < p_2^{(16)} < .368423$$

$$.491227 < p_3^{(16)} < .491230$$

As k increases, the range of $p_j^{(2k)}$ decreases and the probability that the process is in E_j is less dependent on the number of the steps. We reach a kind of quasi-steady state.

The behavior we have illustrated and have been talking about is formalized in the following definition.

8-5b_3 **Definition.** *An $N \times N$ transition matrix is said to* **asymptotically approach** *a steady-state matrix if, given any positive number ϵ, no matter how small, there exists a positive integer v such that for all j and any pair, i, k ($i \neq k$, $i, j, k = 1, 2, \ldots, N$)*

$$|p_{ij}^{(v)} - p_{kj}^{(v)}| < \epsilon.$$

This says that a matrix T asymptotically approaches a steady-state matrix, if, after choosing some $\epsilon > 0$, no matter how small, we can find a power of T, say T^v, such that in any column of T^v the elements, taken pairwise, differ by as little as we please.

For the matrix in our last example above, that is, for

$$T = \begin{pmatrix} .300000 & .000000 & .700000 \\ .000000 & .600000 & .400000 \\ .200000 & .300000 & .500000 \end{pmatrix}$$

we see that if we choose $\epsilon = .01$, $v = 8$, for in T^v, $\Delta_1^{(8)} = .001495 < .01$, $\Delta_2^{(8)} = .002725 < .01$ and $\Delta_3^{(8)} = .001230 < .01$. Were we to choose $\epsilon = .00001 = 10^{-5}$, the integer 8 would not satisfy the requirement of the definition but 16 would, for we computed earlier that $\Delta^{(16)} = .000006 < .00001$. For $\epsilon = .01$ and $\epsilon = .00001$ we have found integers v (namely 8 and 16) such that the elements of T^v satisfy the conditions of Definition 8-5b_3. However, we have not shown that the initial matrix T asymptotically approaches a steady-state matrix, for we have not shown that for every ϵ we can find a v.

We proceed now to prove that every positive matrix asymptotically approaches a steady-state matrix.

c.   Theorem and Proof

Before stating and proving the theorem let us consider briefly what the proof may require. If we single out the jth column of T and look at M_j and m_j, its maximum and minimum elements respectively, we have $0 < m_j \leqslant M_j < 1$. If, $m_j = M_j = \rho_j$ all elements of the jth column are equal. Furthermore, if the entries of the jth column of T are identical and equal to ρ_j, the entries of the jth column of T^v for $v \in I^+$ are also identical and equal to ρ_j. If $m_j < M_j$, we form T^2 and look at $m_j^{(2)}$ and $M_j^{(2)}$, the minimum and maximum of the jth column of T^2, and we show that

$$0 < m_j < m_j^{(2)} \leqslant M_j^{(2)} < M_j < 1.$$

We repeat the procedure until either we find a v_j such that

$$m_j^{(v_j)} = M_j^{(v_j)} = \rho_j$$

and the elements of the jth column of T^{v_j} are equal (and hence for $v > v_j$ all elements of the jth column of T^v equal ρ_j), or until the difference

$$M_j^{(v_j)} - m_j^{(v_j)}$$

is arbitrarily small. We know in this second alternative that

$$0 < m_j < m_j^{(2)} < m_j^{(4)} < \cdots < m_j^{(v_j)} < M_j^{(v_j)} < \cdots < M_j^{(4)} < M_j^{(2)} < M_j < 1$$

and we will have to show that the difference $M_j^{(v_j)} - m_j^{(v_j)}$ while always nonnegative can be made as small as we please. (To show that T asymptotically approaches a steady-state matrix in as few steps as possible we go from T to T^2 to T^4 and so on, so that v_j is some power of two, say $v_j = 2^k$.) To do this we prove first that the smallest element of T cannot be larger than one-half, and then go on to show that $M_j^{2^k} - m_j^{2^k} < a^k$ where $a < 1$. Hence, starting with a positive matrix T and a positive number ϵ, as small as we please, we are able to find for each column of T an integer $v_j = 2^k$ such that in T^{v_j}, $M_j^{(v_j)} - m_j^{(v_j)} < \epsilon$. Furthermore, if $v > v_j$ the elements of the jth column of T^v satisfy the same inequality.

Let $v = \max (v_1, v_2, \ldots, v_N)$, then any two entries in each column of T^v differ by less than ϵ and by 8-5b_1, T asymptotically approaches a steady-state matrix. Let us now begin the formal proof which we will accomplish by a series of lemmas. (In what follows m_j^r and M_j^r will designate respectively the minimum and maximum elements of the jth column of T^v.)

8-5c$_1$ Lemma 1. *Let T be a positive matrix.*

$$m_j \leqslant m_j^{(2)} \leqslant M_j^{(2)} \leqslant M_j$$

Proof: For all i,

$$p_{ij}^{(2)} = \sum_{k=1}^{N} p_{ik} p_{kj} \geqslant \sum_{k=1}^{N} p_{ik} m_j$$

(we replace each p_{kj} by $m_j \leqslant p_{kj}$). But

$$\sum_{k=1}^{N} p_{ik} m_j = m_j$$

(row sums are one), and hence for all i, $p_{ij}^{(2)} \geqslant m_j$. Then

$$m_j^{(2)} = \min {}_i p_{ij}^{(2)} \geqslant m_j.$$

Also

$$p_{ij}^{(2)} = \sum_{k=1}^{N} p_{ik} p_{kj} \leqslant \sum_{k=1}^{N} p_{ik} M_j = M_j,$$

for all i. Hence

$$M_j^{(2)} = \max {}_i p_{ij}^{(2)} \leqslant M_j$$

and we have

$$m_j \leqslant m_j^{(2)} \leqslant M_j^{(2)} \leqslant M_j \qquad \blacksquare$$

8-5c$_2$ Lemma 2. *Let T be a positive $N \times N$ transition matrix ($N \geqslant 2$) and let μ be its smallest entry. Then $\mu \leqslant \dfrac{1}{2}$.*

Proof: Assume $\mu > \dfrac{1}{2}$ and $\mu = p_{ik}$ (that is μ is in the ith row and kth column):

$$\sum_{j=1}^{N} p_{ij} = 1 \quad \text{and} \quad \sum_{\substack{j=1 \\ j \neq k}}^{N} p_{ij} = 1 - \mu < \frac{1}{2}$$

Since T is a positive matrix for all j, $p_{ij} > 0$, and for $j \neq k$, $p_{ij} < \frac{1}{2}$, for if this were not so,

$$\sum_{\substack{j=1 \\ j \neq k}}^{N} p_{ij} > \frac{1}{2} \quad \text{But } p_{ij} < \frac{1}{2} \ (j \neq k) \text{ contradicts the assumption that } \mu > \frac{1}{2} \text{ is the}$$

smallest element of T.   Hence $\mu \leqslant \frac{1}{2}$ ■

    8-5c₃   **Lemma 3.** *Let T be a positive $N \times N$ matrix and μ its smallest element. Then*

$$M_j^{(2)} - m_j^{(2)} \leqslant (1 - 2\mu)\,(M_j - m_j).$$

Proof.   In the jth column let $p_{\zeta j} = M_j$ and $p_{\xi j} = m_j$, then

$$p_{ij}^{(2)} = \sum_{k=1}^{N} p_{ik}p_{kj} \geqslant p_{i\xi}\, M_j + (1 - p_{i\xi})m_j$$

since in each term except the ζ th we replaced p_{kj} by $m_j \leqslant p_{kj}$.   Also,

$$p_{ij}^{(2)} = \sum_{k=1}^{N} p_{ik}p_{kj} \leqslant p_{i\xi}m_j + (1 - p_{i\xi})M_j$$

since in each term except the ξ th we replaced p_{kj} by $M_j \geqslant p_{kj}$.   Together these inequalities yield

$$p_{i\xi}M_j + (1 - p_{i\xi})m_j \leqslant p_{ij}^{(2)} \leqslant p_{i\xi}m_j + (1 - p_{i\xi})M_j$$

$p_{i\xi} \geqslant \mu$ and we write $p_{i\xi} = \mu + (p_{i\xi} - \mu)$ and substitute this in the right-hand inequality:

$$p_{ij}^{(2)} \leqslant \mu m_j + (p_{i\xi} - \mu)m_j + (1 - p_{i\xi})M_j$$

$$\leqslant \mu m_j + (p_{i\xi} - \mu)M_j + (1 - p_{i\xi})M_j$$

where in the middle term we replaced m_j by M_j. Then

$$p_{ij}^{(2)} \leqslant \mu m_j + (1 - \mu)M_j.$$

In an analogous way we work with the left-hand inequality and obtain

$$p_{ij}^{(2)} \geqslant \mu M_j + (1 - \mu)m_j.$$

Since these inequalities hold for all i,

$$\mu M_j + (1 - \mu)m_j \leqslant m_j^{(2)} \leqslant p_j^{(2)} \leqslant M_j^{(2)} \leqslant \mu m_j + (1 - \mu)M_j.$$

We have $M_j^{(2)} \leqslant \mu m + (1 - \mu)M_j$

$$-m_j^{(2)} \leqslant -(1 - \mu)m_j - \mu M_j$$

(We multiply the two left most terms of the inequality by–1). Adding, we have

$$M_j^{(2)} - m_j^{(2)} \leqslant (1 - 2\mu)\,(M_j - m_j).$$ ■

8-5c₄   **Corollary.**   *For* $k \in I^+$,

$$M_j^{(2^k)} - m_j^{(2^k)} \leqslant (1 - 2\mu)^k\,(M_j - m_j).$$

Proof: From Lemma 1 (8-5c₁) and the Principle of Induction we show

$$m_j \leqslant m_j^{(2)} \leqslant m_j^{(4)} \leqslant \cdots \leqslant m_j^{(2^k)} \leqslant M_j^{(2^k)} \leqslant \cdots \leqslant M_j^{(4)} \leqslant M_j^{(2)} \leqslant M_j$$

From Lemma 3 (8-5c₃) and the Principle of Induction we show

$$M_j^{(2^k)} - m_j^{(2^k)} \leqslant (1 - 2\mu)^k\,(M_j - m_j).$$ ■

8-5c₅   **Theorem.**   *If T is a positive $N \times N$ matrix, then T asymptotically approaches a steady-state matrix.*

Proof: Let $\epsilon > 0$ be given. Since $\mu \leqslant \dfrac{1}{2}$, $0 \leqslant 1 - 2\mu < 1$ and there exists an integer v_j such that

$$(1 - 2\mu)^{v_j} < \frac{\epsilon}{M_j - m_j}, \quad (M_j \neq m_j)^{[1]}.$$

This, along with corollary 8-5c₄, yields

$$M_j^{(2^{v_j})} - m_j^{(2^{v_j})} \leqslant (1 - 2\mu)^{v_j}\,(M_j - m_j) < \epsilon,$$

so that the elements of the jth column of $T^{2^{v_j}}$ differ, one from the other, by less than the preassigned number ϵ. For each column j there will be such a v_j. Let

$$v = \max\,(v_1, v_2, \ldots, v_N)$$

Then in

$$T^{2^v} = \begin{pmatrix} p_{11}^{(2^v)} & p_{12}^{(2^v)} & \cdots & p_{1N}^{(2^v)} \\[4pt] p_{21}^{(2^v)} & p_{22}^{(2^v)} & \cdots & p_{1N}^{(2^v)} \\[4pt] \cdot & \cdot & & \cdot \\ \cdot & \cdot & & \cdot \\ \cdot & \cdot & & \cdot \\[4pt] p_{N1}^{(2^v)} & p_{N2}^{(2^v)} & \cdots & p_{NN}^{(2^v)} \end{pmatrix}$$

for every j, and all k and l

$$|p_{kj}^{(2^r)} - p_{ij}^{(2^r)}| < M_j^{(2^r)} - m_j^{(2^r)} < \epsilon.$$

We have shown that if T is a positive matrix, given an arbitrary $\epsilon > 0$, we can find a $2^r \in I^+$ such that the conditions of the definition 8-5b_3 are satisfied Hence T asymptotically approaches a steady-state matrix and the theorem is proved.   ∎

Let T be a transition matrix which asymptotically approaches a steady state, then given any $\epsilon > 0$, there is a $r \in I^+$ such that for all j and any pair i, k $(i, j, k = 1,2,\ldots,N)$ $|p_{ij}^{(r)} - p_{kj}^{(r)}| < \epsilon$. If π_0 is the initial probability vector,

$$\pi_r = \pi_0 T^r = (p_1^{(r)} \ p_2^{(r)} \cdots p_N^{(r)})$$

in which

$$m_j^{(r)} = \min_i p_{ij}^{(r)} \leqslant p_j^{(r)} \leqslant \max_i p_{ij}^{(r)} = M_j^{(r)}.$$

Since $M_j^{(r)} - m_j^{(r)} < \epsilon$, the probability that the Markov process belonging to T is in E_j is made to fall between arbitrarily narrow bounds. Furthermore, since

$$m_j^{(r)} \leqslant m_j^{(r+k)} \leqslant M_j^{(r+k)} \leqslant M_j^{(r)}$$

(see 8-5c_4), for $n > r$, $p_j^{(n)}$ will remain within these bounds. . In many applied problems this quasi-steady state gives useful results.

A construction company classifies its projects at the end of each week as being in one of three categories: ahead of schedule, E_1, behind schedule, E_2, or on schedule, E_3. It has determined that due to weather, labor, and other factors the projects move from one classification to another in a predictable way which is given in the transition matrix

$$T = \begin{pmatrix} .3 & 0 & .7 \\ 0 & .6 & .4 \\ .2 & .3 & .5 \end{pmatrix} *$$

(The entry $p_{12} = 0$ means in a week's time the probability a project moves from being ahead of schedule to being behind schedule is zero. The other entries can be interpreted in a similar manner.) If a project extends over a long period of time, do the fluctuations among the categories approach a steady state so that the probability of being in E_1, E_2, or E_3 varies by less than .01?

*This matrix was investigated in Sec. 8-5b.

$$T^2 = \begin{pmatrix} .23 & .21 & .56 \\ .08 & .48 & .44 \\ .16 & .33 & .51 \end{pmatrix}$$

so T^2 is positive and we know there will be a v such that in any column of T^v the elements taken pairwise will differ by less than .01. In fact, if we go back to 8-5b we find that

$$T^8 = \begin{pmatrix} .141233 & .366813 & .491954 \\ .139738 & .369538 & .490724 \\ .140558 & .368043 & .491399 \end{pmatrix}$$

in which for all j and any pair i, k $(i,j,k = 1,2,3)$

$$|p_{ij}^{(8)} - p_{kj}^{(8)}| < .01.$$

Thus assuming the project is on schedule at its start (i.e., that $\pi_0 = (0\ 0\ 1)$) for $n > 8$

$$\pi_0 T^n = (p_1^{(n)} \quad p_2^{(n)} \quad p_3^{(n)})$$

where

$$.139 < p_1^{(n)} < .142$$

$$.368 < p_2^{(n)} < .370$$

$$.490 < p_3^{(n)} < .492$$

(We certainly could set these limits more narrowly, but since we ask only that the possible fluction be less than .01, there is no reason to do so.)

EXERCISES 8-5c

1. Prove 8-5c_6.
2. Traffic on a certain rail line on any day can be classified as below normal E_1, normal E_2, or above normal E_3. On successive days it moves from one state to another according to the transition matrix

$$T = \begin{pmatrix} .3 & .4 & .3 \\ .2 & .6 & .2 \\ .5 & .4 & .1 \end{pmatrix}$$

If observations extend over a sufficiently long number of days, can we predict to within .001 the probability is below, at, or above normal on a given day? Does it matter whether on the day we began observations traffic was below, at, or above normal levels?

3. The switchboard of a telephone information service is programmed so that up to three incoming calls are held in line. If a fourth call tries to enter the system, a recorded message asks the caller to hang up and try again later. The system has four states: E_0, E_1, E_2 and E_3, where E_i is the state in which there are i calls waiting to be answered. The system has been studied and at regular intervals it has been found to move from one state to another according to the transition matrix:

$$T = \begin{pmatrix} a_0 & a_1 & a_2' & a_3 \\ a_0 & a_1 & a_2 & a_3 \\ 0 & a_0 & a_1 & a_2+a_3 \\ 0 & 0 & a_0 & a_1+a_2+a_3 \end{pmatrix}, \quad (a_0 + a_1 + a_2 + a_3 = 1)$$

(The matrix indicates that during any interval at most one waiting call is answered and one call drops out of the line; $p_{32} > 0$ but $p_{30} = p_{31} = 0$.) If $a_0 = a_1 = a_2 = a_3 = \dfrac{1}{4}$ study T^2, T^4, and T^8 to see whether T will asymptotically approach a steady-state matrix. If so, estimate how much further you will have to go so that $\Delta^{(n)} < .01$. In this case what will be the limits on $p_0^{(n)}$, $p_1^{(n)}$, $p_2^{(n)}$, and $p_3^{(n)}$?

4. Let us look again at the problem of putting balls randomly into cells. As we place the balls in the cells we generate a Markov chain in which the state of the system is the number of occupied cells. If there are six cells, construct the transition matrix. If you let E_i be the state in which i cells are occupied, E_0 is the state in which no cell is occupied and $p_{01} = 1$, $p_{0i} = 0$, $i \neq 1$. Also $p_{i0} = 0$ for all i so that the transition matrix would have a column of zeros. According to our convention 8-1c_6 we drop E_0 since it cannot be entered. Before the first ball is placed no cell is occupied and $\pi_0 = (1\,0\,0\,0\,0\,0\,0)$, and this requires a 7×7 transition matrix—that is the one with the column of zeros. Using this what are π_1 and π_2? What will be the value of $p_0^{(k)}$ for $k > 0$? To avoid these excess computations what is a reasonable choice for π_0 and T? What are π_1 and π_2? Compare these results with Prob. 14, Exercises 7-1a.

5. Let us now look at a common model for the diffusion of molecules of gas through a membrane. Let n balls be distributed between two urns, I and II. At time t a ball is chosen at random and moved from its urn to the other. At any time the state of the system is the number of balls in urn I. Construct the transition matrix. Will this system asymptotically approach a steady state?

In Prob. 5b, Exercises 8-3b we proved $\pi_v T^u = \pi_{u+v}$; for the case $n = 5$ compute T^2 and T^3 then use the formula to find $\pi_1, \pi_2, \pi_3, \pi_5, \pi_8$, assuming that initially there were 3 balls in urn I. Does this indicate anything about the long-range behavior of the chain?

6. Another urn model useful in the study of the mixing of gases starts with N white balls in urn I and N black balls in urn II. At time t a ball is chosen from each urn and placed in the other. The state of the system is the number of white balls in urn I. Construct the transition matrix. Will this approach a steady-state matrix? Again, for the case $n = 5$, compute T^2 and T^3, π_1, π_2, π_3, π_5, and π_8. (At the beginning all the white balls are in urn I.) Does this indicate any trends for the long-range behavior of the chain?

7. Consider the Markov chain with transition matrix

$$T = \begin{pmatrix} \dfrac{1}{3} & \dfrac{2}{3} & 0 & 0 & 0 & 0 \\[2mm] \dfrac{5}{6} & \dfrac{1}{6} & 0 & 0 & 0 & 0 \\[2mm] 0 & 0 & \dfrac{4}{7} & \dfrac{3}{7} & 0 & 0 \\[2mm] 0 & 0 & \dfrac{3}{8} & \dfrac{5}{8} & 0 & 0 \\[2mm] \dfrac{1}{8} & \dfrac{1}{4} & 0 & \dfrac{1}{2} & 0 & \dfrac{1}{8} \\[2mm] 0 & 0 & 0 & \dfrac{1}{2} & \dfrac{1}{2} & 0 \end{pmatrix}$$

If as usual the states are designated by E_1, E_2, E_3, E_4, E_5, E_6, what do we notice about E_1 and E_2, E_3 and E_4, E_5, and E_6? Write T as:

$$\begin{pmatrix} A & 0 & 0 \\ 0 & B & 0 \\ C & D & E \end{pmatrix}$$

where A, B, C, D, and E are all 2×2 matrices and the 0's represent 2×2 matrices all of whose elements are zero. Direct calculation yields:

$$T^2 = \begin{pmatrix} \dfrac{2}{3} & \dfrac{1}{3} & 0 & 0 & 0 & 0 \\[2ex] \dfrac{5}{12} & \dfrac{7}{12} & 0 & 0 & 0 & 0 \\[2ex] 0 & 0 & \dfrac{191}{392} & \dfrac{201}{392} & 0 & 0 \\[2ex] 0 & 0 & \dfrac{201}{448} & \dfrac{247}{448} & 0 & 0 \\[2ex] \dfrac{1}{4} & \dfrac{1}{8} & \dfrac{3}{16} & \dfrac{6}{16} & \dfrac{1}{16} & 0 \\[2ex] \dfrac{1}{16} & \dfrac{1}{8} & \dfrac{3}{16} & \dfrac{9}{16} & 0 & \dfrac{1}{16} \end{pmatrix} = \begin{pmatrix} A^2 & 0 & 0 \\ 0 & B^2 & 0 \\ C' & D' & E^2 \end{pmatrix}$$

Show that

$$T^3 = \begin{pmatrix} A^3 & 0 & 0 \\ 0 & B^3 & 0 \\ C'' & D'' & E^3 \end{pmatrix}$$

Prove that

$$T^n = \begin{pmatrix} A^n & 0 & 0 \\ 0 & B^n & 0 \\ C^{(n)} & D^{(n)} & E^n \end{pmatrix}$$

(A^n, B^n, E^n are the nth powers of A, B, and E but $C^{(n)}$, $D^{(n)}$ are not powers of C and D.)

COMPUTER PROBLEMS

These few problems are suggested for students having knowledge of and access to a computer; many others will probably immediately suggest themselves to students. By observing a Markov process move through many steps the student can see trends that are not always obvious in the first few calculations. (Remember that in transition matrices row sums equal one and this can be a useful check to computations.)

1. In Prob. 5 above use the formula $\pi_{k-1}T = \pi_k$, $k = 1, 2, 3, \dots, 20$, and see what one might expect of this process in the long run. Do this for $n = 8$ and

$n = 9$. (Use smaller dimensions if the speed of computer makes the suggested n too time-consuming.) Will we ever reach an N where $\pi_N T = \pi_N$?

2. Do the same thing for Prob. 6 above. If in either case we do reach an N such that $\pi_N T = \pi_N$, what is the significance of π_N? In your program, what is the accuracy of the entries?

3. In Prob. 3 Exercises 8-1c *Tennis* is played where A wins a point with probability $\frac{3}{7}$ and B wins a point with probability $\frac{4}{7}$. Part b asks: what is the probability A wins the game and the answer is $\frac{9}{25}$. If you again compute $\pi_{k-1} T = \pi_k$, does this give the same result?

4. In 8-5b we used the matrix

$$T = \begin{pmatrix} .3 & 0 & .7 \\ 0 & .6 & .4 \\ .2 & .3 & .5 \end{pmatrix}$$

as an illustration, and speculated that if we continued our calculations to T^{32} or T^{64}, the difference between the rows would be less than 1.0×10^{-6}. Verify this. What is the smallest power of T that will satisfy the condition?

5. Take the numerical example in Prob. 7 above and compute successively higher powers of this T perhaps up to T^{15}. Certainly A and B will approach steady state matrices. Do we see any pattern emerging in the last two rows?

6. Take the nine cell *Maze* introduced in Exercises 8-1c and compute T^n for $n = 1,2,3,\ldots,15$. Can we make any prediction about the location of the mouse as this process continues.

7. Start with

$$T = \begin{pmatrix} \frac{1}{4} & \frac{1}{2} & \frac{1}{4} \\ \frac{1}{2} & 0 & \frac{1}{2} \\ 0 & \frac{1}{3} & \frac{2}{3} \end{pmatrix}$$

(a) Test to see if it approaches a steady-state matrix. (b) Compute $\pi_{k-1} T = \pi_k$, $k = 1,2,3,\ldots,20$ where (*i*) $\pi_0 = (0\ 0\ 1)$, (*ii*) $\pi_0 = (1\ 0\ 0)$, and (*iii*) $\pi_0 = \left(\frac{1}{2}\ \frac{3}{8}\ \frac{1}{8}\right)$. What happens? What is its significance?

chapter **9**

Queuing Theory

9-1. REGULAR ARRIVAL AND SERVICE TIMES

a. Introduction

A queue is a collection of items waiting to be served; cars at a toll booth, supermarket customers in a checkout line, depositors at a bank teller's window and incoming calls at a telephone switchboard are a few such collections. In every case there are "customers" waiting for some "service." This situation is found in so many aspects of twentieth century life that it has been widely studied and analyzed, it being hoped that by better understanding the characteristics of queues, ways of eliminating or of shortening them might be discovered.

While the limited mathematical background assumed here will prevent a full exploration of the subject and its ramifications, we can look at two of the most prevalent types of queues and learn something of the techniques involved and the problems presented. We shall look first at a very simple type of queue which will acquaint us with several of the major characteristics of all queues and with certain questions that they raise.

Our initial problem will look at the single server, first-come, first-served queue; that is, there will be only one line and customers will be serviced in the order in which they arrive. This is the familiar situation experienced at a turnpike exit where there is a single toll booth. Things are not always this simple, of course. There may be several service facilities and customers may be able to move from one line to another as shoppers do at checkout counters. Or, customers may get tired of waiting and leave the line; they cannot do this at a toll booth but they can if waiting for a table is prolonged in a restaurant. Another possibility is a last-come, first-served queue. This would occur in a manufacturing process where one machine feeds another and the second always accepts the most recent output of the first. Then, whenever the second machine is temporarily down and a backlog of items from the first builds up, at the time operation of the second is resumed, there is a queue in which the most recent arrival is serviced first. Other variations exist; those mentioned here will serve to illustrate the complications that must be introduced before queues can be completely analyzed.

So far, the word *probability* has not appeared in our discussion of queues, and in fact it will not in the solution of the introductory problem. However, when we move to the study of a more widely applicable class of queues, the major role of probability in the theory will be evident. The simple situation we now investigate will acquaint us with terminology and concepts peculiar to the field.

b. Single-Server, First-Come, First-Served Queues

Suppose customers arrive at a service facility at a constant rate, say one every T minutes, and that each customer requires S minutes to be serviced. These conditions are very restrictive, for customers usually do not arrive at a constant rate and the services they seek do not require the same amount of time—cars do not arrive at a toll booth at regularly spaced intervals, and they do not take the same amount of time to pay their toll and receive their change. On the other hand, if we study a production line on which a machine at a constant rate stamps out metal disks which are carried to a second station where they are spray-painted at a constant rate, our model is not without application. Let us suppose, in addition, there are n customers in line when the service begins—perhaps the spraying machine was temporarily out of order and a number of unpainted disks had accumulated. How long will it take to dissipate the queue?

We see immediately that if the service time S is greater than the arrival interval T of the customers, i.e., if $\frac{S}{T} > 1$, the queue will grow beyond all bounds, unless the flow of incoming customers is stopped. If the service time and arrival interval are equal ($\frac{S}{T} = 1$), the line will always have n customers; but if $\frac{S}{T} < 1$, we can expect the queue eventually to disappear so that each incoming customer is served without waiting. What can be learned about this type of queue?

Let us first look at the specific example, which we spoke of earlier, of the production line with a stamping operation followed by a spraying operation. Suppose that the stamping machine turns out 60 disks per hour, that the spraying machine can paint 100 disks an hour, and that if all is working smoothly, the second machine will in an hour spray only the 60 items produced by the first. (Spraying is done in the order in which the disks are produced.)

If the spraying machine has been out of operation for two hours, how long will it need to "catch up" on its work—that is, how much time will elapse before it will again be spraying each disk as soon as it comes out of the stamping machine?

In the two hours the spraying machine was out of service, 120 disks piled up. The first hour it is back in service 100 disks are sprayed but 60 additional ones have joined the waiting line so that at the end of the first hour there are 80 "customers" still to be serviced. By the end of the second hour another 100 have been sprayed, the eighty that were waiting at the beginning of the hour plus the first 20, of those produced during the hour, so that there are 40 in the queue after 120 minutes. In the third hour another 100 are sprayed, the 40 which were ready and the 60 disks stamped out during the hour. Hence, after three hours the operation is back to its steady state in which disks are spray-painted as soon as they are stamped out.

Can we now construct a formula which will give the length of time it takes a queue with parameters S, T, and n to vanish? The method we shall use can be easily illustrated; let $S = 3$, $T = 5$, and $n = 4$. The original four customers, s_1, s_2, s_3, s_4, are finished service at time $t = 12$, the first to arrive c_1 after the service facility opened is finished at $t = 15$, the second c_2 at $t = 18$ and so forth. If the first to arrive after the facility opened, arrives at $t = 2$ and those who follow arrive at five-unit intervals ($T = 5$), c_5 will have been served at $t = 27$, the time when c_6 arrives; hence c_6 and those who follow him enter and are serviced immediately and the queue no longer exists.

9-1b_1 Figure

Progress of a Queue with Parameters $S = 3$, $T = 5$, $n = 4$.

9-1b_2 Figure

Process of Queue Dissipation (n customers waiting at start of service).

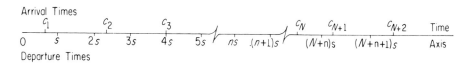

On the basis of this specific case we see that the queue will disappear if there is an integer N such that $N + n$ customers can be served no later than the time the $(N + 1)$st customer arrives at the facility: that is, we seek an N such that

$$(N + n)S \leqslant NT + c_1$$

where c_1 is the time of the arrival of the first customer after the facility opens and $0 \leqslant c_1 \leqslant T$. Then

$$nS - c_1 \leqslant N(T - S)$$

$$\frac{nS - c_1}{T - S} \leqslant N.$$

The N we want is the smallest integer satisfying the inequality. If we let $<x>$ stand for the largest integer less than x, so that $<3.14> = 3$ and $<3> = 2$, then

$$N = \left\langle \frac{nS - c_1}{T - S} \right\rangle + 1.$$

In the example above where $S = 3$, $T = 5$, $n = 4$ and $c_1 = 2$,

$$N = \left\langle \frac{(4)(3) - 2}{5 - 3} \right\rangle + 1 = <5> + 1 = 5$$

and we found that nine customers—the four who were there when service began plus the first five additional customers to arrive—had received service by the time the sixth additional customer—the $(N + 1)$st—arrived at the facility.

We remark that the formula for N involves not only the parameters S, T, and n, but also c_1, the arrival time of the first customer after the service facility opened.

c. Total Elapsed Time

The total amount of time τ elapsed from the beginning of service to the disappearance of the waiting line is the time required to serve the n customers waiting at the time the service facility opened plus the N additional customers who join the existing queue. Thus

$$\tau = (n + N)S = \left(n + \left\langle \frac{nS - c_1}{T - S} \right\rangle + 1 \right)S = \left\langle \frac{(n + 1)T - S - c_1}{T - S} \right\rangle S$$

where the second form of τ is derived from the first by induction on the equality $<x> + 1 = <x + 1>$. In our first example this formula yields

$$\tau = \left\langle \frac{(5)(5) - 2 - 3}{5 - 3} \right\rangle 3 = \left\langle \frac{20}{2} \right\rangle 3 = 27$$

and this, as it should, agrees with our earlier calculation that the queue disappears after the ninth customer was served at $t = 27$.

We can compute how long a customer waits for completion of his service. If he arrives at time $t > \tau$, his waiting time is the time required to perform the service, namely S units. If he arrives before the service facility opens, we consider that he arrives at $t = 0$ and his waiting time is iS, if he is the ith customer served. If he is the kth customer to arrive after service has started,

$$0 < t = c_1 + (k-1)T < \tau$$

he waits

$$(n + k)S - (c_1 + (k-1)T)$$

time units. The $(n + k)S$ is the time needed to serve all the customers who arrived before he did plus himself and the $(c_1 + (k-1)T)$ is subtracted since he did not enter the line until that long after service had begun.

The time spent in waiting for and receiving service by a customer who comes to a given service facility at time t can be summarized as follows:

$$Q(t) = \begin{cases} S & \text{where } \tau \leqslant t \\ (n + k)S - t & 0 < t = c_1 + (k-1)T < \tau \\ iS & t = 0, \end{cases}$$

where $Q(0) = iS$ is the time spent by the ith one to be served of the n customers waiting at the time service was begun. In the numerical example the fourth customer to arrive after service was begun arrived at $t = 2 + 3(5) = 17$. He finished service at $t = (n + k)S = (4 + 4)(3) = 24$ so that he waited $24 - 17 = 7$ time units, a fact that is clearly illustrated in Figure 9-1b_1.

As we mentioned earlier, this model of a single-server facility with constant service time and regular arrival times is limited in its application. To achieve greater applicability we now remove the restrictions of regular arrival times and uniform service time.

EXERCISES 9-1c

1. If $n = 4$, $S = 3$, $T = 5$ and $c_1 = 2$ (the situation given in our numerical examples), and if the last customer to arrive is served first, how long will it be before the queue disappears.

2. Find a general formula in terms of n, S, T, and c_1 for the time it takes a queue to vanish if the service plan is last-come, first-served. How long will a customer wait from the time he enters the queue until he finishes being served?

3. Suppose there are two servers in a facility and that customers are served on a first-come, first-served plan. If service time S is uniform and arrival times T are regular with $S = 5$ and $T = 3$, and if $c_1 = 2$ and $n = 6$, how long will it be before the waiting line disappears? Can you generalize this situation for any n, S, T, and c_i? What relationship must exist between S and T if after some fixed period of time a customer can enter service immediately on joining the queue?

4. Suppose there are three serving facilities and customers are served on a first-come, first-served basis. If $S = 7$, $T = 3$, $c_i = 3$ and $n = 6$, when will the queue disappear? For a three-service facility what relationship must exist between S and T if the queue is to disappear?

5. Develop a formula for the length of time a customer waits for his turn to be served in a single-server, first-come, first-served queue with regular service and arrival times.

6. Let $0 < t < \tau$. Develop a formula for l, the length of the waiting line a time t. (The customer being served is not in the waiting line.)

7. In the text we showed that if $0 \leqslant c_1 < T$ the queue will disappear before the $(N + 1)$st customer arrives where $N = \left\langle \dfrac{nS - c_1}{T - S} \right\rangle + 1$. What will be the situation if $c_1 = T$?, $c_i > T$?

9-2. THE POISSON PROCESS

a. Basic Assumptions

Unlike the queue just studied, most waiting lines are not added to at regular intervals nor is the time to service a customer constant. In many common phenomena, such as activity on a telephone switchboard or the work load of a crew of repairmen charged with keeping a set of machines in operation, we do not know when a customer (an incoming call, a machine breakdown) will arrive at the service facility nor how long it will take to service it (duration of call, length of repair job). However, if we make certain assumptions concerning the probabilities that in a given interval of time a customer will join the queue and that one leaves the queue, we can consider the queue—the totality of those waiting for service and those being served—as being in some state E_n, $n = 0,1,2,\ldots$, during each time interval. Our study of the queue will yield the probabilities p_n that the system is in the state E_n.

Serious mathematical analysis of queues began with the publication in 1917 of A. K. Erlang's paper "Solutions of Some Probability Problems of Significance for Automatic Telephone Exchanges."[*] The original work was in Danish but was soon translated into English, German, and French. Erlang and his

---

[*]*Electroteknikeren*, Vol. 13 (1917), pp. 5-13.

contemporaries were concerned with the "jamming" of an exchange and whether calls arriving at a fully used switchboard would get a busy signal (these would join the queue) or whether they would be lost. Today problems are more sophisticated, but the basic techniques developed by these early investigators are still used in the study of queues.

Queues, of course, change continually with time and in particular they have to start—the switchboard must be installed or the repair crew organized. We shall not look at this initial period in which the queue is forming, rather we shall wait until the system has been in operation sufficiently long so that the state of the system at any time is no longer influenced by its initial state.

The first assumption we make is that the time axis, which is indeed continuous, can be looked at as a sequence of disjoint intervals and that these are sufficiently short so that with positive probability at most one change can occur in the system in a single interval. Using the notation p_{ij} for the probability the system will move from E_i to E_j in one time interval (the same notation we used in the preceding chapter) we have $p_{nn} > 0$, $p_{n\,n\pm 1} > 0$ but $p_{n\,n\pm k} = 0$, $k \neq 0,1$. On the switchboard, for example, the time interval might be a second, and if n calls are in the system, either in progress or waiting for a connection, in any given second the system may remain unchanged, increase to $n + 1$ calls by having another call join the waiting line or decrease to $n - 1$ when one of the calls in progress finishes and leaves the system. By our assumption the probability of two or more changes is negligible. We assume, too, that the customers act independently of each other.

We are all aware of the fact that whenever customers come to a service facility, they tend to seek service at special times—a switchboard is little used at three in the morning and a restaurant is crowded at meal times. In limiting ourselves to the steady state of queues we have removed the time dependent factor and actually look at them during their busy times. In a restaurant lunchtime customers, which may be parties of people, begin to trickle in at 11:00 a.m. By 12:15 p.m. the state of that system—persons seated and being served plus those waiting for tables—is steady; variations exist but the probability that it is in the state E_n is constant. After 1:30 p.m. the incoming traffic slows down appreciably and the pressure on the facility is off. A firm's switchboard has little or no activity during nonbusiness hours; but repeated observations have shown that usage during the business day is predictable. Thus we shall be concerned with "peak" hours. Studies of the behavior of the queue outside these times are also useful but will not be treated here.

In terms of the queue we can look either at the sample space

$$\Omega = \{E_0, E_1, E_2, \ldots\}$$

whose elementary events are the states of the queue or at the random variable X whose range values are the nonnegative integers which count the number of customers in the queue; specifically $X(E_n) = n$. We are looking for

$$p_n = P(E_n) = P(X = n) = P_X(n).$$

We shall speak of the set of p_n as the probability distribution of the queue, and so we tacitly work with the random variable X rather than the sample space. The distinction is obviously not important, for our interest is centered on the number of individuals in the queue.

Before looking at several variations of this type of queue, let us introduce some notations and outline briefly our assumptions. We set:

9-2a_1 (i) p_n = probability there are n customers in the queue,
(ii) λ = probability a customer enters the system in one time interval, and
(iii) μ = probability a customer leaves the queue in one time interval.

We assume:

9-2a_2 (i) the probability of two or more changes is negligible,
(ii) the customers act independently of each other, and
(iii) the queue has reached a steady state and all probabilities are independent of time.

From 9-2a_1, (ii) and (iii) we see that the probability no customer enters the queue is $1 - \lambda$ and the probability no one leaves the queue is $1 - \mu$. Since the customers act independently of each other (9-2a_2,(ii)) the probability of no change in the queue is the product of the probability no customer enters the queue and the probability no customer leaves. We have

$$p \text{ (no change)} = (1 - \lambda)(1 - \mu)$$
$$= 1 - (\lambda + \mu) + \lambda\mu.$$

Since $\lambda\mu$ is the probability of the independent events, a customer enters and a customer leaves, it is the probability of two changes and by 9-2a_2, (i), this is negligible. Hence,

9-2a_3 The probability of no change in a queue is

$$1 - (\lambda + \mu).$$

Suppose it has been determined that, correct to two decimal places, $\lambda = \mu = .01$: then $\lambda\mu = .0001 = .00$ in calculations limited to two-place accuracy. This illustrates what we mean when we speak of the probability of two changes being negligible. The probability of no change, $1 - (\lambda + \mu) = 1 - .02 = .98$, is certainly not negligible.

b. Single-Service Facility

The simplest situation to consider is that in which there is only one service facility—perhaps the single window in a small post office. We are not concerned with the beginning of the process but rather after it has "settled down." Then at any interval of time the queue can be in the state E_n (the queue can contain n customers) if

    (1)   it was in E_n during the preceding interval and no changes occured, or

    (2)   it was in E_{n-1} and a customer joined the queue, or

    (3)   it was in E_{n+1} and a customer completed service. Thus

9-2b$_1$
$$p_n = (1 - (\lambda + \mu))p_n + \lambda p_{n-1} + \mu p_{n+1}. \quad n \geqslant 1$$

$$p_0 = (1 - \lambda)p_0 + \mu p_1.$$

The second equation becomes

$$\lambda p_0 = \mu p_1$$

and, setting $\dfrac{\lambda}{\mu} = r$, we have

9-2b$_2$
$$p_1 = \frac{\lambda}{\mu} p_0 = r p_0.$$

(We call $\dfrac{\lambda}{\mu}$ the service ratio.) In the first equation of 9-2b$_1$, set $n = 1$, then

$$p_1 = (1 - (\lambda + \mu))p_1 + \lambda p_0 + \mu p_2$$

and

$$\mu p_2 = (\lambda + \mu)r p_0 - \lambda p_0 \quad \text{(from 9-2b}_2\text{)}$$

$$p_2 = (r + 1)r p_0 - r p_0 \quad \text{(dividing by } \mu\text{)}$$

$$p_2 = r^2 p_0.$$

If $n = 2$,

$$p_2 = (1 - (\lambda + \mu))p_2 + \lambda p_1 + \mu p_3$$

$$\mu p_3 = (\lambda + \mu)r^2 p_0 - \lambda r p_0$$

$$p_3 = (r + 1)r^2 p_0 - r^2 p_0$$

$$p_3 = r^3 p_0.$$

For $n = 1, 2, 3$, $p_n = r^n p_0$ and we want to see whether all p_n can be so simply expressed in terms of p_0. If we look at what we did for the cases $n = 2$ and $n = 3$, we see we needed the relationship to hold for both $n - 1$ and $n - 2$. Thus we require a variation on our principle of induction. Let us suppose the equality, $p_n = r^n p_0$, holds for $n = k - 1$ and $n = k$; then

$$p_k = (1 - (\lambda + \mu))p_k + \lambda p_{k-1} + \mu p_{k+1}$$

$$\mu p_{k+1} = (\lambda + \mu)r^k p_0 - \lambda r^{k-1} p_0$$

$$p_{k+1} = (r + 1)r^k p_0 - (r)r^{k-1} p_0$$

$$= r^{k+1} p_0.$$

Hence, whenever $p_n = r^n p_0$ holds for $n = k - 1$ and $n = k$, it holds for $n = k + 1$. Since it holds for $n = 2$ and $n = 3$, it holds for $n = 4$, and since it holds for $n = 3$ and $n = 4$, it holds for $n = 5$, and so on. Thus for all $n \in I^+$,

9-2b_3
$$p_n = r^n p_0.$$

Since at every interval of time the system must be in some state,

$$\sum_{n=0} p_n = 1$$

so that

9-2b_4
$$\sum_{n=0} r^n p_0 = p_0 \sum_{n=0} r^n = 1.$$

This is possible only if $r < 1$,—that is, only if $\dfrac{\lambda}{\mu} < 1$. We immediately see the analogue between this and the requirement in the preceding section that $\dfrac{S}{T} < 1$.

If customers are more likely to join the queue than to leave it, the queue will soon grow beyond all bounds and the system will move from E_n to E_{n+1} to $E_{n+2}...$ without ever easing off, so that for any n, $p_n = 0$. Mathematically, if $r > 1$, $1 < r < r^2 < r^3 < \cdots$, and consequently $p_0 < p_1 < p_2 < p_3 < \cdots$. Thus no matter how small p_0 is, if it is positive, there will be an n such that $p_n = r^n p_0 > 1$. This contradicts $\sum_{n=0} p_n = 1$ and so $p_0 = 0$ and for all n, $p_n = 0$.

Thus if the queue is to be in the state E_n with positive probability for at least some n, we must have $r < 1$. Then from Eqs. 9-2b_4 and 4-2b_2, we obtain

$$p_0 \sum_{n=0} r^n = \frac{p_0}{1 - r} = 1$$

and

9-2b_5
$$p_0 = 1 - r$$

so that for any n,

9-2b₆ $$p_n = r^n(1 - r).$$

Thus the distribution function of the random variable that counts the number of customers in the queue is geometric (see 6-1e_3).

For this case of a single-server facility with known probabilities λ and μ and with $\dfrac{\lambda}{\mu} = r < 1$, we observe that

$$p_0 > p_1 > p_2 > p_3 > \cdots$$

Now if r is "large," that is, close to one, $p_0 = 1 - r$ is small and although $p_n = r^n(1 - r)$ is less than p_0, there is not too much difference between them, and the queue will at times have some length. On the other hand if r is "small," p_0 is large and $p_n = r^n(1 - r)$ is much smaller than p_0 so the queue will never become too great. If X is the random variable which counts the number of customers in the queue, $P_X = \{(n, r^n(1 - r)) | n = 0,1,2,\ldots\}$ is geometric, and extrapolating from 6-3a_4 $E(X) = \dfrac{r}{1 - r}.$* Hence, the expected number of customers in the queue is

9-2b₇ $$E(X) = \frac{r}{1 - r}.$$

If $r = .9$, $E(X) = \dfrac{.9}{.1} = 9$, but if $r = .5$, $E(X) = \dfrac{.5}{.5} = 1$ supporting our earlier remark concerning the relationship between r and the length of the queue. The variation of the system is

$$V(X) = E(X^2) - (E(X))^2$$

$$-\sum_{n=0} n^2 r^n(1 - r) - \left(\frac{r}{1 - r}\right)^2$$

$$= \frac{2r^2}{(1 - r)^2} + \frac{r}{1 - r} - \frac{r^2}{(1 - r)^2} \quad \text{[7]}$$

$$= \frac{r}{(1 - r)^2}.$$

In the case where $r = .9$, $E(X) = 9$ and $V(X) = 90$ and large deviations from 9 can be anticipated, but where $r = .5$, $E(X) = 1$ and $V(X) = 2$, and the length of the queue will remain fairly constant.

---

*In the derivation of Eq. 6-3a_4, $P_X = \{(n, p(1 - p)^{n-1}) | n \in I^+\}$ and this gave $E(X) = \dfrac{1}{p}$

EXERCISES 9-2b

1. Consider a queue at a toll booth. If in any minute a car arrives with probability $\frac{1}{3}$ and car departs with probability $\frac{1}{2}$, what is the probability there is no one in the queue? Five cars in the queue? What is the expected length of the queue? Its variation? (Assume the probability of two or more changes in any minute is negligible.)

2. If cars arrive with probability $\frac{1}{2}$ and leave with probability $\frac{1}{3}$, what is the expected length of the queue? If the probability the queue is in E_0 is $\frac{1}{10}$, what is the probability it is in E_5? E_6?

3. Suppose the service rendered by a facility takes between θ and $\theta + k$ time units to complete, $\theta \leqslant S \leqslant \theta + k$. Let us assume that service always begins at the start of a time interval and ends at the completion of one so that k is always a nonnegative integer. If the time intervals are sufficiently small, this does not impose a serious restriction on our problem. Set $p(S = \theta + v) = p_v$ where S is the time of service; $p_0 + p_1 + \cdots + p_k = 1$.

(a) If a customer enters a queue as the only individual already in the queue begins to be serviced, what are the limits on the length of time T the arriving customer will remain in the queue?

(b) What is the probability the arriving customer spends $2\theta + \mu$, $0 \leqslant \mu \leqslant 2k$, time units in the queue?

(c) How are you certain it is a probability?

(d) What is the arriving customer's expected time in the queue?

4. If a customer enters a queue as the first of four waiting customers begins service, what are the limits on the length of time T he will spend in the queue? What is the probability he spends $5\theta + v$, $0 \leqslant v \leqslant 5k$, time units in the queue?

c. Unlimited Number Of Service Channels

We now look at the case where a customer is served as soon as he joins the queue; the system can accommodate everyone that comes to it. There will never be a line and so the service ratio $\frac{\lambda}{\mu}$ will not play a central role. The equations which describe this type of queue differ from those of the single-server queue in one important detail. If the system is in the state E_n, the probability that a change will occur because of one customer's leaving the queue is $n\mu$, since any one of the n individuals could finish service and they all act independently. Thus the equation giving the probabilities the system is in E_n are

9-2c$_1$ $p_n = (1 - (\lambda + n\mu))p_n + \lambda p_{n-1} + (n + 1)\mu p_{n+1}, \quad n \geqslant 1$

$p_0 = (1 - \lambda)p_0 + \mu p_1.$

(The similarity between these equations and those in Eq. 9-2b_1 is readily seen.) As before, we set $\dfrac{\lambda}{\mu} = r$, and in this case we also have

9-2c$_2$
$$p_1 = rp_0.$$

Let us solve for p_2 and p_3.

$$p_1 = (1 - (\lambda + \mu))p_1 + \lambda p_0 + 2\mu p_2$$

$$2p_2 = (r + 1)p_1 - rp_0$$

$$2p_2 = (r + 1)rp_0 - rp_0$$

$$p_2 = \frac{r^2}{2} p_0 = \frac{r^2}{2!} p_0.$$

From this we derive

$$p_2 = (1 - (\lambda + 2\mu))p_2 + \lambda p_1 + 3\mu p_3$$

$$3p_3 = (r + 2)\frac{r^2 p_0}{2!} - r(rp_0)$$

$$3p_3 = \frac{r^3 p_0}{2!}$$

$$p_3 = \frac{r^3}{3 \cdot 2!} p_0 = \frac{r^3}{3!} p_0.$$

These results suggest that for all n, $p_n = \dfrac{r^n}{n!} p_0$. If, following the method we used in studying the single-server queue, we assume that for two consecutive integers, $k-1$ and k the result does hold; then using the same variation on the principle of induction we appealed to in the previous case, we have for all $n \in I^+$

9-2c$_3$
$$p_n = \frac{r^n}{n!} p_0.$$

Since the system must be in some state E_n,

9-2c$_4$
$$\sum_{n=0} p_0 = \sum_{n=0} \frac{r^n}{n!} p_0 = p_0 \sum_{n=0} \frac{r^n}{n!} = 1.$$

The sum above, although we have not met it before, is a familiar one to mathematicians and we know*

9-2c$_5$
$$\sum_{n=0} \frac{r^n}{n!} = e^r$$

*See Tom M., Apostol, *Calculus,* Vol. 1 (Waltham, Mass.: Blaisdell Publishing Company, 1961). p. 452.

where e is Euler's constant. (An approximate value for the irrational e is 2.7183.) Putting this into 9-2c_4 we obtain

9-2c_6 $$p_0 = e^{-r}.$$

We remark that since $e > 1$ and $r > 0$ (it is the ratio of two nonzero probabilities), $e^r > 1$ and so $0 < e^{-r} < 1$. Substituting this last result into Eq. 9-2c_3, we have

9-2c_7 $$p_n = \frac{r^n}{n!} e^{-r}.$$

For every n, $p_n \geqslant 0$ and $\Sigma_{n=0} \, p_n = 1$, so the set p_n indeed forms a probability distribution for the queue. This distribution, $\{(n,p_n) | p_n = \dfrac{r^n}{n!} e^{-r}\}$, is known as the *Poisson distribution* and has the sole parameter r. The Poisson distribution has many applications other than this, but the calculus is required for a complete study of its properties.

The expectation of this distribution

$$\Sigma_{n=0} \, n \frac{r^n}{n!} e^{-r} = re^{-r} \Sigma_{n=1} \frac{r^{n-1}}{(n-1)!} = re^{-r}e^r = r$$

where we use 9-2c_5 to evaluate the sum. We also have

$$\sigma^2 = \Sigma_{n=0} \, n^2 \frac{r^n}{n!} e^{-r} - r^2 = re^{-r} \Sigma_{n=1} n \frac{r^{n-1}}{(n-1)!} - r^2$$

$$= re^{-r} \left(\Sigma_{n=1} (n-1) \frac{r^{n-1}}{(n-1)!} + \Sigma_{n=1} \frac{r^{n-1}}{(n-1)!} \right) - r^2$$

$$= re^{-r} (re^r + e^r) - r^2$$

$$= r^2 + r - r^2 = r.$$

So the Poisson distribution with parameter r has the interesting property that its expectation and variation both equal the parameter r.

EXERCISES 9-2c

1. Prove that if $p_n = \dfrac{r^n}{n!} p_0$ for two consecutive integers $k-1$ and k, $p_n = \dfrac{r^n}{n!} p_0$ for all integers. (At the beginning of Sec. 9-2c we proved the equality for $k = 2$ and $k = 3$.)

2. In preparing for future growth a firm installs a telephone switchboard much larger than required to accommodate its current needs. Hence at this time no call coming into it has to wait for service.

(a) If in any second a call comes in with probability $\frac{1}{4}$ and one leaves with probability $\frac{1}{2}$, what is the probability there are no incoming calls on the switchboard?

(b) What is the expected number of calls?

(c) If calls come in with probability .6 and leave with probability .1, what is the expected number of calls?

(d) Answer the same three questions for the case of a single-service facility.

3. If length of service s satisfies $\theta \leqslant s \leqslant \theta + k$ and $p(s = \theta + v) = p_v$, what is the expected length of time spent in a queue with an unlimited number of service channels? What role does the state of the system play?

d.   Fixed Number Of Service Channels

We now modify the last case by considering a fixed number of service channels. This increases measurably the realism. Let us consider the case of c service channels. If $n \leqslant c$, the equations defining p_n are like those in the previous case, but when n becomes greater than c, a waiting line will form and a third equation must be introduced. Again, returning to our basic assumptions ($9\text{-}2a_1$ and $9\text{-}2a_2$), we have

9-2d₁
$$p_0 = (1 - \lambda)p_0 + \mu p_1$$

$$p_n = (1 - (\lambda + n\mu))p_n + \lambda p_{n-1} + (n + 1)\mu p_{n+1}, \qquad n < c$$

9-2d₂ $$p_n = (1 - (\lambda + c\mu))p_n + \lambda p_{n-1} + c\mu p_{n+1}, \qquad n \geqslant c$$

As in the case of an unlimited number of service channels,

9-2d₃
$$p_1 = r p_0$$

$$p_n = \frac{r^n}{n!} p_0, \qquad n < c.$$

Now if $n = c - 1$,

$$p_{c-1} = (1 - (\lambda + (c - 1)\mu))p_{c-1} + \lambda p_{c-2} + c\mu p_c$$

$$c p_c = (r - (c - 1))p_{c-1} - r p_{c-2}$$

$$c p_c = (r + (c - 1)) \frac{r^{c-1}}{(c-1)!} p_0 - \frac{r^{c-1}}{(c-2)!} p_0$$

$$p_c = \frac{r^c}{c!} p_0.$$

And so,

9-2d_4
$$p_n = \frac{r^n}{n!} p_0, \quad n \leqslant c.$$

Starting with $n = c$, we can compute

$$p_c = (1 - (\lambda + c\mu))p_c + \lambda p_{c-1} + c\mu p_{c+1}$$

$$cp_{c+1} = (r + c)p_c - rp_{c-1}$$

$$cp_{c+1} = (r + c)\frac{r^c}{c!} p_0 - \frac{r^c}{(c-1)!} p_0$$

$$p_{c+1} = \frac{r^{c+1}}{c!c} p_0 = \left(\frac{r^c}{c!}\right)\left(\frac{r}{c}\right) p_0.$$

Continuing, we obtain

$$p_{c+1} = (1 - (\lambda + c\mu))p_{c+1} + \lambda p_c + c\mu p_{c+2}$$

$$cp_{c+2} = (r + c)p_{c+1} - rp_c$$

$$cp_{c+2} = (r + c)\frac{r^{c+1}}{c! \, c} p_0 - \frac{r^{c+1}}{c!} p_0$$

$$p_{c+2} = \frac{r^{c+2}}{c!c^2} p_0 = \left(\frac{r^c}{c!}\right)\left(\frac{r}{c}\right)^2 p_0.$$

Again using the induction variation we employed in the previous two cases we can prove for $k \in I^+$

$$p_{c+k} = \left(\frac{r^c}{c!}\right)\left(\frac{r}{c}\right)^k.$$

Writing this in terms of n, we have

9-2d_5
$$p_n = \left(\frac{r^c}{c!}\right)\left(\frac{r}{c}\right)^{n-c} p_0, \quad n > c.$$

Bringing 9-2d_3, 9-2d_4, and 9-2d_5 together, the probability distribution for a queue with c service channels and service ratio r,

$$p_n = \frac{r^n}{n!} p_0, \qquad\qquad 0 \leqslant n \leqslant c$$

9-2d_6

$$= \left(\frac{r^c}{c!}\right)\left(\frac{r}{c}\right)^{n-c} p_0, \quad c < n.$$

Since the system must be in some state E_n,

$$\sum_{n=0} p_n = 1,$$

or

9-2d$_7$ $\sum_{n=0} p_n = \sum_{n=0}^{c} \frac{r^n}{n!} p_0 + \sum_{n=c+1} \left(\frac{r^c}{c!}\right)\left(\frac{r}{c}\right)^{n-c} p_0$

$$= p_0 \left(\sum_{n=0}^{c} \frac{r^n}{n!} + \frac{r^c}{c!} \sum_{k=1} \left(\frac{r}{c}\right)^k \right) = 1.$$

Since, if $\frac{r}{c} > 1$, the second sum above increases beyond all bounds, this is possible

only if $\frac{r}{c} < 1$, or, what is the same thing, only if $\frac{\lambda}{\mu} < c$. We recognize this as a

generalization of the single-server case, where $c = 1$, and here, as there, the
system becomes "jammed" (the lines increase beyond all limits, and $p_n = 0$ for
all n) if the ratio of the arrival rate λ to the service rate μ is greater than the
number of service channels available.

Our next step is to compute the expected length of this queue.   Here

$$\sum_{n=0} n p_n = \sum_{n=0}^{c} n \frac{r^n}{n!} p_0 + \sum_{n=c+1} n \frac{r^c}{c!} \left(\frac{r}{c}\right)^{n-c} p_0$$

$$= p_0 \left(r \sum_{n=1}^{c} \frac{r^{n-1}}{(n-1)!} + \frac{r^c}{c!} \sum_{k=1} (c + k)\left(\frac{r}{c}\right)^k \right)$$

$$= p_0 \left(r \sum_{n=1}^{c} \frac{r^{n-1}}{(n-1)!} + \frac{r^c}{c!} \left\{ c \sum_{k=1} \left(\frac{r}{c}\right)^k + \sum_{k=1} k \left(\frac{r}{c}\right)^k \right\} \right)$$

$$= p_0 \left(r \sum_{n=1}^{c} \frac{r^{n-1}}{(n-1)!} + \frac{r^c}{c!} \left\{ \frac{rc}{c-r} + \frac{rc}{(c-r)^2} \right\} \right)$$

where the middle sum, being geometric, is evaluated by 4-2b$_2$ and the last sum is
given in 6-3a$_4$.   Although additional simplification is possible it soon becomes
evident that no simple expression for the $\sum_{n=0} n p_n$ can be found.   We remark
that if r, the ratio of λ to μ, is fixed and c, the number of service channels, increases, the contribution of $\frac{r^c}{c!} \left\{ \frac{rc}{c-r} + \frac{rc}{(c-r)^2} \right\}$ to the expected value decreases
and the situation begins to approximate the previous case where the number of
channels was unlimited.   There the expected number of customers in the queue
was r, and here, if c is large in comparison to r, the expected number of customers
in the queue will be close to r.

This type of queue with $c \geqslant 1$ is quite familiar. If a customer calls an airline information office, he sometimes gets a clerk immediately, and at other times he hears a recorded voice asking him to be patient and promising to connect him to a line, in turn, as one becomes available. The information service has a fixed number of lines, and when they are all being used to service customers, additional customers coming in have to wait. Engineers study the flow of incoming calls and duration of service time to determine reasonable values for the parameters λ and μ. They then balance this against the cost of each line and expected length of the queue to determine the optimum number of lines for the service they want performed. The service must be optimum in terms of economy for the airline and convenience for the customer—if customers have to wait too long, they may turn to a competitor.

Fig. $9\text{-}2d_8$ displays the situation in which there are four service channels and a service ratio of three. The values of p_n are computed from Eq. $9\text{-}2d_6$.

$9\text{-}2d_8$ Figure

Probability Distribution of a Queue with c = 4 *and* r = 3.

n	0	1	2	3	4	5	6	7
Channels in use	0	1	2	3	4	4	4	4
Customers waiting	0	0	0	0	0	1	2	3
p_n	$\dfrac{2}{53}$	$\dfrac{6}{53}$	$\dfrac{9}{53}$	$\dfrac{9}{53}$	$\dfrac{27}{4\cdot53}$	$\dfrac{81}{16\cdot53}$	$\dfrac{243}{64\cdot53}$	$\dfrac{729}{256\cdot53}$
p_n (decimal form)	.0377	.1132	.1698	.1698	.1273	.0955	.0716	.0537

$$\left(p_n = \frac{27}{4 \cdot 53}\left(\frac{3}{4}\right)^{n-4}, \; n \geqslant 5\right).$$

The expectation of this distribution is

$$\sum_{n=0} np_n = 0\frac{2}{53} + 1\frac{6}{53} + 2\frac{9}{53} + 3\frac{9}{53} + \frac{27}{4.53}\left(4 + 5\frac{3}{4} + 6\left(\frac{3}{4}\right)^2 + \cdots\right)$$

$$= \frac{51}{53} + \frac{27}{4.53}\left(1 + 2\left(\frac{3}{4}\right) + 3\left(\frac{3}{4}\right)^2 + \cdots + 3\left(1 + \left(\frac{3}{4}\right) + \left(\frac{3}{4}\right)^2 + \cdots\right)\right)$$

$$= \frac{51}{53} + \frac{27}{4.53}\left(\frac{1}{\left(1 - \frac{3}{4}\right)^2} + \frac{3}{1 - \frac{3}{4}}\right) = \frac{51}{53} + \frac{27}{4.53}(16 + 12)$$

$$= \frac{240}{53} = 4.53.$$

At first glance it might seem that if $r = 3$, four service channels would keep the queue quite manageable since we would expect to find no more than one person waiting to be served. But how great is the variation? What are the probabilities of its becoming so great at times that customers will leave the line? On the other hand, if there were only three lines (this would cut costs), could we still adequately serve the customers? Questions such as these, which we are not prepared to answer here, must be answered in optimization problems.

EXERCISES 9-2d

1. In the last illustrative example what would be the distribution and expectation of the queue if $c = 5$ and $r = 4$?

2. In a certain supermarket the manager has determined that at peak hours customers join the checkout lines at the rate of one a minute and that, on the average, it takes five minutes to check out a customer. What is the minimum number of lines he must provide if the number of waiting customers will not increase beyond all bounds? If the manager provides 7 checkout counters, what is the expected number of people waiting to be checked out? (A desk calculator will be necessary.)

3. Suppose a service facility can only accommodate N persons ($N - 1$ waiting for service, 1 in service). What are the p_i, $i = 0,1,\ldots,N$? What is the expected number in the queue? Again let $r = \dfrac{\lambda}{\mu}$ be the service ratio.

4. In Prob. 3 suppose there are $1 < c < N$ service counters but still only N customers can be accommodated at any one time. What are the p_i, $i = 0, 1,2,\ldots,N$?

5. Consider again a single-service facility where as many as wish can join the queue, but now let us take into account the impatience of customers. If there are n persons in the queue, a newcomer will join the queue with probability $e^{\frac{-\alpha}{\mu} n}$ (where again we have a service ratio of $\dfrac{\lambda}{\mu}$). What are the p_n?

9-3. POSTSCRIPT

Our introduction to queueing theory has been concerned with service given on a first-come, first-served basis and with arrival and service times described parametrically. Many queues behave very differently; some serve the most recent arrival before those already waiting, in others the patterns of arrival and service times are more complex. We have not considered the problems of customers who come for service and seeing the length of the queue refuse to join it, or that of those who wait for a time but leave before receiving service. In any analysis of the costs and profits associated with a queueing system these problems cannot be ignored.

An exhaustive study of the field is found in Tomas L. Saaty, *Elements of Queueing Theory* (New York: McGraw-Hill Book Company, 1961). A comprehensive bibliography is included in this work, a bibliography which contains not only references in the general theory but an extensive list of papers concerned with applications.

chapter **10**

Branching Processes

In the last quarter of the nineteenth century two British scientists, W. H. Watson and Frances Galton, published a paper in the *Journal of the Anthropological Institute of Great Britain and Ireland** in which they constructed a model for predicting whether a family name would become extinct after a certain number of generations. The problem initially arose in discussions of the continuance of the royal family and the early papers were as much concerned with demography as with mathematics.

However, the model they devised—known as the Galton-Watson process—has found application in many areas of scientific research. In genetics it has been used to study the survival of mutant genes; in physics and chemistry, a variety of common particle-splitting processes. Our interest in the model, quite expectedly, will center on its use of probability. In our study we shall employ the descriptive terms of "generation," "offspring," "extinction," and others which will be useful in fixing the major ideas.

10-1. THE GALTON-WATSON PROCESS

a. Generation Probabilities

Consider a species of individuals—men, amoeba, neutrons—which can generate additional individuals of its own kind. We start with one individual and consider him to be the zeroth generation; he produces offspring which we call the first generation. The members of this first generation reproduce and their progeny comprise the second generation and continuing in this manner we create a sequence of generations. We assume that each individual produces offspring independently of the other members of his generation. We remind the reader that what we are discussing here is a model—that it fits no "real" situation perfectly, but it embodies the common characteristics of many

*Vol. 4 (1874), pp. 138-144.

branching processes. The assumption of independence limits the applicability of the model to biological processes but reflects realistically the splitting activity of certain atomic particles.

Let X be a random variable which counts the number of offspring of an arbitrary individual. The range of X is a subset of the nonnegative integers and its probability distribution is given by

10-1a$_1$ $$P_X = \{(i,p_i) | i \in R_X, p_i \geqslant 0, \sum_{i \in R_X} p_i = 1\}.$$

In a nuclear chain reaction a neutron hits another neutron and splits into m neutrons with probability p or has no descendants with probability $1 - p$. For this neutron the distribution of the random variable which counts the number of offspring is the two-element function $\{(0, 1 - p), (m,p)\}$. In biological systems the number of offspring of a given individual is always finite (although in certain forms of insect and plant life it can be very large), and hence the range of X is finite and the sum $\sum_{i \in R_X} p_i = 1$ has only a finite number of nonzero terms.

Let G_μ be a random variable which counts the number of individuals in the μth generation. Its range is contained in the nonnegative integers, and our first problem is to determine the probability distribution. We let

$$P(G_\mu = k) = P_\mu(k)$$

be the probability that the μth generation contains k individuals and we want to show P_μ is a probability distribution. We assume

10-1a$_2$ $$P(G_0 = 1) = P_0(1) = 1.$$

Since X counts the number of offspring of any individual,

10-1a$_3$ $$P(G_1 = k) = P_X(k) = p_k,$$

that is, P_X is the distribution function of G_1.

We write the conditional probability that $G_{\mu+1} = j$, given that $G_\mu = k$ as

$$P(G_{\mu+1} = j | G_\mu = k).$$

Since the sequence of generations will stop whenever one generation has zero members, we have

10-1a$_4$ $P(G_{\mu+1} = j \neq 0 | G_\mu = 0) = 0$ and $P(G_{\mu+1} = 0 | G_\mu = 0) = 1.$

If the μth generation has k members, let the random variable $X_{\mu v}$, $v = 1,2,\ldots,k$, count the number of offspring of the νth individual in the μth generation. For fixed μ the $X_{\mu\nu}$ are independent (see 7-3a_1), identically distributed* random variables with $R_{X_{\mu\nu}} = R_X$ and $P_{X_{\mu\nu}} = P_X$, where P_X is given in 10-1a_1. We can now write

10-1a_5 $P(G_{\mu+1} = j | G_\mu = k) = P(X_{\mu 1} + X_{\mu 2} + \cdots + X_{\mu k} = j)$

In the Appendix to this chapter we derive in detail the formula for the sum of N independent, identically distributed random variables, with $R_X = \{0,1,2,\ldots,n\}$ and

$$P_X = \{(i, p_i) | i \in R_X, \, p_i \geq 0, \quad \sum_{i \in R_X} p_i = 1\}$$

Using the major result (10-1A_2) derived in the Appendix, the last equation becomes

10-1a_6 $P(G_{\mu+1} = j | G_\mu = k) = P(X_{\mu 1} + X_{\mu 2} + \cdots + X_{\mu k} = j)$

$$= \sum_{x_1 + 2x_2 + \cdots + nx_n = j} \binom{k}{x_0, x_1, \ldots, x_n} p_0^{x_0} p_1^{x_1} \cdots p_n^{x_n}$$

where x_i is the number of individuals of the μth generation that produced i offspring ($0 \leq x_i \leq k$, $i \in R_X$). The events $\{G_\mu = k\}$ form a partition of $\{G_{\mu+1} = j\}$, and applying Bayes' rule (7-1b_3) we have

10-1a_7

$$P_{\mu+1}(j) = \sum_{k=0}^{n^\mu} P_\mu(k) P(G_{\mu+1} = j | G_\mu = k)$$

$$= \sum_{k=0}^{n^\mu} P_\mu(k) \sum_{x_1 + 2x_2 + \cdots + nx_n = j} \binom{k}{x_0, x_1, \ldots, x_n} p_0^{x_0} p_1^{x_1} \cdots p_n^{x_n}.$$

Since this notation is rather staggering, let us look at an example before going any further, for this should help to clarify the notation and, in addition, it may give a "feel" for the process.

b.   An Example

Suppose individuals of a certain species produce one or two offspring with probabilities $\dfrac{1}{2}$ and $\dfrac{3}{8}$, so that there is a $1 - \left(\dfrac{1}{2} + \dfrac{3}{8}\right) = \dfrac{1}{8}$ probability an individual will produce no offspring. If X is a random variable which counts the offspring of an individual of this species, $R_X = \{0,1,2\}$ and $P_X = \{(0, \dfrac{1}{8})$,

---

*Two random variables, X and Y, are said to be *identically distributed* if $P_X(i) = P_Y(i)$ for all $i \in R_X = R_Y$.

$(1, \frac{1}{2})$, $(2, \frac{3}{8})\}$. Using the notation introduced in the preceding paragraph, G_μ is a random variable which counts the number of individuals in the μth generation, P_μ its associated probability distribution and $X_{\mu\nu}$ the random variable which counts the number of offspring of the νth individual of the μth generation. The $X_{\mu\nu}$ are independent, identically distributed random variables with range R_X probability distribution P_X. We start with a single individual, so that

10-1b₁

G_0	1
P_0	1

10-1b₂

G_1	0	1	2
P_1	$\dfrac{1}{8}$	$\dfrac{1}{2}$	$\dfrac{3}{8}$
	.1250	.5000	.3750

Our next step is to create a table with entries G_2 and P_2. We know the range of G_2 will be $\{0,1,2,3,4\}$ and we want to find its probability distribution P_2. In the calculations which follow we make repeated use of 10-1a₇.

$$P_2(0) = \sum_{k=0}^{2} P_1(k)P(G_2 = 0|G_1 = k)$$

$$= \sum_{k=0}^{2} P_1(k) \sum_{x_1+2x_2=0} \binom{k}{x_0,x_1,x_2} p_0^{x_0} p_1^{x_1} p_2^{x_2}$$

$$= P_1(0) + P_1(1) \binom{1}{1,0,0} p_0 + P_1(2) \binom{2}{2,0,0} p_0^2$$

$$= \frac{1}{8} + \frac{1}{2}\left(\frac{1}{8}\right) + \frac{3}{8}\left(\frac{1}{8}\right)^2 = \frac{99}{8^3}.$$

$$P_2(1) = \sum_{k=0}^{2} P_1(k)P(G_2 = 1|G_1 = k)$$

$$= \sum_{k=0}^{2} P_1(k) \sum_{x_1+2x_2=1} \binom{k}{x_0,x_1,x_2} p_0^{x_0} p_1^{x_1} p_2^{x_2}$$

$$= (P_1(0))0 + P_1(1) \binom{1}{0,1,0} p_1 + P_1(2) \binom{2}{1,1,0} p_0 p_1$$

$$= 0 + \left(\frac{1}{2}\right)\left(\frac{1}{2}\right) + \left(\frac{3}{8}\right)(2)\left(\frac{1}{8}\right)\left(\frac{1}{2}\right) = \frac{152}{8^3}.$$

Hereafter we shall suppress the index $k = 0$, since $P(G_2 = j \neq 0 | G = 0) = 0$
(see 10-1a_4) and also the first expression of the sum.

$$P_2(2) = \sum_{k=1}^{2} P_1(k) \sum_{x_1+2x_2=2} \binom{k}{x_0,x_1,x_2} p_0^{x_0} p_1^{x_1} p_2^{x_2}$$

$$= P_1(1) \binom{1}{0,0,1} p_2 + P_1(2)\left[\binom{2}{0,2,0} p_1^2 + \binom{2}{1,0,1} p_0 p_2 \right]$$

$$= \left(\frac{1}{2}\right)\left(\frac{3}{8}\right) + \frac{3}{8}\left[\left(\frac{1}{2}\right)^2 + 2\left(\frac{1}{8}\right)\left(\frac{3}{8}\right)\right] = \frac{162}{8^3}$$

$$P_2(3) = \sum_{k=1}^{2} P_1(k) \sum_{x_1+2x_2=3} \binom{k}{x_0,x_1,x_2} p_0^{x_0} p_1^{x_1} p_2^{x_2}$$

$$= P_1(1)(0) + P_1(2) \binom{2}{0,1,1} p_1 p_2$$

$$= \frac{3}{8}(2)\left(\frac{1}{2}\right)\left(\frac{3}{8}\right) = \frac{72}{8^3}.$$

Finally

$$P_2(4) = \sum_{k=1}^{2} P_1(k) \sum_{x_1+2x_2=4} \binom{k}{x_0,x_1,x_2} p_0^{x_0} p_1^{x_1} p_2^{x_2}$$

$$= P_1(1)(0) + P_2(2) \binom{2}{0,0,2} p_2^2$$

$$= \left(\frac{3}{8}\right)\left(\frac{3}{8}\right)^2 = \frac{27}{8^3}.$$

We have

G_2	0	1	2	3	4
P_2	$\dfrac{99}{8^3}$	$\dfrac{152}{8^3}$	$\dfrac{162}{8^3}$	$\dfrac{72}{8^3}$	$\dfrac{27}{8^3}$
	.1933	.2968	.3164	.1406	.0527

10-1b_3

$$(P_2(0) + P_2(1) + P_2(2) + P_2(3) + P_2(4) = 1)$$

We repeat the analysis to find P_3. The range of G_3 is $\{0,1,2,3,4,5,6,7,8\}$ and

$$P_3(j) = \sum_{k=0}^{4} P_2(k) \sum_{x_1+2x_2=j} \binom{k}{x_0,x_1,x_2} p_0^{x_0} p_1^{x_1} p_2^{x_2}.$$

We shall demonstrate only the computation of $P_3(4)$ and $P_3(6)$.

$$P_3(4) = \sum_{k=0}^{4} P_2(k) \sum_{x_1+2x_2=4} \binom{k}{x_0,x_1,x_2} p_0^{x_0} p_1^{x_1} p_2^{x_2}$$

$$= P_2(2) \binom{2}{0,0,2} p_2^2 + P_2(3) \left[\binom{3}{0,2,1} p_1^2 p_2 + \binom{3}{1,0,2} p_0 p_2^2 \right]$$

$$+ P_2(4) \left[\binom{4}{2,0,2} p_0^2 p_2^2 + \binom{4}{0,4,0} p_1^4 + \binom{4}{1,2,1} p_0 p_1^2 p_2 \right]$$

$$= \frac{162}{8^3} \left(\frac{3}{8}\right)^2 + \frac{72}{8^3} \left[3\left(\frac{1}{2}\right)^2\left(\frac{3}{8}\right) + 3\left(\frac{1}{8}\right)\left(\frac{3}{8}\right)^2 \right]$$

$$+ \frac{27}{8^3} \left[6\left(\frac{1}{8}\right)^2\left(\frac{3}{8}\right)^2 + \left(\frac{1}{2}\right)^4 + 12\left(\frac{1}{8}\right)\left(\frac{1}{2}\right)^2\left(\frac{3}{8}\right) \right]$$

$$= \frac{215730}{8^7}.$$

$$P_3(6) = \sum_{k=1}^{4} P_2(k) \sum_{x_1+2x_2=6} \binom{k}{x,0,x_1,x_2} p_0^{x_0} p_1^{x_1} p_2^{x_2}$$

$$= P_2(3) \binom{3}{0,0,3} p_2^3 + P_1(4) \left[\binom{4}{1,0,3} p_0 p_2^3 + \binom{4}{0,2,2} p_1^2 p_2^2 \right]$$

$$= \frac{72}{8^3} \left(\frac{3}{8}\right)^3 + \frac{27}{8^3} \left[4\left(\frac{1}{8}\right)\left(\frac{3}{8}\right)^3 + 6\left(\frac{1}{2}\right)^2\left(\frac{3}{8}\right)^2 \right]$$

$$= \frac{41796}{8^7}.$$

In the same way we can compute the other values of the distribution and we have

10-1b_4

G_3	0	1	2	3	4	5	6	7	8
P_3	$\frac{494299}{8^7}$	$\frac{401584}{8^7}$	$\frac{497316}{8^7}$	$\frac{337968}{8^7}$	$\frac{215730}{8^7}$	$\frac{94608}{8^7}$	$\frac{41796}{8^7}$	$\frac{11664}{8^7}$	$\frac{2187}{8^7}$
	.2357	.1920	.2371	.1612	.1029	.0451	.0199	.0056	.0010

We remark that $8^7 = 2{,}097{,}152$ and $\sum_{j=0}^{8} P_3(j) = 1$.

These few calculations illustrate the method of computing the P_i's.

EXERCISES 10-1b

1. What is the range of P_4 for the example discussed above? $P_4(0)$? $P_4(13)$?

2. At regular intervals a neutron splits into 3 neutrons with probability p and does not split with probability $1 - p$. What are the probability distributions of the number of neutrons "descended" from the initial one after one, two and three intervals?

3. Suppose the neutron in Prob. 2 two splits into m neutrons with probability p and does not split with probability $1 - p$. What are the distributions of the first and second generations? What are $P_3(0)$ and $P_3(m)$?

4. A certain species of fish bears its first offspring when it is two months old and thereafter bears offspring every month. If each female bears $0, 1, \ldots, n$ females with probabilities $p_0, p_1, \ldots, p_n \left(\sum_{j=0}^{n} p_j = 1 \right)$, what are the probability distributions of the total number of female descendants of a single female at two and three months?

c. Generating Functions

Let X be a random variable with range $R_X = \{0, 1, 2, \ldots, n\}$ and probability distribution $P_X = \{(i, p_i) | i \in R_X\}$. The polynomial

$$\sum_{k=0}^{n} p_k s^k = p_0 + p_1 s + p_2 s^2 + \cdots + p_n s^n$$

is called the generating function of the distribution P_X. The variable s has no significance. The generating function for P_1 in the example of the preceding paragraph is

$$\sum_{k=0}^{2} p_k s^k = \frac{1}{8} + \frac{1}{2} s + \frac{3}{8} s^2,$$

and that for P_2 is

$$\sum_{k=0}^{2^2} P_2(k) s^k = \frac{1}{8^3} (99 + 152s + 162s^2 + 72s^3 + 27s^4).$$

Similarly the generating function for P_3 is

$$\sum_{j=0}^{8} P_3(j) s^j$$

where we can read the values of $P_3(j)$ from Table 10-1b_4.

If X counts the number of offspring of one individual of a given species and $R_X = \{0, 1, \ldots, n\}$ and $P_X = \{(i, p_i) \mid i \in R_X\}$ we define:

10-1c$_1$
$$g(s) = \sum_{k=0}^{n} p_k s^k = \sum_{k=0}^{n} P_X(k) s^k,$$

so that $g(s)$ is the generating function of $P_1 = P_X$ (since we assume the first generation consists of a single individual). If we set $s = 1$, we have

10-1c$_2$
$$g(1) = \sum_{k=0}^{n} p_k = \sum_{k=0}^{n} P_X(k) = 1.$$

The generating function of P_2 is, by definition

$$g_2(s) = \sum_{j=0}^{n^2} P_2(j) s^j.$$

Putting 10-1a_7 in this equation,

$$g_2(s) = \sum_{j=0}^{n^2} \left(\sum_{k=0}^{n} P_1(k) P(G_2 = j \mid G_1 = k) \right) s^j$$

$$= \sum_{j=0}^{n^2} \left(\sum_{k=0}^{n} P_1(k) \sum_{x_1 + 2x_2 + \cdots + nx_n = j} \binom{k}{x_0, x_1, \ldots, x_n} p_0^{x_0} p_1^{x_1} \cdots p_n^{x_n} \right) s^j.$$

Since all sums involved are finite, we can interchange the order of summation and

$$g_2(s) = \sum_{k=0}^{n} P_1(k) \left(\sum_{j=0}^{n^2} \sum_{x_1 + 2x_2 + \cdots + nx_n = j} \binom{k}{x_0, x_1, \ldots, x_n} p_0^{x_0} p_1^{x_1} \cdots p_n^{x_n} s^j \right).$$

Since $x_1 + 2x_2 + \cdots + nx_n = j$, s^j can be associated with the $p_i^{x_i}$ by rewriting $p_0^{x_0} p_1^{x_1} \cdots p_n^{x_n} s^j$ as $p_0^{x_0} (p_1 s)^{x_1} (p_2 s^2)^{x_2} \ldots (p_n s^n)^{x_n}$. This yields

$$g_2(s) = \sum_{k=0}^{n} P_1(k) \left(\sum_{j=0}^{n^2} \sum_{x_0 + 2x_2 + \cdots + nx_n = j} \binom{k}{x_0, x_1, \ldots, x_n} p_0^{x_0} (p_1 s)^{x_1} \ldots (p_n s^n)^{x_n} \right).$$

Having removed the index j from our addends, the double sum is meaningful only where the coefficients are meaningful, that is only when $x_0 + x_1 + \cdots + x_n = k$. Hence

$$g_2(s) = \sum_{k=0}^{n} P_1(k) \sum_{x_0 + x_1 + \cdots + x_n = k} \binom{k}{x_0, x_1, \ldots, x_n} p_0^{x_0} (p_1 s)^{x_1} \ldots (p_n s^n)^{x_n}.$$

Now the second summation is the well-known multinomial formula (see Prob. 8, Exercises 4-2a) and so

$$g_2(s) = \sum_{k=0}^{n} P_1(k)(p_0 + p_1 s + \cdots + p_n s^n)^k.$$

Applying 10-1c_1 to this we obtain

10-1c_3
$$g_2(s) = \sum_{k=0}^{n} P_1(k)(g(s))^k = g(g(s)).$$

We have, using 10-1c_2,

10-1c_4
$$g_2(1) = \sum_{j=0}^{n^2} P_2(j) = \sum_{k=0}^{n} P_1(k)(g(1))^k = \sum_{k=0}^{n} P_1(k) = 1$$

and thus we show for the first time that P_2 is the probability distribution of G_2.

(The P_2 is our example was a probability distribution since $\sum_{k=0}^{4} P_2(k) = 1$, but that was a special case and we could not generalize from it. Here we have demonstrated P_3 defined by 10-1a_7 is a probability distribution).

We can now prove

10-1c_5
$$g_\mu(s) = g_{\mu-1}(g(s)).$$

Proof:

$$g_\mu(s) = \sum_{j=0}^{n^\mu} P_\mu(j)s^j$$

$$= \sum_{j=0}^{n^\mu} \left(\sum_{k=0}^{n^{\mu-1}} P_{\mu-1}(k) P(G_\mu = j \mid G_{\mu-1} = k) \right) s^j$$

$$= \sum_{k=0}^{n^{\mu-1}} P_{\mu-1}(k) \left(\sum_{j=0}^{n^\mu} \sum_{x_1+2x_2+\cdots+nx_n=j} \binom{k}{x_0, x_1, \ldots, x_n} p_0^{x_0} p_1^{x_1} \cdots p_n^{x_n} \right) s^j$$

$$= \sum_{k=0}^{n^{\mu-1}} P_{\mu-1}(k) \sum_{x_1+x_2+\cdots+x_n=k} \binom{k}{x_0, x_1, \ldots, x_n} p_0^{x_0} (p_1 s)^{x_1} \cdots (p_n s^n)^{x_n}$$

$$= \sum_{k=0}^{n^{\mu-1}} P_{\mu-1}(k)(p_0 + p_1 s + \cdots + p_n s^n)^k$$

$$= \sum_{k=0}^{n^\mu} P_{\mu-1}(k)(g(s))^k$$

$$= g_{\mu-1}(g(s)).$$

where we have reemployed the methods used in showing $g_2(s) = g(g(s))$. ∎

Since 10-1c_5 holds for all μ we can write

$$g_\mu(s) = g_{\mu-1}(g(s)) = g_{\mu-2}(g(g(s))) = g_{\mu-2}(g_2(s)) = g_{\mu-3}(g(g_2(s)))$$
$$= g_{\mu-3}(g(g(g(s)))) = g_{\mu-3}(g_2(g(s))) = g_{\mu-3}(g_3(s)).$$

By repetition of this argument suppose that we have shown for some $k < \mu$ that

$$g_\mu(s) = g_{\mu-k}(g_k(s)).$$

Working from this we have

$$g_\mu(s) = g_{\mu-k}(g_k(s))$$
$$= g_{\mu-k-1}(g(g_k(s)))$$
$$= g_{\mu-(k+1)}(g(g(g_{k-1}(s)))) = g_{\mu-(k+1)}(g_2(g_{k-1}(s)))$$
$$= g_{\mu-(k+1)}(g_2(g(g_{k-2}(s)))) = g_{\mu-(k+1)}(g_3(g_{k-2}(s))).$$

After $k - 2$ additional repetitions of this argument we come to:

$$g_\mu(s) = g_{\mu-(k+1)}(g_k(g(s)))$$
$$= g_{\mu-(k+1)}(g_{(k+1)}(s)).$$

Thus for all $k < \mu$,

$$g_\mu(s) = g_{\mu-k}(g_k(s));$$

and in particular, if $k = \mu - 1$, we have

10-1c_6 $\qquad\qquad g_\mu(s) = g(g_{\mu-1}(s)).$

This last equation can be expressed as the sum

10-1c_7 $\qquad\qquad g_\mu(s) = \sum_{k=0}^{n} P_1(k)(g_{\mu-1}(s))^k.$

We know $g(1) = g_2(1) = 1$ (10-1c_2 and 10-1c_4) and we use induction to prove that for all μ, $g_\mu(1) = 1$. Suppose there is an integer j such that $g_j(1) = 1$; then

$$g_{j+1}(1) = \sum_{k=0}^{n} P_1(k)(g_j(1))^k = \sum_{k=0}^{n} P_1(k) = g(1) = 1.$$

Hence

10-1c_8 $\qquad\qquad$ for all $\mu \in I^+$, $g_\mu(1) = 1.$

This yields to the important result

10-1c₉ $P_\mu(k)$ is a probability distribution for all $\mu \in I^+$.

We have from 10-1c₈,

$$\sum_{k=0}^{n^\mu} P_\mu(k) = g_\mu(1) = 1.$$

EXERCISES 10-1c

1. The generating function for the example in Sec. 10-1b is $g(s) = \dfrac{1}{8} + \dfrac{1}{2}s + \dfrac{3}{8}s^2$. Use this and Eq. 10-1c₃ to find $g_2(s)$ and compare the results with Table 10-1b₃. Similarly find $g_3(s)$ and compare the results with the Table 10-1b₄.

2. Let $g(s) = (1 - p) + ps^3$ (the generating function for the nuclear splitting described in Prob. 2, Exercises 10-1b). Find $g_2(s)$ and $g_3(s)$ and compare the result with the earlier calculations.

3. What is the generating function for the case where a neutron splits into m neutrons with probability p and does not split with probability $1 - p$? Also find $g_2(s)$. Find the first two nonzero coefficients of $g_3(s)$ and compare your results with your calculations in Prob. 3, Exercises 10-1b.

4. If $g(s) = \dfrac{3}{8} + \dfrac{1}{2}s + \dfrac{1}{16}s^2 + \dfrac{1}{16}s^3$, find $g_2(s)$.

*d. Expectations and Variances of the G_μ's

The expectation of X, the random variable which counts the number of offspring of an individual, we shall denote by m. Hence

10-1d₁ $$E(X) = \sum_{k=0}^{n} k p_k = m.$$

This is also the expectation of G_1, and for the species considered in 10-1b,

$$m = E(G_1) = \frac{1}{8} \cdot 0 + \frac{1}{2} \cdot 1 + \frac{3}{8} \cdot 2 = \frac{5}{4}.$$

The derivative with respect to s of the generating function $g(s) = \sum_{k=0}^{n} p_k s^k$ is

$$g'(s) = \sum_{k=0}^{n} k p_k s^{k-1}$$

*This paragraph uses notions of elementary calculus and can be omitted without loss of continuity. It is included for the interest of the results 10-1d₃ and 10-1d₁₁.

and setting $s = 1$, we have from 10-1d_1,

10-1d_2 $m = g'(1).$

Whereas straightforward calculation of the expectations of $G_2, G_3, \ldots$, from the definition of E would be very tedious, we remark

$$g_2(s) = \sum_{k=0}^{n^2} P_2(k)s^k$$

$$g_2'(s) = \sum_{k=0}^{n^2} kP_2(k)s^{k-1}$$

and when $s = 1$, this yields

$$g_2'(1) = \sum_{k=0}^{n^2} kP_2(k) = E(G_2).$$

Fortunately we also have (10-1c_3),

$$g_2(s) = g(g(s))$$

and using the chain rule for differentiation,

$$g_2'(s) = g'(g(s))g'(s).$$

Hence

$$g_2'(1) = g'(g(1))g'(1) = (g'(1))^2$$

and

$$E(G_2) = g_2'(1) = (g'(1))^2 = m^2.$$

For any μ,

$$g_\mu(s) = \sum_{k=0}^{n^\mu} P_\mu(k)s_k$$

$$g_\mu'(s) = \sum_{k=0}^{n^\mu} kP_\mu(k)s^{k-1}$$

and

$$E(G_\mu) = g_\mu'(1) = \sum_{k=0}^{n^\mu} kP_\mu(k).$$

Since $E(G_1) = m$ and $E(G_2) = m^2$ we use Induction to prove

10-1d_3 $\qquad\qquad E(G_\mu) = m^\mu, \quad \mu \in I^+.$

Proof: The statement is true for $\mu = 1$ and $\mu = 2$.

Suppose there is an integer j such that $E(G_j) = g_j'(1) = m^j$. Then

$$g_{j+1}(s) = \sum_{k=0}^{n^{j+1}} P_{j+1}(k)s^k = g(g_j(s))$$

$$g_{j+1}'(s) = \sum_{k=0}^{n^{j+1}} kP_{j+1}(k)s^{k-1} = g'(g_j(s))g_j'(s).$$

Now

$$E(G_{j+1}) = g_{j+1}'(1) = \sum_{k=0}^{n^{j+1}} kP_{j+1}(k)$$

$$= g'(g_j(1))g_j'(1) = g'(1)g_j'(1)$$

$$= m(m^j) = m^{j+1}.$$

Hence the statement holds for all $\mu \in I^+$. ∎

Thus we have shown that the expectations of successive generations are successive powers of the expectation of the first generation. We remark that if $m < 1$, $m^j < m < 1$ so that whenever $E(G_1) < 1$, the expected number of members in successive generations decreases. If $E(G_1) = m > 1$, for $j \geqslant 2$, $E(G_j) = m^j > m > 1$ and the expected size of successive generations increase. Again returning to the example in Sec. 10-1b, we have computed that

$$E(G_1) = \frac{5}{4}.$$

Hence

$$E(G_2) = \frac{25}{16}, \quad E(G_3) = \frac{125}{64}, \quad E(G_4) = \frac{625}{256},$$

and for any n,

$$E(G_n) = \left(\frac{5}{4}\right)^n.$$

The variance of a random variable X satisfies, by 6-4a_2,

$$V(X) = E(X^2) - (E(X))^2.$$

Hence

10-1d_4 $\qquad V(G_1) = E(G_1^2) = (E(G_1))^2 = E(G_1^2) - m^2$

where by definition

10-1d$_5$
$$E(G_1^2) = \sum_{k=0}^{n} k^2 P_1(k).$$

Since the derivative of the generating function assisted greatly in computing values of $E(G_1)$, let us try using it for $E(G_1^2)$. We have

$$g'(s) = \sum_{k=0}^{n} kP_1(k)s^{k-1}$$

and differentiating again,

$$g''(s) = \sum_{k=0}^{n} k(k-1)P_1(k)s^{k-2}$$

and

$$= \sum_{k=0}^{n} k^2 P_1(k)s^{k-2} - \sum_{k=0}^{n} kP_1(k)s^{k-2}$$

$$g''(1) = \sum_{k=0}^{n} k^2 P_1(k) - \sum_{k=0}^{n} kP_1(k) = \sum_{k=0}^{n} k^2 P_1(k) - m.$$

Substituting these results first into 10-1d$_5$ then into 10-1d$_4$, we have

$$E(G_1^2) = g''(1) + m$$

and

10-1d$_6$ $\quad V(G_1) = g''(1) + m - m^2 = g''(1) - m(m - 1).$

For future use let us set

10-1d$_7$ $$V(G_1) = \sigma^2$$

and see if we can find expression for the $V(G_\mu)$ in terms of μ, σ^2, and m. We start with

$$V(G_\mu) = E(G_\mu^2) - (E(G_\mu))^2 = \sum_{k=0}^{n^\mu} k^2 p_\mu(k) - (m^\mu)^2.$$

Repeating the analysis we used to find $E(G_1^2)$,

$$g_\mu(s) = \sum_{k=0}^{n^\mu} P_\mu(k)s^k$$

$$g'_\mu(s) = \sum_{k=0}^{n^\mu} kP_\mu(k)s^{k-1}$$

$$g''_\mu(s) = \sum_{k=0}^{n^\mu} k(k-1)P_\mu(k)s^{k-2} = \sum_{k=0}^{n^\mu} k^2 P_\mu(k)s^{k-2} - \sum_{k=0}^{n} kP_\mu(k)s^{k-2}.$$

Then

$$g''_\mu(1) = E(G_{\bar{\mu}}^2) - g'_\mu(1)$$

and

$$E(G_{\bar{\mu}}^2) = g''(1) + m^\mu.$$

Thus

10-1d$_8$ $V(G_\mu) = g''_\mu(1) + m^\mu - (m^\mu)^2 = g''_\mu(1) - m^\mu(m^\mu - 1).$

Can we eliminate $g''_\mu(1)$ by expressing it in terms of m, μ and σ? From 10-1c$_6$,

$$g_{\mu+1}(s) = g(g_\mu(s))$$

$$g'_{\mu+1}(s) = g'(g_\mu(s))g'_\mu(s)$$

$$g''_{\mu+1}(s) = g''(g_\mu(s))(g'_\mu(s))^2 + g'(g_\mu(s))g''_\mu(s).$$

Letting $s = 1$,

10-1d$_9$ $g''_{\mu+1}(1) = g''(1)(m^\mu)^2 + mg''_\mu(1).$

We also have (10-1c$_5$)

$$g_{\mu+1}(s) = g_\mu(g(s))$$

$$g'_{\mu+1}(s) = g'_\mu(g(s))g'(s)$$

$$g''_{\mu+1}(s) = g''_\mu(g(s))(g'(s))^2 + g'_\mu(g(s))g''(s)$$

and

10-1d$_{10}$ $g''_{\mu+1}(1) = g''_\mu(1)m^2 + m^\mu g''(1).$

We eliminate $g''_{\mu+1}(1)$ from the last two equations (10-1d$_9$ and 10-1d$_{10}$),

$$g''(1)(m^\mu)^2 + mg''_\mu(1) = g''_\mu(1)m^2 + m^\mu g''(1).$$

After rearranging terms, we obtain

$$g''_\mu(1)(m^2 - m) = g''(1)m^\mu(m^\mu - 1).$$

Combining 10-1d$_6$ and 10-1d$_7$, we have

$$\sigma^2 = g''(1) - m(m - 1)$$

and putting this into the equation directly above,

$$g''_\mu(1)m(m - 1) = [\sigma^2 + m(m - 1)]m^\mu(m^\mu - 1).$$

If $m \neq 1$,

$$g''_\mu(1) = \sigma^2 m^\mu \frac{m^\mu - 1}{m(m - 1)} + m^\mu(m^\mu - 1)$$

Then from 10-1d_8,

10-1d_{11} $$V(G_\mu) = \sigma^2 m^\mu \frac{m^\mu - 1}{m(m - 1)}, \quad m \neq 1;$$

and if $m = 1$, we use L'Hospital's rule and obtain

10-1d_{12} $$V(G_\mu) = \mu\sigma^2.$$

For the example in Sec. 10-1b we have already computed $E(G_1) = \dfrac{5}{4}$; furthermore

$$E(G_1^2) = 0^2 \frac{1}{8} + 1^2 \cdot \frac{1}{2} + 2^2 \cdot \frac{3}{8} = 2$$

$$V(G_1) = E(G_1^2) - (E(G_1))^2 = 2 - \left(\frac{5}{4}\right)^2 = \frac{7}{16} = \sigma^2$$

$$V(G_2) = \frac{7}{16} \left(\frac{5}{4}\right)^2 \frac{\left(\frac{5}{4}\right)^2 - 1}{\frac{5}{4}\left(\frac{5}{4} - 1\right)} = \frac{315}{256}$$

$$V(G_3) = \frac{7}{16} \left(\frac{5}{4}\right)^3 \frac{\left(\frac{5}{4}\right)^3 - 1}{\frac{5}{16}} = \frac{3800}{64^2}$$

EXERCISES 10-1d

1. In Probs. 2 and 4 of Exercises 10-1c find the expectations and variances of the first, second and third generations.

2. In Prob. 3 of Exercises 10-1c find the expectation and variances of the first four generations.

3. For a certain species P_X is given by

i	0	1	2	3	4
p_i	$\dfrac{23}{48}$	$\dfrac{1}{4}$	$\dfrac{1}{8}$	$\dfrac{1}{12}$	$\dfrac{1}{16}$

For any r, what are the expectations and variations of G_r?

10-2. PROBABILITIES OF EXTINCTION

a. Some Necessary Inequalities

We remark that if $0 < s_1 < s_2$, $s_1^k < s_2^k$ for $k \in I^+$.
Then since the coefficients of a generating function are probabilities, and hence nonnegative, if $0 \leqslant s_1 < s_2$, for any $\mu \in I^+$,

$$g_\mu(s_2) - g_\mu(s_1) = \sum_{k=0}^{n''} P_\mu(k)\,(s_2^k - s_1^k) > 0.$$

From this we can conclude:

10-2a_1   if $0 \leqslant s_1 < s_2$, $g_\mu(s_1) < g_\mu(s_2)$

and, in particular,

10-2a_2   if $0 < s < 1, g_\mu(s) < g_\mu(1) = 1.$

From elementary algebra, we know

$$\frac{s_2^k - s_1^k}{s_2 - s_1} = s_2^{k-1} + s_2^{k-2}\,s_1 + s_2^{k-3}\,s_1^2 + \cdots + s_1^{k-1}$$

and if $s_1 < s_2$,

10-2a_3 $\dfrac{g(s_2) - g(s_1)}{s_2 - s_1} = \displaystyle\sum_{k=0}^{n} P_1(k)\frac{s_2^k - s_1^k}{s_2 - s_1}$

$$= \sum_{k=0}^{n} P_1(k)\,(s_2^{k-1} + s_2^{k-2}\,s_1 + \cdots + s_1^{k-1})$$

$$\leqslant \sum_{k=0}^{n} kP_1(k)s_2^{k-1}.$$

(Here, since $s_1 < s_2$, $s_2^{k-j}\,s_1^{j-1} < s_2^{k-1}$, for we have replaced each s_1 factor by s_2.)

b. Extinction

The event that a species becomes extinct we designate by

$$\{G_n \to 0\}.$$

Then

$$P(G_n \to 0) = P(G_n = 0, \text{ for some } n)$$

$$= P(\{G_1 = 0\} \cup \{G_2 = 0\} \cup \{G_3 = 0\} \cup \cdots) \leqslant 1.$$

By definition (10-1c_1),

$$g(0) = P_1(0) = p_0.$$

Let us assume

10-2b_1 $\qquad\qquad\qquad 0 < p_0 < 1.$

For if $p_1 = 0$, the species will never become extinct, and if $p_1 = 1$, the sole individual in the first generation does not reproduce and so the species surely becomes extinct. Now

$$g_2(0) = P_2(0) = \sum_{k=0}^{n} P_1(k)P(G_2 = 0 | G_1 = k)$$

$$= p_0 + \sum_{k=1}^{n} P_1(k)\, p_0^k > p_0 = g(0);$$

$$g_3(0) = P_3(0) = \sum_{k=0}^{n^2} P_2(k)P(G_3 = 0 | G_2 = k)$$

$$= P_2(0) + \sum_{k=1}^{n^2} P_2(k)\, p_0^k > P_2(0) = g_2(0).$$

Similarly, for every μ

10-2b_2 $\qquad\qquad\qquad g_\mu(0) > g_{\mu-1}(0)$

since

$$g_\mu(0) = P_\mu(0) = \sum_{k=0}^{n^\mu} P_{\mu-1}(k)P(G_\mu = 0 | G_{\mu-1} = k)$$

$$= P_{\mu-1}(0) + \sum_{k=1}^{n^\mu} P_{\mu-1}(k) p_0^k > P_{\mu-1}(0) = g_{\mu-1}(0).$$

Thus we have

10-2b_3 $\qquad\qquad\qquad g(0) < g_2(0) < g_3(0) < \cdots \leqslant 1.$

(This property can be observed in the example of Sec. 10-1b.)

Since the $g_\mu(0)$ increase as μ increases and since the $g_\mu(0)$ never surpass one, there is a number $\rho \leqslant 1$, such that: (1) $g_\mu(0) \leqslant \rho$ for all μ and (2) given any positive number ϵ, no matter how small, there is an integer M_ϵ such that if $\mu > M_\epsilon$, $\rho - g_\mu(0) < \epsilon$.* If, for example,

$$g_\mu(0) = 1 - \frac{1}{\mu}, \quad g_{\mu+1}(0) = 1 - \frac{1}{\mu+1} > 1 - \frac{1}{\mu} = g_\mu(0),$$

---

*Readers familiar with elementary mathematical analysis will recognize that we are using here the fact that every bounded monotone sequence has a limit. See for example, Tom M., Apostle, *Calculus* (Waltham, Mass.: Blaisdell Publishing Company, 1961), Vol. 1, p. 417.

and for all μ, $g_\mu(0) \leqslant 1$. Now $1 - g_\mu(0) = \dfrac{1}{\mu}$, so for any $\epsilon > 0$, no matter how small, $1 - g_\mu(0)$ will be less than ϵ whenever $\mu > \dfrac{1}{\epsilon}$. Thus if we insist that

$$\mu > M_\epsilon = \left[\frac{1}{\epsilon}\right] + 1, \ 1 - g_\mu(0) < \epsilon. \quad \text{The smaller } \epsilon, \text{ the larger } M_\epsilon, \text{ but given } \epsilon,$$

there is always an M_ϵ, and so by choosing μ sufficiently large, we can find $g_\mu(0)$'s as close to one as we please.

10-2b_4 Figure

Relation of ρ to the $g_\mu(0)$'s.

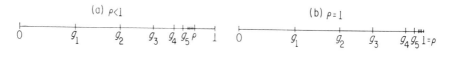

Let us choose ϵ so small that the difference between ρ, whatever it happens to be, and $g_\mu(0)$ is negligible for $\mu > M_\epsilon$; then for all practical purposes we can set

$$g_\mu(0) = \rho.$$

This number ρ is the probability of extinction; it is the probability that from some point onward all generations have the same probability of producing no offspring and thus terminating the species. Employing the notation introduced at the beginning of this section,

$$P(G_n \to 0) = \rho.$$

Whenever μ is so large that we can set

$$g_\mu(0) = \rho,$$

since $g_\mu(0) < g_{\mu+1}(0) < \rho$, we also can write

$$g_{\mu+1}(0) = \rho.$$

Using 10-1c_6, we obtain

$$g_{\mu+1}(0) = g(g_\mu(0)) = \rho$$

which leads us to

10-2b_5 $$g(\rho) = \rho.$$

Hence the probability of extinction is a solution of Eq. 10-2b_5. Let $x > 0$ be another solution of the same equation, so that

Since $$g(x) = x.$$

$$g_\mu(x) = g_{\mu-1}(g(x)) = g_{\mu-1}(x) = g_{\mu-2}(g(x)) = g_{\mu-2}(x) = \cdots = g(x),$$

and since, from 10-2a_1,

$$g_\mu(0) < g_\mu(x)$$

we have, for sufficiently large μ,

$$\rho = g_\mu(0) < g_\mu(x) = g(x) = x.$$

Hence

$$P(G_n \to 0) = \rho$$

is the smallest solution of the equation

$$g(\rho) = \rho.$$

One is a solution of this equation, since

10-2b_6 $$g(1) = \sum_{k=0}^{n} P_1(k) = 1.$$

From 10-2a_3, if $s_1 < s_2 \leqslant 1$,

$$\frac{g(s_2) - g(s_1)}{s_2 - s_1} \leqslant \sum_{k=0}^{n} kP_1(k)s_2^{k-1} \leqslant m$$

where m is the expectation of X, the random variable which counts the number of offspring (see 10-1d_1). Then, if $s_2 = 1$, and $0 < s < 1$, from 10-2a_2 and the preceding inequality we obtain

10-2b_7 $$\frac{1 - g(s)}{1 - s} \leqslant m$$

or

$$1 - g(s) \leqslant m(1 - s).$$

Now if $m \leqslant 1$,

$$1 - g(s) \leqslant (1 - s)$$

$$-g(s) \leqslant -s$$

and

10-2b_8 $$g(s) \geqslant s.$$

Hence if $m \geqslant 1$, $s = 1$ is the only solution, $0 \leqslant s \leqslant 1$, of the equation. If in addition $g(x) = x$ for some $x < 1$, we would have

$$\frac{1 - g(x)}{1 - x} = 1 \leqslant m$$

which is a contradiction whenever $m < 1$ (and the inequality 10-2b_7 must hold for $m < 1$ as well as for $m = 1$). Using

$$g_\mu(s_1) < g_\mu(s_2), \quad 0 < s_1 < s_2 \tag{10-2a_1}$$

$$g_\mu(s) < g_n(1), \quad 0 < s < 1 \tag{10-2a_2}$$

$$g(0) = p_0 > 0 \tag{10-2b_1}$$

$$g(1) = 1 \tag{10-2b_6}$$

$$\frac{1 - g(s)}{1 - s} \leqslant m, \quad 0 < s < 1 \tag{10-2b_7}$$

without knowing the precise values of $g(s)$ we can illustrate the relationship between $g(s)$ and s by the graph in Fig. 10-2b_9. Since whenever $m \leqslant 1$, the smallest solution of $g(\rho) = \rho$ is $\rho = 1$, we have just shown that for those species in which the expected number of offspring of an individual is less than or equal to one the species becomes extinct with probability one.

10-2b_9 Figure
   Graphical representation of $g(s)$ and s for the case $m \leqslant 1$

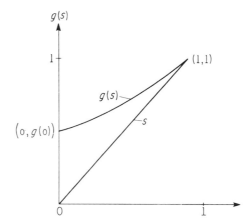

If $m > 1$, 10-2b_7 still holds, so for $0 < s < 1$,

$$\frac{1 - g(s)}{1 - s} \leqslant m.$$

From 10-2a_3,

$$\frac{1 - g(s)}{1 - s} = \sum_{k=0}^{n} P_1(k)(1 + s + s^2 + \cdots + s^{k-1}) \leqslant \sum_{k=0}^{n} kP_1(k) = m.$$

As s comes closer to one, all the s^j's come closer to one, and $\dfrac{1 - g(s)}{1 - s}$ comes

closer to $\sum\limits_{k=0}^{n} kP_1(k) = m > 1$.   If we choose s close enough to one,

$$1 < \frac{1 - g(s)}{1 - s} \leqslant m.$$

Hence, for such an s, one that is sufficiently close to one,

$$1 - g(s) > 1 - s$$

and

10-2b$_{10}$ $\qquad\qquad\qquad\qquad g(s) < s.$

Thus, since

$$g(0) > 0 \qquad \text{and} \qquad g(1) = 1$$

and for some s, $0 < s < 1$,

$$g(s) < s,$$

there must be a $\rho < 1$ such that

$$g(\rho) = \rho.$$

Suppose there are two such numbers, $\rho_1 < \rho_2 < 1$, such that

$$g(\rho_1) = \rho_1 \qquad \text{and} \qquad g(\rho_2) = \rho_2.$$

We have

$$g(\rho_1) = \rho_1 < \rho_2 = g(\rho_2) < g(1) = 1.$$

From 10-2a_3,

$$1 = \frac{g(\rho_2) - g(\rho_1)}{\rho_2 - \rho_1} = \sum_{k=1}^{n} P_1(k)(\rho_2^{k-1} + \rho_2^{k-2}\,\rho_1 + \rho_2^{k-3}\,\rho_1^2 + \cdots + \rho_1^{k-1})$$

and since $\rho_1 < \rho_2$,

$$\sum_{k=1}^{n} kP_1(k)\rho_1^{k-1} < \sum_{k=1}^{n} P_1(k)(\rho_2^{k-1} + \rho_2^{k-2}\rho_1 + \cdots + \rho_1^{k-1}) < \sum_{k=1}^{n} kP_1(k)\rho_2^{k-1}$$

Hence there is a $\bar{\rho}$, $\rho_1 < \bar{\rho} < \rho_2$, such that

$$\sum_{k=1}^{n} P_1(k)(\rho_2^{k-1} + \rho_2^{k-2}\,\rho_1 + \cdots + \rho_1^{k-1}) = \sum_{k=1}^{n} kP_1(k)\bar{\rho}^{k-1} = 1.^*$$

---

*Readers familiar with elementary calculus will recognize this as an application of the Mean Value Theorem of calculus. See, for example, Tom M., Apostle, Calculus (Waltham, Mass.: Blaisdell Publishing Company, 1961), Vol. I, p. 355.

Also,
$$1 = \frac{1 - g(p_2)}{1 - p_2} = \sum_{k=0}^{n} P_1(k)(1 + p_2 + p_2^2 + \cdots + p_2^{k-1}).$$

and there is a $\bar{\bar{p}}$, $p_2 < \bar{\bar{p}} < 1$ such that

$$1 = \sum_{k=0}^{n} kP_1(k)\bar{\bar{p}}^{k-1}.$$

Since $\bar{p} < p_2 < \bar{\bar{p}}$, we have

$$1 = \sum_{k=0}^{n} kP_k(1)\bar{p}^{k-1} < \sum_{k=0}^{n} kP_k(1)\bar{\bar{p}}^{k-1} = 1$$

but this is a contradiction. Hence if $m > 1$, there is only one p less than one such that $g(p) = p$. Again using 10-2a_1, 10-2a_2, 10-2b_1, and 10-2b_{10}, without knowledge of the exact values of $g(s)$, we illustrate the relative values of $g(s)$ and s in the case where $m > 1$ in Figure 10-2b_{11}.

10-2b_{11}   **Figure**

        Graphical representation of $g(s)$ and s for the case $m > 1$.

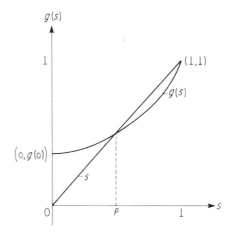

We have just shown that whenever the expected number of offspring of the individuals of a species is greater than one, the probability of the extinction of that strain, though positive, is less than one.

To illustrate the theory we have just described, let us apply it to the example in 10-1b. There $X = G_1$ was the random variable with range and distribution

G_1	0	1	2
P_1	$\frac{1}{8}$	$\frac{1}{2}$	$\frac{3}{8}$

The generating function of the species is

$$g(s) = \frac{1}{8} + \frac{1}{2}s + \frac{3}{8}s^2$$

and by straightforward calculations we found

$$g(0) = .1250$$
$$g_2(0) = .1933$$
$$g_3(0) = .2357$$

which as far as it goes agrees with 10-2b_3. We calculated that

$$E(G_1) = \frac{5}{4} = m$$

which means the probability of extinction should be less than one. We set

$$g(\rho) = \frac{1}{8} + \frac{1}{2}\rho + \frac{3}{8}\rho^2 = \rho$$

$$1 + 4\rho + 3\rho^2 = 8\rho$$

$$1 - 4\rho + 3\rho^2 = 0$$

$$(1 - \rho)(1 - 3\rho) = 0$$

The solutions of this equation are $\rho = 1$ and $\rho = \frac{1}{3}$, and the probability of extinction is the minimum solution, that is,

$$\rho = P(G_n \to 0) = \frac{1}{3}.$$

If instead we consider a species in which G_1 and P_1 are given by

G_1	0	1	2	3
P_1	$\frac{3}{8}$	$\frac{1}{2}$	$\frac{1}{16}$	$\frac{1}{16}$

we would have

$$m = \sum_{k=0}^{3} kP_1(k) = \frac{1}{2} + \frac{2}{16} + \frac{3}{16} = \frac{13}{16} < 1.$$

Then $g(\rho) = \rho$ becomes

$$\frac{1}{16}\rho^3 + \frac{1}{16}\rho^2 + \frac{1}{2}\rho + \frac{3}{8} = \rho$$

or

$$p^3 + p^2 - 8p + 6 = 0.$$

This factors into

$$(p - 1)(p^2 + 2p - 6) = 0$$

so that $p = 1$ is one solution and the other two are

$$p = \frac{-2 \pm \sqrt{4 + 24}}{2} = -1 \pm \sqrt{7}$$

The minimum positive root is $p = 1$ (as it should be) and hence this species becomes extinct with probability one.

EXERCISES 10-2b

1. Individuals of a certain species have no offspring with probability p, $0 < p < 1$, and one offspring with probability $q = 1 - p$. What are the generating functions $g(s)$, $g_2(s)$ and $g_3(s)$? What is $g_n(s)$? What are the expectations of G_1 and the probability of extinction? For $p = \frac{1}{3}$, graph $g(s)$ and s on the same pair of axes, $0 \leqslant s \leqslant 2$.

2. Certain amoeba die with probability p, $0 < p < 1$, are split into two with probability $q = 1 - p$, —that is,

G_1	0	1	2
P_1	p	0	q

What are $g(s)$, $g_2(s)$, $g_3(s)$? What are the expectations of G_1 and the probability of extinction. For $p = \frac{1}{3}$ and $p = \frac{2}{3}$, graph $g(s)$ and s on the same pair of axes.

3. Let G_1 and P_1 be given by

G_1	0	1	2	3
P_1	$1 - p$	0	0	p

(see Prob. 2, Exercises 10-1c). What are the probabilities of extinction for $p = \frac{1}{4}, p = \frac{1}{3}, p = \frac{1}{2}, p = \frac{3}{4}$?

4. If G_1 and P_1 are given by

G_1	0	1	2	...	m
P_1	$1 - p$	0	0	...	p

(see Prob. 3, Exercises 10-1c), how is the probability of extinction related to m?

10-A. APPENDIX

Probability of the sum of identically distributed random variables

Let X_1, X_2,...,X_N be independent identically distributed random variables with range $R_X = \{0, 1, 2,...,n\}$ and with the common distribution

$$P_X = \{(i, p_i) | i \in R_X, p_i \geqslant 0, \sum_{i=1}^{n} p_i = 1\}.$$

Consider the set

$$\{X_1 = v_1, X_2 = v_2,..., X_N = v_N\}$$

where $v_i \in R_X$ for every i but where we do not assume that the v_i are distinct. Since the X_i are independent,

$$P\{X_1 = v_1, X_2 = v_2,...,X_N = v_N\} = P(X_1 = v_1)P(X_2 = v_2)\cdots P(X_N = v_N),$$

(see 7-3a_1), and since they are identically distributed,

10-A_1 $\qquad P(X_1 = v_1, X_2 = v_2,...,X_N = v_N) = p_{v_1} p_{v_2} \cdots p_{v_N}$

Let x_r of the v_i equal $r \in R_X$ ($r = 0,1,2,...,n$, $i = 1,2,...,N$), and for these factors $P(X_i = v_i) = P(X_i = r) = p_r$. We remark that some of the x_r may be zero. With this Eq. 10-A_1 becomes

$$P(X_1 = v_1, X_2 = v_2,...,X_N = v_N) = p_0^{x_0} p_1^{x_1} \cdots p_n^{x_n}$$

where $x_0 + x_1 + \cdots + x_n = N$ and $x_1 + 2x_2 + \cdots + nx_n = v_1 + v_2 + \cdots + v_N$.

To illustrate what we have just set up, let X_1, X_2, and X_3 be defined on $\Omega = \{(d_1, d_2, d_3) | d_i \in \{1, 2, 3, 4\}\}$ by the equations $X_1((d_1, d_2, d_3)) = d_1$, $X_2((d_1, d_2, d_3)) = d_2$, and $X_3((d_1, d_2, d_3)) = d_3$. X_1, X_2 and X_3 are independent identically distributed random variables with $R_X = \{1, 2, 3, 4\}$ and

$$P_X = \{(i, p_i) | i \in R_X, \sum_{i=1}^{4} p_i = 1\}.$$

$$P(X_1 = 1, X_2 = 4, X_3 = 1) = P(X_1 = 1)P(X_2 = 4)P(X_3 = 1)$$

$$= p_1 p_4 p_1 = p_1^2 p_4.$$

In this example $x_1 = 2$, $x_4 = 1$, $x_2 = x_3 = 0$ so that $x_1 + x_2 + x_3 + x_4 = 3$ and $x_1 + 2x_2 + 3x_3 + 4x_4 = 2 + 4(1) = 1 + 4 + 1$. We shall return to these random variables to illustrate further points in our discussion.

Coming back to the general case, if $(v_1', v_2', \ldots, v_N')$ is a permutation of $(v_1, v_2, \ldots, v_N)$,

$$P(X_1 = v_1, X_2 = v_2, \ldots, X_N = v_N) = P(X_1 = v_1', X_2 = v_2', \ldots, X_N = v_N')$$

$$= p_0^{x_0} p_1^{x_1} \cdots p_n^{x_n}$$

since a permutation in no way alters the number of zeros, ones, twos, et cetera, that appear in $(v_1, v_2, \ldots, v_N)$. In our example $(1, 1, 4)$ is a permutation of $(1, 4, 1)$ and

$$P(X_1 = 1, X_2 = 1, X_3 = 4) = P(X_1 = 1, X_2 = 4, X_3 = 1) = p_1^2 p$$

We recall that there are $\begin{pmatrix} n \\ r_1, r_2, \ldots, r_k \end{pmatrix}$ number of ways a population of n elements can be subdivided into k parts containing respectively $r_1, r_2, \ldots, r_k$ elements (see 4-1d_1). Hence there are $\begin{pmatrix} n \\ x_0, x_1, \ldots, x_n \end{pmatrix}$ permutations of the original set $(v_1, v_2, \ldots, v_N)$. With this in mind, let us look at the set

$$\{X_1 + X_2 + \cdots + X_N = v_1 + v_2 + \cdots + v_N\}.$$

This set is the disjoint union of the sets

$$\{X_1 = v_1', X_2 = v_2', \ldots, X_N = v_N'\}$$

where $(v_1', v_2', \ldots, v_N')$ is a permutation of $(v_1, v_2, \ldots, v_N)^*$; so that

$$\{X_1 + X_2 + \cdots + X_N = v_1 + v_2 + \cdots + v_N\} = \sum \{X_1 = v_1', X_2 = v_2', \ldots, X_N = v_N'\}$$

where the summation is over all permutations of $(v_1, v_2, \ldots, v_N)$. Since all the sets on the right-hand side of the equation have the same probability,

$$P(X_1 + X_2 + \cdots + X_N = v_1 + v_2 + \cdots + v_N) = \begin{pmatrix} N \\ x_0, x_1, \ldots, x_n \end{pmatrix} p_0^{x_0} p_1^{x_1} \cdots p_n^{x_n}$$

where $x_1 + 2x_2 + \cdots + nx_n = v_1 + v_2 + \cdots + v_N$.

Again going back to our example,

$$P(X_1 + X_2 + X_3 = 1 + 4 + 1) = \begin{pmatrix} 3 \\ 2, 0, 0, 1 \end{pmatrix} p_1^2 p_4$$

---

*The union is disjoint, for if $(v_1', v_2', \ldots, v_N')$ is a proper permutation of $(v_1, v_2, \ldots, v_N)$ there are at least two i's such that $v_i' \neq v_i$, and if $\omega \in \{X_1 = v_1, X_2 = v_2, \ldots, X_N = v_N\} = \{X_1 = v_1\} \{X_2 = v_2\} \{X_N = v_N\}$ and $\omega \in \{X_1 = v_1', X_2 = v_2', \ldots, X_N = v_N'\}$, $X_i(\omega) = v_i$ and $X_i(\omega) = v_i' \neq v_i$ which is a contradiction since X_i is a random variable.

for

$$\{X_1 + X_2 + X_3 = 1 + 4 + 1\} = \{X_1 = 1, X_2 = 4, X_3 = 1\}$$
$$= \{X_1 = 1, X_2 = 1, X_3 = 4\}$$
$$= \{X_1 = 4, X_2 = 1, X_3 = 1\}.$$

Instead of specifying the addends we now look only at their sum; that is, we consider the set

$$\{X_1 + X_2 + \cdots + X_N = \mu\}$$

for some nonnegative integer μ. There may be two or more distinct sequences of N numbers from R_X whose sums are μ. That is there may exist collections $(v_1, v_2, \ldots, v_N)$, $(\bar{v}_1, \bar{v}_2, \ldots, \bar{v}_N)$, $(\bar{\bar{v}}_1, \bar{\bar{v}}_2, \ldots, \bar{\bar{v}}_N)$, ... such that

$$v_1 + v_2 + \cdots + v_N = \bar{v}_1 + \bar{v}_2 + \cdots + \bar{v}_N = \bar{\bar{v}}_1 + \bar{\bar{v}}_2 + \cdots + \bar{\bar{v}}_N = \cdots = \mu$$

where no group is a permutation of the other. In our example $(1, 1, 4)$, $(1, 2, 3)$ and $(2, 2, 2)$ are the three distinct sequences from R_X whose elements sum to six. Hence the set

$$\{X_1 + X_2 + \cdots + X_N = \mu\} = \sum_{v_1 + v_2 + \cdots + v_N = \mu} \{X_1 = v_1, X_2 = v_2, \ldots, X_N = v_N\}$$

and

10-A_2

$$P\{(X_1 + X_2 + \cdots + X_N = \mu\} = \sum_{v_1 + v_2 + \cdots + v_N = \mu} P(X_1 = v_1)P(X_2 = v_2) \cdots P(X_N = v_N)$$

$$= \sum_{x_1 + 2x_2 + \cdots + nx_n = \mu} \binom{N}{x_0, x_1, \ldots, x_n} p_0^{x_0} p_1^{x_1} \cdots p_n^{x_n}$$

(where, of course, each $v_i \in R_X$ since otherwise $X_i = v_i$ would be meaningless) In our example,

$$P(X_1 + X_2 + X_3 = 6) = \sum_{x_1 + 2x_2 + 3x_3 + 4x_4 = 6} \binom{3}{x_1, x_2, x_3, x_4} p_1^{x_1} p_2^{x_2} p_3^{x_3} p_4^{x_4}$$

$$= \binom{3}{2,0,0,1} p_1^2 p_4 + \binom{3}{1,1,1,0} p_1 p_2 p_3 + \binom{3}{0,3,0,0} p_2^3.$$

chapter **11**

On Information Theory

About twenty years ago Claude E. Shannon published an article in the *Bell System Technical Journal** in which he presented a mathematical model of a communication system. The paper gave rise to Information Theory which, in the brief period since its introduction, has become a major field of mathematical research. Upon early acquaintance with the field the Theory developed seems both disappointing and bizarre—disappointing because it has nothing to do with meaning and bizarre because it deals not with a single message but rather with the statistical character of whole ensembles of messages. Thus the word "information" takes on new meaning in terms of the mathematical theory we investigate. That a word has different meaning in different fields is not unusual; "angel" means one thing to a theologian and quite another to a theatrical producer.

Though it may be quite unnecessary to do so, we point out that Information Theory, like Queuing Theory, had its origins in the research departments of telephone companies.

11-1. A COMMUNICATION SYSTEM

A basic communication system is made up of five components:

a *source* of messages,
a *transmitter* which sends the messages through
a *channel* to
a *receiver* which delivers the messages to
a *destination.*

*Vol. 27 (1948), pp. 379-423, 623-656.

11-1   Figure

A Communication System

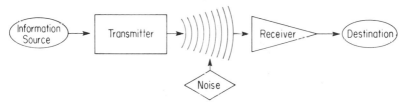

In most communication systems there is in addition to the system source a *noise* source which feeds messages into the channel, messages which tend to distort and garble the messages intentionally transmitted through the channel.   In human speech the brain of one person is the information source, his voice the transmitter; the air channels the message to the ear of another; this ear is the receiver which delivers it to the listener's consciousness, the destination.   If this message is delivered in a crowded, noisy room, the hearer often does not catch what has been said or he misunderstands because the transmitted message was interfered with by the other conversations in the area.   In television the image of a newscaster is picked up by a camera and carried by radio waves to a TV picture tube where it (the image) reappears on the screen.   In outline form we see that we can equate:

newscaster	to	source
camera	to	transmitter
radio waves	to	channel
TV tube	to	receiver
screen	to	destination

Another important phase of information enters into the television example, it is coding.   The newscaster's image after being picked up by the camera is coded into electrical impulses which are carried by the radio waves to the receiver where they are decoded and the picture reappears.   Picture distortion results from interference with the channel—airplanes overhead cause modifications in the transmitted waves, other electrical waves in the atmosphere may merge with the transmitted waves to produce unwanted results.

Some of the major areas of investigation in the theory are:

(a) Can information be quantified?   Can we measure the amount of information transmitted?

(b) How efficient is the channel?   Do we get as much information at the destination as was given at the source?

(c) How much information can a channel carry?

(d) What are the characteristics of an efficient coding process?

(e) Can noise be eliminated or at least minimized?

Our probabilistic background enables us to understand how the first two questions can be answered, but the others require more sophisticated analysis than we assume, so we shall not concern ourselves with them. Although the mathematical theory we study has nothing to do with the meaning of messages, applications of the results have enabled engineers concerned with the development of communications equipment (radar, computing machines, television— to mention only a few) to greatly increase the efficiency and reliability of their apparatus.

11-2. A MEASURE OF INFORMATION

a. Entropy

Every information source produces a sequence of messages; they may, for example, be letters of the alphabet, musical notes, electrical impulses or digits. The total array of different messages a source can transmit is called its *alphabet;* if the source sends out English letters, it has an alphabet consisting of the letters *a* through *z* and a space, if the source feeds a computer ten-digit binary numbers, its alphabet has 2^{10} elements. The various characters in any alphabet come out of the source with a given probability so that with any source is associated a probability space. As a very simple example, let the information source be a die which sends out a message each time it is rolled. Its alphabet is the sequence 1, 2, 3, 4, 5, 6; and assuming the die to be fair, the probability space associated with this source is the usual one which assigns the probability $\frac{1}{6}$ to each of the digits.

Information is given when the source emits a message and the first question is "how much?" Can we actually quantify information? In the probability space $\{\Omega, A_\Omega, P\}$ associated with a source let $\{A_i\}$, $i = 1, 2, \ldots, n$ be a set of events which partition Ω. Writing p_i for $P(A_i)$, the array

$$\begin{pmatrix} A_1 & A_2 & \cdots & A_n \\ p_1 & p_2 & \cdots & p_n \end{pmatrix}$$

will be called an *uncertainty scheme* and when the source emits a message (which necessarily belongs to one of the A_i's), the uncertainty is removed and information is given. Let $I(A_i)$ stand for the information given when A_i occurs; hence, I is a random variable on $\{\Omega, A_\Omega, P\}$.

$$E(I) = p_1 I(A_1) + p_2 I(A_2) + \cdots + p_n I(A_n)$$

is the expected amount of information contained in the uncertainty scheme, or, it is the average amount of information conveyed in learning which of the A_i actually does occur. We intuitively feel information should increase with uncertainty—the less likely an event the more information given when it does occur. The coupling of "information" and "uncertainty" that we have just introduced will keep reappearing in our work.

Returning to the example of a die as an information source, two possible uncertainty schemes are

$$\begin{pmatrix} \{1\} & \{2\} & \{3\} & \{4\} & \{5\} & \{6\} \\ \dfrac{1}{6} & \dfrac{1}{6} & \dfrac{1}{6} & \dfrac{1}{6} & \dfrac{1}{6} & \dfrac{1}{6} \end{pmatrix} \quad \text{and} \quad \begin{pmatrix} \{1,2,3\} & \{4,5\} & \{6\} \\ \dfrac{1}{2} & \dfrac{1}{3} & \dfrac{1}{6} \end{pmatrix}$$

The average information contained in the first is

$$E(I) = \sum_{n=1}^{6} \frac{1}{6} I(\{n\})$$

and in the second

$$E(I) = \frac{1}{2} I(\{1,2,3\}) + \frac{1}{3} I(\{4,5\}) + \frac{1}{6} I(\{6\}).$$

In none of this, of course, has the function I been explicitly defined. In an intuitive sense, the more uncertain a scheme is, the more information it should convey. Whatever our I function is to be, we expect more information from the first scheme than from the second; we learn more if we know $\{1\}$ occurs than if we know $\{1,2,3\}$ occurs. This idea of greater uncertainty yielding greater information is also apparent in the following two schemes, the first associated with a true coin and the second with a weighted one:

$$\begin{pmatrix} H & T \\ \dfrac{1}{2} & \dfrac{1}{2} \end{pmatrix} \quad \text{and} \quad \begin{pmatrix} H & T \\ \dfrac{9}{10} & \dfrac{1}{10} \end{pmatrix}$$

Here we expect more information if H occurs in the first scheme than if H occurs in the second.

$$\begin{pmatrix} \{1\} & \{2\} & \{3\} & \{4\} & \{5\} & \{6\} \\ \dfrac{1}{6} & \dfrac{1}{6} & \dfrac{1}{6} & \dfrac{1}{6} & \dfrac{1}{6} & \dfrac{1}{6} \end{pmatrix} \quad \text{and} \quad \begin{pmatrix} H & T \\ \dfrac{1}{2} & \dfrac{1}{2} \end{pmatrix}$$

are illustrations of *equiprobable uncertainty schemes* in which each event in the scheme occurs with the same probability. For notational purposes only, whenever we work with an equiprobable scheme, we shall denote the events of the partition by E_i (instead of A_i) so that our scheme has the form

$$\begin{pmatrix} E_1 & E_2 & \cdots & E_n \\ \dfrac{1}{n} & \dfrac{1}{n} & \cdots & \dfrac{1}{n} \end{pmatrix}$$

Although the random variable I is defined on the sample space Ω in $\{\Omega, A_\Omega, P\}$, its use in information theory will be limited to considerations of its expectations on uncertainty schemes. For this reason we introduce some new terminology. Let $\{\Omega, A_\Omega, P\}$ be associated with an information source and let I be a random variable on the probability space. Consider the scheme

$$\begin{pmatrix} A_1 & A_2 & \cdots & A_n \\ p_1 & p_2 & \cdots & p_n \end{pmatrix}.$$

The function

11-2a₁ $H(A_1, A_2, \ldots, A_n) = p_1 I(A_1) + p_2 I(A_2) \cdots + p_n I(A_n)$

is called the *entropy* of the scheme and is, as we indicated earlier, the average amount of information conveyed when uncertainty is removed, that is, when we know which of the A_i occur. The term "entropy" results from a physical analogy which we have no need to investigate here. We emphasize that H is a function ot uncertainty schemes, and hence it is meaningful if, and only if, $A_i \in A_\Omega$ for each i, $A_i A_j = \varnothing$, $i \neq j$, $A_1 + A_2 + \cdots + A_n = \Omega$ and $p_1 + p_2 + \cdots + p_n = 1$.

b.   Development of I and H

In the preceding section we introduced the random variable I and the function H; now by putting some "reasonable" restraints on them, we hope to obtain explicit expressions for both.

Consider two uncertainty schemes

$$\begin{pmatrix} A & A' \\ p(A) & 1 - p(A) \end{pmatrix} \text{ and } \begin{pmatrix} B & B' \\ p(B) & 1 - p(B) \end{pmatrix}$$

where $0 < p(A) < p(B) \leq \dfrac{1}{2}$. The second scheme contains more uncertainty than the first, for it is less easy to predict which of the messages B and B' will be sent than to predict which of A and A' will be sent; hence we want $H(B,B')$ to be greater than $H(A,A')$. In the first scheme where $p(A) < \dfrac{1}{2} < p(A') = 1 - p(A)$

we would expect to gain more information by learning that A occurred than by learning that A' occurred. Thus our first requirements on I and H are

$$R_1: \quad \begin{array}{l} I(A) > I(B) \text{ if } p(A) < p(B) \\ \\ H(A,A') < H(B,B') \text{ if } 0 < p(A) < p(B) \leqslant \dfrac{1}{2} \end{array}$$

As a sort of corollary to the first part of R_1 if $p(A) = p(B)$ then we should get the same amount of information on learning which occurs; that is,

$$I(A) = I(B) \text{ whenever } p(A) = p(B).$$

What we have done here is to require that $I(A)$ be in fact a function of $p(A)$, so that in the uncertainty scheme

$$\begin{pmatrix} A_1, & A_2, & \ldots, & A_n \\ \\ p_1, & p_2, & \ldots, & p_n \end{pmatrix}$$

$I(A_i) = I(p_i)$ and $H(A_1,A_2,\ldots,A_n) = \displaystyle\sum_{i=1}^{n} p_i I(p_i) = H(p_1,p_2,\ldots,p_n)$.

To illustrate the significance of the requirement that the information conveyed by a message depends on the probability of the message's being sent, consider the random variables X and I on the sample space associated with the roll of two distinguishable dice. If $D = \{1,2,3,4,5,6\}$,

$$\Omega = D \times D = \{(r,g)|r,g \in D\}.$$

Let $X(r,g) = r + g$ and $I(r,g) = I(p(r + g))$. The elementary events of Ω and their probabilities are given in the uncertainty scheme:

$$\begin{pmatrix} 2 & 3 & 4 & 5 & 6 & 7 & 8 & 9 & 10 & 11 & 12 \\ \\ \dfrac{1}{36} & \dfrac{2}{36} & \dfrac{3}{36} & \dfrac{4}{36} & \dfrac{5}{36} & \dfrac{6}{36} & \dfrac{5}{36} & \dfrac{4}{36} & \dfrac{3}{36} & \dfrac{2}{36} & \dfrac{1}{36} \end{pmatrix}$$

$$X(2,2) = 4 < X(3,4) = 7 < X(5,5) = 10$$

$$I(2,2) = I(p(4)) = I\left(\frac{3}{36}\right); \; I(3,4) = I(p(7)) = I\left(\frac{1}{6}\right)$$

$$I(5,5) = I(p(10)) = I\left(\frac{3}{36}\right) = I(2,2).$$

Since we want information to increase as probability decreases, we want

$$I(2,2) = I(5,5) = I\left(\frac{3}{36}\right) > I\left(\frac{1}{6}\right) = I(3,4).$$

Since I and H are to depend on the probabilities rather than the messages themselves it will be convenient to express them as functions of the p_i, so we rewrite our first set of requirements as:

11-2b_1 R_1: $I(p) > I(q)$ if $p < q$

$$H(P, 1 - p) < H(q, 1 - q) \text{ if } 0 < p < q \leqslant \frac{1}{2}$$

and when $p = q$, we shall have $I(p) = I(q)$.

The entropy, H, by its definition (11-2a_1) is symmetric in all its arguments and, in particular, if $A + A' = \Omega$,

$$H(p, 1 - p) = H(1 - p, p) = pI(p) + (1 - p)I(1 - p).$$

It is for this reason that the second part of R_1 in 11-2b_1 is conditioned upon $p < q \leqslant \frac{1}{2}$. (Without the condition, if $p < \frac{1}{2}$, $1 - p > \frac{1}{2}$ so that $p < 1 - p$, and we might be tempted to write $H(p, 1 - p) < H(1 - p, p)$.) The uncertainty in any scheme

$$\begin{pmatrix} A_1 & A_2 & \dots & A_n \\ p_1 & p_2 & \dots & p_n \end{pmatrix}$$

depends not on the order in which the A_i's are listed but rather in their corresponding probabilities.

Consider the special case of the two-message scheme in which the message A cannot be sent, so that A' must be sent. Since A cannot occur, $I(A) = I(0)$ is meaningless and to avoid a meaningless entropy we set

$$pI(p) = 0 \quad \text{if} \quad p = 0.$$

Furthermore, if A' is the only message which can be sent, $p(A') = 1$, and we get no information by knowing it was sent, so we must have

$$I(1) = 0.$$

Hence, the entropy of the scheme

$$\begin{pmatrix} A & A' \\ 0 & 1 \end{pmatrix}$$

is

$$H(0, 1) = 0 \cdot I(0) + 1 \cdot I(1) = 0.$$

We combine these into our second set of requirements for I and H:

11-2b₂ R_2: $pI(p) = 0$ if $p = 0$

$$H(0,1) = 0$$

R_1 and R_2 coupled with the symmetry of $H(p,1 - p)$ indicate that were we to graph $H(p,1 - p)$ as a function of p we would expect a curve like that in Fig. 11-2b₃. The precise nature of the curve is not of concern here. What is of importance is its maximum at $p = \dfrac{1}{2} = 1 - p$, its symmetry about the point $p = \dfrac{1}{2}$, and that it increases from $p = 0$ to $p = \dfrac{1}{2}$, embodying the requirements imposed by R_1 and R_2.

11-2b₃ Figure

The graph of $H(p,1 - p)$ as a function of p

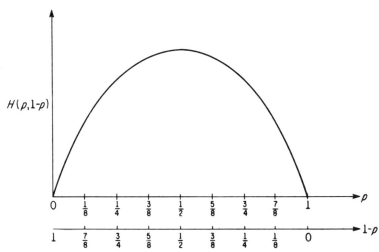

To learn more about H, let us consider an uncertainty scheme of three events:

$$\begin{pmatrix} A_1 & A_2 & A_3 \\ p_1 & p_2 & p_3 \end{pmatrix}$$

$$H(p_1,p_2,p_3) = p_1I(p_1) + p_2I(p_2) + p_3I(p_3)$$

is the average amount of information conveyed when we learn which of A_1, A_2, and A_3 is sent. We can look at this from another angle. In finding out whether A_1, A_2, or A_3 is sent, we ask first whether A_1 was sent and if not, which of A_2 and A_3 was sent. Thus H should satisfy

$$H(A_1,A_2,A_3) = p_1I(A_1) + (1 - p_1)\Big(I(A_1') + p(A_2|A_1')I(A_2|A_1') + p(A_3|A_1')I(A_3|A_1')\Big).$$

From 7-1a_1 and the fact that $A_j \subset A'_1, j = 2, 3$,

$$p(A_j|A'_1) = \frac{p(A_jA'_1)}{p(A'_1)} = \frac{p(A_j)}{p(A'_1)} = \frac{p_j}{1 - p_1}.$$

Substituting this into the equation for $H(A_1,A_2,A_3)$ and writing it in terms of p_1, p_2 and p_3 we have

11-2b_4

$$H(p_1,p_2,p_3) = p_1I(p_1) + (1 - p_1)I(1 - p_1) + p_2I\left(\frac{p_2}{1 - p_1}\right) + p_3I\left(\frac{p_3}{1 - p_1}\right).$$

This can be grouped into

$$H(p_1,p_2,p_3)$$
$$= p_1I(p_1) + (1 - p_1)I(1 - p_1) + (1 - p_1)\left\{\frac{p_2}{1 - p_1}I\left(\frac{p_2}{1 - p_1}\right) + \frac{p_3}{1 - p_1}I\left(\frac{p_3}{1 - p_1}\right)\right\}$$

which can be written

11-2b_5 $$H(p_1,p_2,p_3) = H(p_1, 1 - p_1) + (1 - p_1)H\left(\frac{p_2}{1 - p_1}, \frac{p_3}{1 - p_1}\right).$$

Does this last equation tell us anything about I? We have

$$p_1I(p_1) + p_2I(p_2) + p_3I(p_3)$$

$$= p_1I(p_1) + (1 - p_1)I(1 - p_1) + p_2I\left(\frac{p_2}{1 - p_1}\right) + p_3I\left(\frac{p_3}{1 - p_1}\right)$$

Since $p_2 + p_3 = 1 - p_1$, this becomes

$$p_2\left[I(p_2) - I(1 - p_1) - I\left(\frac{p_2}{1 - p_1}\right)\right] + p_3\left[I(p_3) - I(1 - p_1) - I\left(\frac{p_3}{1 - p_1}\right)\right] = 0.$$

This equation is surely satisfied if for two events A and B with $A \subset B$ (so that $p = p(A) < p(B) = q$),

11-2b_6 $$I(p) - I(q) = I\left(\frac{p}{q}\right).$$

One common function in which the function of the quotient of two numbers is the difference of the functions of the two numbers is the logarithm. Therefore the logarithm is a possible choice for I.

To get a final requirement on H we extend 11-2b_5 to more than three events. In the scheme

$$\begin{pmatrix} A_1 & A_2 & \dots & A_n \\ p_1 & p_2 & \dots & p_n \end{pmatrix}$$

set $A_1 + A_2 + \dots + A_{n-2} = A$ and $p_1 + p_2 + \dots + p_{n-2} = p$ and form

$$\begin{pmatrix} A & A_{n-1} & A_n \\ p & p_{n-1} & p_n \end{pmatrix}.$$

We require that the entropies of the two schemes be equal since they both contain the same set of messages. Hence we want

$$H(p_1, p_2, \dots, p_n) = H(p, p_{n-1}, p_n).$$

Using 11-2b_5 to express $II(p, p_{n-1}, p_n)$, we have

$$H(p_1, p_2, \dots, p_n) = H(p, 1-p) + (1-p)H\left(\frac{p_{n-1}}{1-p}, \frac{p_n}{1-p}\right),$$

or, in terms of the I function,

11-2b$_7$

$$\sum_{i=1}^{n} p_i I(p_i) = pI(p) + (1-p)I(1-p) + p_{n-1}I\left(\frac{p_{n-1}}{1-p}\right) + p_n I\left(\frac{p_n}{1-p}\right).$$

If, as in 11-2b_6, $I(p) - I(q) = I\left(\dfrac{p}{q}\right)$ whenever $p < q$, since $p_{n-1} + p_n = 1 - p$,

$$p_{n-1} I(p_{n-1}) + p_n I(p_n) = (1-p) I(1-p) + p_{n-1}I\left(\frac{p_{n-1}}{1-p}\right) + p_n I\left(\frac{p_n}{1-p}\right)$$

Putting this into 11-2b_7, we obtain

$$\sum_{i=1}^{n} p_i I(p_i) = \sum_{i=1}^{n-2} p_i I(p_i) + (1-p)I(1-p) + p_{n-1}I\left(\frac{p_{n-1}}{1-p}\right) + p_n I\left(\frac{p_n}{1-p}\right),$$

and writing this in the H notation, we have

11-2b$_8$ $H(p_1, p_2, \dots, p_n) = H(p_1, p_2, \dots, p_{n-2}, 1-p) + (1-p)H\left(\dfrac{p_{n-1}}{1-p}, \dfrac{p_n}{1-p}\right).$

Using this, the entropy of the scheme

$$\begin{pmatrix} \{1\} & \{2\} & \{3\} & \{4\} & \{5\} & \{6\} \\ \dfrac{1}{6} & \dfrac{1}{6} & \dfrac{1}{6} & \dfrac{1}{6} & \dfrac{1}{6} & \dfrac{1}{6} \end{pmatrix}$$

can be computed as

$$H\left(\frac{1}{6},\frac{1}{6},\frac{1}{6},\frac{1}{6},\frac{1}{6},\frac{1}{6}\right) = H\left(\frac{1}{6},\frac{1}{6},\frac{1}{6},\frac{1}{6},\frac{1}{3}\right) + \frac{1}{3}H\left(\frac{1}{2},\frac{1}{2}\right).$$

We now collect 11-2b_6 and 11-2b_8 into a third set of requirements to be imposed on I and H.

11-2b_9 R_3: $I(p) - I(q) = I\left(\dfrac{p}{q}\right)$ whenever $p < q$

$$H(p_1,p_2,\ldots,p_n) = H(p_1,p_2,\ldots,p_{n-2},1-p) + (1-p)H\left(\frac{p_{n-1}}{1-p},\frac{p_n}{1-p}\right)$$

where $p_1 + p_2 + \cdots + p_{n-2} = p$.

Let us briefly pull together our definitions of I and H and the properties we require of them.   Given a source and its associated probability space, $\{\Omega, A_\Omega, P\}$, we construct an uncertainty scheme

$$\begin{pmatrix} A_1 & A_2 & \ldots & A_n \\ p_1 & p_2 & \ldots & p_n \end{pmatrix}$$

where the set of events $\{A_i\}$, $i = 1,2,\ldots,n$ is a partition of Ω and $\sum_{i=1}^{n} p_i = 1$. We introduce a random variable I, which we call the information unit, and define the entropy H as a function on uncertainty schemes.   By definition H is the expectation of the random variable I, the information on the uncertainty scheme—that is, $H(p_1,p_2,\ldots,p_n) = p_1 I(p_1) + p_2 I(p_2) + \cdots + p_n I(p_n)$.   We remark that H is symmetric in its arguments, p_i.   In order to find explicit expressions for I and H we put three "reasonable" conditions on them—"reasonable" in the sense that they are in keeping with the layman's concept of information and uncertainty.   These are:

$$I(p) > I(q) \text{ if } p < q$$

R_1:   (11-2b_1)

$$H(p,1-p) < H(q,1-q) \text{ if } 0 < p < q \leqslant \frac{1}{2}$$

$$pI(p) = 0 \text{ if } p = 0$$

R_2:   (11-2b_2)

$$H(0,1) = 0$$

$$I(p) - I(q) = I\left(\frac{p}{q}\right) \quad \text{whenever } p < q$$

R_3:

$$H(p_1,p_2,\ldots,p_n) = H(p_1,p_2,\ldots,p_{n-2},1-p) + (1-p)H\left(\frac{p_{n-1}}{1-p}, \frac{p_n}{1-p}\right)$$

where
$$\sum_{i=1}^{n-2} p_i = p. \qquad\qquad (11\text{-}2b_9)$$

EXERCISES 11-2b

1. Demonstrate why the following functions do or do not satisfy the conditions R_1 (Eq. $11\text{-}2b_1$): (a) $I(p) = k$ where k is a constant; (b) $I(p) = 1 - \frac{p}{2}$; (c) $I(p) = \frac{1}{p}$; (d) $I(p) = p^2$; (e) $I(p) = k \log \frac{1}{p} = -k \log p$, where k is a constant.

2. Of those functions in Prob. 1 which do satisfy R_1, which also satisfy R_2 ($11\text{-}2b_2$)? (Assume $x \log \frac{1}{x} = 0$ whenever $x = 0$.)

3. Of those functions which satisfy both R_1 and R_2, are any a possible choice for $I(p)$; i.e., do any also satisfy R_3?

The remaining problems are based on the function which emerges from Prob. 3.

4. (a) If the 26 letters of the English alphabet occur with equal probability, how much information does one letter give? (b) An approximation to the frequency of occurrence of the various letters in common English usage is given by the number of times each letter appears in the game *Scrabble*. The frequency of the letters is given in the following list:

A – 9	E – 12	I – 9	M– 2	Q– 1	U– 4	Y – 2
B – 2	F – 2	J – 1	N– 6	R– 6	V – 2	Z – 1
C – 2	G– 3	K– 1	O– 8	S – 4	W– 2	
D– 4	H – 2	L – 4	P – 2	T– 6	X – 1	

How much information does an A give? An F, a Q, a S?

5. Forty percent of all automobiles are bought for personal use; of these 30 percent are four-door sedans while only 20 percent of all cars are four-door sedans. If a dealer knows a certain car was purchased for personal use how much more information does he get if he also is told it was a four-door sedan?

6. North knows that his opponents (East and West) have three trump between them. How much additional information does he have if he learns these trump are divided between the two hands?

7. Consider the uncertainty schemes associated with one true and two weighted coins:

(a) $\begin{pmatrix} H & T \\ \frac{1}{2} & \frac{1}{2} \end{pmatrix}$, (b) $\begin{pmatrix} H & T \\ \frac{1}{3} & \frac{2}{3} \end{pmatrix}$ and (c) $\begin{pmatrix} H & T \\ \frac{9}{10} & \frac{1}{10} \end{pmatrix}$

What are their entropies? [Note: We have yet said nothing about an appropriate base for the logarithm defining $I(p)$. Choose base 2, and use the table of $-p \log_2 p$ (page 326) to compute the entropies.]

8. What is the average amount of information conveyed by the following uncertainty schemes:

(a) $\begin{pmatrix} A & B & C & D \\ \frac{1}{4} & \frac{1}{4} & \frac{1}{4} & \frac{1}{4} \end{pmatrix}$, (b) $\begin{pmatrix} \{A,B\} & C & D \\ \frac{1}{2} & \frac{1}{4} & \frac{1}{4} \end{pmatrix}$

(c) $\begin{pmatrix} \{A,B\} & \{C,D\} \\ \frac{1}{2} & \frac{1}{2} \end{pmatrix}$, (d) $\begin{pmatrix} \{A,B,C\} & \{D\} \\ \frac{3}{4} & \frac{1}{4} \end{pmatrix}$

9. What are the entropies of the letters of the English alphabet when their probabilities are given in Probs. 4a and 4b?

c.   Properties of *I* and *H*

11-2c₁ Theorem.   *The entropy of the scheme*

$$\begin{pmatrix} A_1 & A_2 & \cdots & A_n & B_1 & B_2 & \cdots & B_m \\ p_1 & p_2 & \cdots & p_n & q_1 & q_2 & \cdots & q_m \end{pmatrix}$$

where

$$\sum_{i=1}^{n} p_i = p \text{ and } \sum_{j=1}^{m} q_j = q = 1 - p$$

satisfies the equation

$$H(p_1,\ldots,p_n,q_1,\ldots,q_m) = H(p_1,\ldots,p_n,q) + qH\left(\frac{q_1}{q}, \frac{q_2}{q},\ldots,\frac{q_m}{q}\right)$$

Proof: If $m = 2$, the theorem is the same as R_3. We proceed using the Principle of Induction. If the theorem is true for $m = k$,

$$H(p_1,p_2,\ldots,p_n,q_1,q_2,\ldots,q_k) = H(p_1,p_2,\ldots,p_n,q) + qH\left(\frac{q_1}{q},\frac{q_2}{q},\ldots,\frac{q_k}{q}\right)$$

Let $m = k + 1$ and set $q_2 + q_3 + \cdots + q_{k+1} = \bar{q} = q - q_1$ and since the theorem holds for $m = k$,

$$H(p_1,\ldots,p_n,q_1,\ldots,q_{k+1}) = H(p_1,\ldots,p_n,q_1,\bar{q}) + \bar{q}H\left(\frac{q_2}{\bar{q}},\ldots,\frac{q_{k+1}}{\bar{q}}\right)$$

$$= H(p_1,p_2,\ldots,p_n,q) + qH\left(\frac{q_1}{q},\frac{\bar{q}}{q}\right) + \bar{q}H\left(\frac{q_2}{\bar{q}},\ldots,\frac{q_{k+1}}{\bar{q}}\right)$$

where the last equality comes from R_3 (11-2b_9).

Again, since the theorem is true for $m = k$,

$$H\left(\frac{q_1}{q},\frac{q_2}{q},\ldots,\frac{q_{k+1}}{q}\right) = H\left(\frac{q_1}{q},\frac{\bar{q}}{q}\right) + \frac{\bar{q}}{q}H\left(\frac{q_2}{\bar{q}},\ldots,\frac{q_{k+1}}{\bar{q}}\right).$$

Hence

$$H(p_1,\ldots,p_n,q) + qH\left(\frac{q_1}{q},\ldots,\frac{q_{k+1}}{q}\right) = H(p_1,\ldots,p_n,q_1,\ldots,q_{k+1})$$

and so the theorem is true for all $m \in I^+$.   ∎

As a numerical example of the theorem start with the scheme

$$\begin{pmatrix} \{1\} & \{2\} & \{3\} & \{4\} & \{5\} & \{6\} \\ \dfrac{1}{6} & \dfrac{1}{6} & \dfrac{1}{6} & \dfrac{1}{6} & \dfrac{1}{6} & \dfrac{1}{6} \end{pmatrix}$$

Then, by the theorem

$$H\left(\frac{1}{6},\frac{1}{6},\frac{1}{6},\frac{1}{6},\frac{1}{6},\frac{1}{6}\right) = H\left(\frac{1}{6},\frac{1}{6},\frac{2}{3}\right) + \frac{2}{3}H\left(\frac{1}{4},\frac{1}{4},\frac{1}{4},\frac{1}{4}\right).$$

11-2c_2 Theorem. *In the scheme*

$$\begin{pmatrix} A_{11} & A_{12} & ,\ldots, & A_{1m_1} & A_{21} & A_{22} & \cdots & A_{2m_2} & \cdots & A_{n1} & A_{n2} & \cdots & A_{nm_n} \\ p_{11} & p_{12} & & p_{1m_1} & p_{21} & p_{22} & \cdots & p_{2m_2} & \cdots & p_{n1} & p_{n2} & \cdots & p_{nm_n} \end{pmatrix}$$

set $p_i = p_{i1} + p_{i2} + \cdots + p_{im_i}$ *so that* $p_1 + p_2 + \cdots + p_n = 1$.

Then

$$H(p_{11},p_{12},\ldots,p_{1m_1},\ldots,p_{n1},p_{n2},\ldots,p_{nm_n})$$

$$= H(p_1,p_2,\ldots,p_n) + \sum_{i=1}^{n} p_i H\left(\frac{p_{i1}}{p_i},\frac{p_{i2}}{p_i},\ldots,\frac{p_{im_i}}{p_i}\right).$$

Proof: From Theorem 11-2c_1

$$H(p_{11},p_{12},\ldots,p_{nm_n}) = H(p_{11},\ldots,p_{n-1\,m_{n-1}},p_n) + p_n H\left(\frac{p_{n1}}{p_n},\ldots,\frac{p_{nm_n}}{p_n}\right)$$

Since H is symmetric in all its arguments we can rotate p_n to the leftmost position, then

$$H(p_{11},\ldots,p_{nm_n}) = H(p_n,p_{11},\ldots,p_{n-2\,m_{n-2}},p_{n-1})$$

$$+ p_{n-1}H\left(\frac{p_{n-1\,1}}{p_{n-1}},\frac{p_{n-1\,2}}{p_{n-1}},\ldots,\frac{p_{n-1\,m_{n-1}}}{p_{n-1}}\right) + p_n H\left(\frac{p_{n1}}{p_n},\ldots,\frac{p_{nm_n}}{p_n}\right)$$

$$= H(p_{n-1},p_n,p_{11},\ldots,p_{n-3\,m_{n-3}},p_{n-2})$$

$$+ \sum_{j=0}^{2} p_{n-j}H\left(\frac{p_{n-j\,1}}{p_{n-j}},\ldots,\frac{p_{n-j\,m_{n-j}}}{p_{n-j}}\right).$$

Continuing this process for all i,

$$H(p_{11},p_{12},\ldots,p_{nm_n}) = H(p_1,p_2,\ldots,p_n) + \sum_{i=1}^{n} p_i H\left(\frac{p_{i1}}{p_i},\frac{p_{i2}}{p_i},\ldots,\frac{p_{im_i}}{p_i}\right)$$

and our theorem is proved.   ∎

Again starting with the scheme

$$\begin{pmatrix} \{1\} & \{2\} & \{3\} & \{4\} & \{5\} & \{6\} \\[2mm] \dfrac{1}{6} & \dfrac{1}{6} & \dfrac{1}{6} & \dfrac{1}{6} & \dfrac{1}{6} & \dfrac{1}{6} \end{pmatrix}$$

using Theorem 11-2c_2, its entropy is

$$H\left(\frac{1}{6},\frac{1}{6},\frac{1}{6},\frac{1}{6},\frac{1}{6},\frac{1}{6}\right) = H\left(\frac{1}{3},\frac{1}{3},\frac{1}{3}\right) + \left[\frac{1}{3}H\left(\frac{1}{2},\frac{1}{2}\right) + \frac{1}{3}H\left(\frac{1}{2},\frac{1}{2}\right) + \frac{1}{3}H\left(\frac{1}{2},\frac{1}{2}\right)\right]$$

$$= H\left(\frac{1}{3},\frac{1}{3},\frac{1}{3}\right) + 3\left[\frac{1}{3}H\left(\frac{1}{2},\frac{1}{2}\right)\right] = H\left(\frac{1}{3},\frac{1}{3},\frac{1}{3}\right) + H\left(\frac{1}{2},\frac{1}{2}\right).$$

We now have what we need to choose a numerical function for I. Given an equiprobable uncertainty scheme

$$\begin{pmatrix} E_1 & E_2 & \ldots & E_n \\[2mm] \dfrac{1}{n} & \dfrac{1}{n} & \ldots & \dfrac{1}{n} \end{pmatrix}$$

set

11-2c_3 $$H\left(\frac{1}{n}, \frac{1}{n}, \dots, \frac{1}{n}\right) = F(n).$$

If in turn each E_i is expressed as the sum of m equiprobable events, that is, if $E_i = E_{i1} + E_{i2} + \dots + E_{im}$ with $P(E_{ij}) = \frac{1}{nm}$ for all j, then from Theorem 11-2c_2

11-2c_4 $$H\left(\frac{1}{nm}, \frac{1}{nm}, \dots, \frac{1}{nm}\right) = H\left(\frac{1}{n}, \frac{1}{n}, \dots, \frac{1}{n}\right) + \frac{1}{n}\sum_{i=1}^{n} H\left(\frac{\frac{1}{nm}}{\frac{1}{n}}, \frac{\frac{1}{nm}}{\frac{1}{n}}, \dots, \frac{\frac{1}{nm}}{\frac{1}{n}}\right)$$

$$= H\left(\frac{1}{n}, \frac{1}{n}, \dots, \frac{1}{n}\right) + \frac{1}{n}\left(nH\left(\frac{1}{m}, \frac{1}{m}, \dots, \frac{1}{m}\right)\right)$$

$$= H\left(\frac{1}{n}, \frac{1}{n}, \dots, \frac{1}{n}\right) + H\left(\frac{1}{m}, \frac{1}{m}, \dots, \frac{1}{m}\right).$$

From Eq. 11-2c_3, this becomes

11-2c_5 $$F(mm) = F(n) + F(m).$$

One very common function which satisfies this functional equation is the logarithm

$$\log mn = \log m + \log n.$$

Recalling that

$$F(n) = H\left(\frac{1}{n}, \frac{1}{n}, \dots, \frac{1}{n}\right) = \frac{1}{n}\sum_{j=1}^{n} I\left(\frac{1}{n}\right) = \frac{1}{n}\left(n I\left(\frac{1}{n}\right)\right) = I\left(\frac{1}{n}\right).$$

setting $F(n) = \log n$ yields

$$I\left(\frac{1}{n}\right) = \log n.$$

Thus the choice of $\log n$ for $F(n)$ gives a useful measure of information for an equiprobable uncertainty scheme. Does it provide any lead in finding an appropriate quantifier for information if the messages do not all occur with the same probability?

Since $\log n = -\log \frac{1}{n}$, the information conveyed by learning which of the equiprobable messages E_i is sent equals the negative of the probability of the message's being sent—that is,

$$I(E_i) = I\left(\frac{1}{n}\right) = \log n = -\log \frac{1}{n} = -\log p(E_i).$$

This suggests that a workable definition of information might be

11-2c$_6$ $$I(p) = -\log p$$

and with this we would have

11-2c$_7$ $$H(p_1, p_2, \ldots, p_n) = -\sum_{i=1}^{n} p_i \log p_i.$$

(We remark that since $p < 1$, $\log p < 0$ and $I(p) = -\log p > 0$). In particular

11-2c$_8$ $$H\left(\frac{1}{n}, \frac{1}{n}, \ldots, \frac{1}{n}\right) = -\sum_{i=1}^{n} \frac{1}{n} \log \frac{1}{n} = -n\left(\frac{1}{n} \log \frac{1}{n}\right) = \log n.$$

Now, does this choice of a function for I satisfy our requirements R_1, R_2, and R_3? Let us first look at I. If $p < q$,

$$I(p) = -\log p > -\log q = I(q).$$

Also, if $p < q$,

$$I(p) - I(q) = -\log p + \log q = \log \frac{q}{p} = -\log \frac{p}{q} = I\left(\frac{p}{q}\right).$$

Furthermore, leaning on methods of analysis that we do not assume here, the requirement that for $p = 0$, $pI(p) = 0$ is a logical consequence of the choice $I(p) = -\log p$.* Thus the choice $I(p) = -\log p$ satisfies all requirements placed on I.

We turn now to the requirements on H. Since we have $pI(p) = 0$ for $p = 0$, and since $\log 1 = 0$, $H(1.0) = 1 \cdot \log 1 = 0$ and so R_2(11-2b$_2$) is satisfied. Also, since the choice $I(p) = -\log p$ yields $I\left(\frac{p}{q}\right) = I(p) - I(q)$,

$$H(p_1, p_2, \ldots, p_n) = -\sum_{i=1}^{n} p_i I(p_i) = -\sum_{i=1}^{n-2} p_i \log (p_i) - (1-p) \log (1-p)$$

$$= H(p_1, p_2, \ldots, p_{n-2}, 1-p) + (1-p)H\left(\frac{p_{n-1}}{1-p}, \frac{p_n}{1-p}\right)$$

*For from elementary calculus we have $\lim_{x \to 0} x \log x = 0$.

where $\sum_{i=1}^{n-2} p_i = p$, and the condition R_3 (11-2b_9) is fulfilled. To show that R_1 is satisfied, that is, to show that $H(p,1-p) < H(q,1-q)$ whenever $p < q \leqslant \frac{1}{2}$

requires some elementary calculus[8] and so must be assumed here without proof. With this assumption the 'reasonable" requirements R_1, R_2, and R_3 which we imposed on I and H are satisfied.

Two other important properties of the entropy H must also be accepted without formal demonstration. The first is that

11-2c_9 $$H(p_1,p_2,\ldots,p_n) \leqslant H\left(\frac{1}{n},\ \frac{1}{n},\ldots,\frac{1}{n}\right).^{[9]}$$

This says the scheme with the greatest uncertainty yields the greatest amount of information. Secondly,

11-2c_{10}   *log p is the only function which simultaneously satisfies R_1, R_2, and R_3.*

Hence, as a measure of information, it is unique. Thus for any event A with probability p and any uncertainty scheme

$$\begin{pmatrix} A_1 & A_2 & \ldots & A_n \\ p_1 & p_2 & \ldots & p_n \end{pmatrix}$$

we have

11-2c_{11}   **Definition**

$$I(A) = I(p) = -\log p;$$

$$H(p_1,p_2,\ldots,p_n) = -\sum_{i=1}^{n} p_i \log p_i.$$

In our discussion so far nothing has been said about the base of the logarithm used to quantify information. Should we choose the common logarithm with base 10 or the natural logarithm with base e or make some other choice. Actually in the development of the Theory the particular logarithm is not important and no particular reference is made to it. In applications the logarithm to the base 2 is most frequently chosen and we speak of "bits" of information. Thus the scheme

$$\begin{pmatrix} A & B & C & D \\ \dfrac{1}{4} & \dfrac{1}{2} & \dfrac{1}{8} & \dfrac{1}{8} \end{pmatrix}$$

yields

$$H\left(\frac{1}{4}, \frac{1}{2}, \frac{1}{8}, \frac{1}{8}\right) = -\frac{1}{4}\log_2\frac{1}{4} - \frac{1}{2}\log_2\frac{1}{8} - \frac{2}{8}\log_2\frac{1}{8} = \frac{1}{2} + \frac{1}{2} + \frac{6}{8} = \frac{7}{4} \text{ bits}$$

of information.

EXERCISES 11-2c

1. Use the table of $-p\log_2 p$ on page 326 to verify the two statements in this section relating to the schemes:

$$\begin{pmatrix} \{1\} & \{2\} & \{3\} & \{4\} & \{5\} & \{6\} \\ \dfrac{1}{6} & \dfrac{1}{6} & \dfrac{1}{6} & \dfrac{1}{6} & \dfrac{1}{6} & \dfrac{1}{6} \end{pmatrix}$$

The statements are:

(1) $H\left(\dfrac{1}{6},\dfrac{1}{6},\dfrac{1}{6},\dfrac{1}{6},\dfrac{1}{6},\dfrac{1}{6}\right) = H\left(\dfrac{1}{6},\dfrac{1}{6},\dfrac{2}{3}\right) + \dfrac{2}{3}H\left(\dfrac{1}{4},\dfrac{1}{4},\dfrac{1}{4},\dfrac{1}{4}\right)$; and

(2) $H\left(\dfrac{1}{6},\dfrac{1}{6},\dfrac{1}{6},\dfrac{1}{6},\dfrac{1}{6},\dfrac{1}{6}\right) = H\left(\dfrac{1}{3},\dfrac{1}{3},\dfrac{1}{3}\right) + H\left(\dfrac{1}{2},\dfrac{1}{2}\right)$

2. Rank from smallest to largest:

(a) $H\left(\dfrac{1}{n},\dfrac{1}{n},\dots,\dfrac{1}{n}\right)$, (b) $H\left(\dfrac{1}{nm},\dfrac{1}{nm},\dots,\dfrac{1}{nm}\right)$, (c) $H\left(\dfrac{1}{nm},\dfrac{1}{nm},\dots,\dfrac{1}{nm},\dfrac{1}{n}\right)$,

(d) $H\left(\dfrac{1}{n},\dfrac{1}{n},\dots,\dfrac{1}{n},\ \dfrac{1}{nm},\dfrac{1}{nm},\dots,\dfrac{1}{nm}\right)$. (Recall that the symbol $H(p_1,p_2,\dots,p_n)$

has meaning only if $\displaystyle\sum_{i=1}^{n} p_i = 1$; therefore in the third symbol there are $m(n-1)$ terms $\dfrac{1}{nm}$.)

3. If $N > n$ but $N \neq nm$ for $m \in I^+$, which is larger, $H\left(\dfrac{1}{n},\dfrac{1}{n},\dots,\dfrac{1}{n}\right)$ or $H\left(\dfrac{1}{N},\dfrac{1}{N},\dots,\dfrac{1}{N}\right)$? Why?

4. A bridge hand may contain 0, 1, 2, 3, or 4 aces. What is the expected amount of information in learning how many aces a given hand contains?

5. An eight-character binary number contains zero to eight ones (the other characters are zeros). What is the expected amount of information in learning how many ones such a number contains? In learning exactly which number is sent?

11-3. THE CHANNEL

a. Conditional Entropy

Given two schemes

$$A = \begin{pmatrix} A_1 & A_2 & \dots & A_n \\ p_1 & p_2 & \dots & p_n \end{pmatrix} \quad \text{and} \quad B = \begin{pmatrix} B_1 & B_2 & \dots & B_m \\ q_1 & q_2 & \dots & q_m \end{pmatrix}$$

we can construct a third scheme AB whose events are the ordered pairs (A_i,B_j), $A_i \in A$, $B_j \in B$, occurring with probabilities $p(A_i,B_j) = p_{A_i}(B_j)p(A_i)$. For simplicity let us write,

$$H(p_1,p_2,\dots,p_n) = H(A)$$

$$H(q_1,q_2,\dots,q_m) = H(B)$$

$$H(p(A_1,B_1),p(A_1,B_2),\dots,p(A_n,B_m)) = H(A,B).$$

Then from 11-2c$_{11}$

$$H(A,B) = -\sum_{i=1}^{n}\sum_{j=1}^{m} p(A_i,B_j) \log p(A_i,B_j)$$

$$= -\sum_{i=1}^{n}\sum_{j=1}^{m} p_{A_i}(B_j)p(A_i)\log p_{A_i}(B_j)p(A_i)$$

$$= -\sum_{i=1}^{n}\sum_{j=1}^{m} p_{A_i}(B_j)p(A_i)\left[\log p_{A_i}(B_j) + \log p(A_i)\right]$$

$$= -\sum_{i=1}^{n}p(A_i)\sum_{j=1}^{m} p_{A_i}(B_j)\log p_{A_i}(B_j) - \sum_{i=1}^{n} p(A_i)\log p(A_i)\sum_{j=1}^{m} p_{A_i}(B_j).$$

Since for fixed A_i, the events (A_i,B_j) form a partition of A_i,

$$\sum_{j=1}^{m}p_{A_i}(B_j) = \sum_{j=1}^{m}\frac{p(A_i,B_j)}{p(A_i)} = \frac{p(A_i)}{p(A_i)} = 1$$

(see 7-1b_1) and

$$H(A,B) = -\sum_{i=1}^{n} p(A_i)\sum_{j=1}^{m} p_{A_i}(B_j)\log p_{A_i}(B_j) + H(A).$$

The quantity

$$- \sum_{j=1}^{m} p_{A_i}(B_j) \log p_{A_i}(B_j)$$

has the form of an entropy and is taken as the uncertainty in B if we know A_i has occurred. Thus we have the *conditional entropy*

11-3a₁ $$H_{A_i}(B) = - \sum_{j=1}^{m} p_{A_i}(B_j) \log p_{A_i}(B_j).$$

We see immediately that $H_{A_i}(B)$ is a random variable on the uncertainty scheme A and we set its expectation

11-3a₂ $$\sum_{i=1}^{n} p(A_i)\, H_{A_i}(B) = H_A(B).$$

$H_A(B)$ is the conditional entropy of B averaged over A or the average amount of uncertainty in B when we know which event in A occurred. Putting this last result in our expression for $H(A,B)$ we have

11-3a₃ $$H(A,B) = H_A(B) + H(A)$$

and since the expression is symmetric in A and B we can similarly have

$$H(A,B) = H_B(A) + H(B).$$

EXERCISES 11-3a

1. Prove $H_A(B) \leqslant H(B)$; that is on the average the amount of information in a scheme never increases if it is known an event occurred in some other scheme.

2. If A and B are mutually independent, prove $H(A,B) = H(A) + H(B)$.

3. Let

$$A = \begin{pmatrix} \{1,2\} & \{3\} & \{4,5,6\} \\ \dfrac{1}{3} & \dfrac{1}{6} & \dfrac{1}{2} \end{pmatrix} \quad \text{and} \quad B = \begin{pmatrix} \{1\} & \{2\} & \{3\} & \{4\} & \{5\} & \{6\} \\ \dfrac{1}{6} & \dfrac{1}{6} & \dfrac{1}{6} & \dfrac{1}{6} & \dfrac{1}{6} & \dfrac{1}{6} \end{pmatrix}$$

If $\{4,5,6\}$ occurs what is the conditional entropy of B?

4. Let A be a sample space consisting of 10 equiprobable elementary events, $A_0, A_1, \ldots, A_9$, and let B be another sample space with ten elementary events (not necessarily equiprobable). The accompanying table gives the probabilities $P_{A_i}(B_j)$. Find $H(A)$, $H_A(B)$ and $H(A,B)$.

B \ A	0	1	2	3	4	5	6	7	8	9
0	.09	.01	.005	0	0	0	.0	0	0	0
1	.005	.08	.005	.005	0	0	0	0	0	0
2	.005	.005	.08	.005	.005	0	0	0	0	0
3	0	.005	.005	.08	.005	.005	0	0	0	0
4	0	0	.005	.005	.08	.005	.005	0	0	0
5	0	0	0	.005	.005	.08	.005	.005	0	0
6	0	0	0	0	.005	.005	.08	.005	.005	0
7	0	0	0	0	0	.005	.005	.08	.005	.005
8	0	0	0	0	0	0	.005	.005	.08	.005
9	0	0	0	0	0	0	0	.005	.01	.09

b.   Channels Without Memory

We now turn our attention to the channel and some elementary problems associated with it.   In an information system we have a source, a transmitter which takes the information given by the source and emits signals which are carried by the channel to a receiver which converts them to a form intelligible at the intended destination.   In a computing system the source may be a length of punched paper tape, the transmitter the electronic reader that converts the information on the tape to electrical impulses carried to the storage unit by the computer's circuitry.   In this system the channel is the circuitry; the transmitted signals are received at the storage unit where they are converted to a form which the computer can store. A phonograph record is a channel by which musical sounds converted to cuts in a plastic disk are carried to an instrument which reconverts the transmitted signals to musical sounds for the listener.

In order to characterize a channel we need an input alphabet which the channel will accept.   In the computer it is a set of electrical impulses, in telegraphy it is the dots and dashes of the Morse code.   We need also the output alphabet, the set of signals after conversion by the receiver—the output of a phonograph record is sound, the output alphabet in telegraphy is the English alphabet.   Since channels are subject to noise which to a greater or less degree distort the emitted signal, we need to know the probabilities that a given letter of the input alphabet after transmission will emerge as various letters of the output alphabet.   For example, suppose an electronic printer reads a digit from a punched card and prints it on a tally sheet.   Here the input and output alphabets are both the set of digits 0, 1, 2,...,9.   If the printer reads the digit 3 from a card we want to know $p_3(0)$, $p_3(1)$,...,$p_3(9)$ where the notation $p_3(i)$ means the probability an i, $i = 0,1,...,9$, was printed if a 3 were read.   We shall assume that both alphabets are finite, but not necessarily of the same length; in some cases the input and output alphabets may be identical.

Denote the input alphabet by

$$A = (a_1,a_2,...,a_n)$$

and the output alphabet by

$$B = (b_1,b_2,...,b_m)$$

If each input signal $a_i \in A$ produces one and only one output signal $b_j \in B$, the channel is said to be *noiseless*.   Generally there is noise, so that if $a_i \in A$ is sent different $b_j \in B$ will be received at different times.   Each time $a_i \in A$ is sent, only one $b_j \in B$ will be received, but because of the noise it will not always be the same b_j.   This is essentially a random phenomenon, so we speak in terms of the probability of receiving $b_j \in B$ when $a_i \in A$ is transmitted.

We consider first the case of channels *without memory;* that is channels in which the received signals depend only on the signal which is sent.   Thus if...$a_i a_j a_k$ is a sequence of transmitted signals, the signal received after a_k has traveled throught the channel is due to a_k alone.   In many systems the output in response to a_k is affected by some finite set of signals immediately preceding a_k. To complete the characterization of a memoryless channel we define on the space $A \times B$ the function $p_{B|A}$ so that for each pair $(a_i,b_j) \in A \times B$,

$$p_{B|A}(a_i,b_j) = p_{a_i}(b_j)$$

is the probability that b_j is received if a_i is sent.   Since some $b_j \in B$ must be received when $a_i \in A$ is sent,

11-3b_1 $$\sum_{j=1}^{m} p_{a_i}(b_j) = 1$$

for all $i = 1,2,...,n$.   Given the input alphabet A, the output alphabet B and the set function $p_{B|A}$ on $A \times B$ a channel is described by the triplet $[A,p_{B|A},B]$.

As an example of a channel consider a self-service elevator which serves a building with a basement and four floors.   When a rider pushes a button inside the elevator, the signal causes the elevator to move and it will stop at one of the floors.   Here we have a channel with input alphabet corresponding to the buttons in the elevator, say

$$b = (B, 1,2,3,4)$$

and output alphabet corresponding to the floors of the building, say

$$f = (\beta,I,II,III,IV)$$

The set function $p_{f|b}$ is displayed in Figure 11-3b_2 and the triplet $[b,p_{f|b},f]$ describes the channel.

11-3b_2 Figure

The range values of the set function $p_{f|b}$ on $b \times f$

b f	B	1	2	3	4
β	.980	.010	0	0	0
I	.020	.980	.010	0	0
II	0	.010	.980	.010	0
III	0	0	.010	.980	.020
IV	0	0	0	.010	.980

[Each entry of the table is the probability the elevator will go to certain floor when a given button is pushed; $p_{(b=3)}$ ($f = $ III) $= p_3($III$) = .980$, $p_1($II$) = .010$]

We observe that the column sums are one, thus illustrating the property given in 11-3b_1,—that is, when a signal is sent, there will be a response. We say such channels are *lossless* and we limit our considerations to such channels.

In addition to characterizing the channel by the triplet $[A,p_{B|A},B]$ we also need to know the frequency with which the letters in the input alphabet are transmitted. Hence we can look upon the source as the uncertainty scheme

$$[A,p_A] = \begin{pmatrix} a_1 & a_2 & \cdots & a_n \\ p_A(a_1) & p_A(a_2) & & p_A(a_n) \end{pmatrix}$$

where A is the input alphabet and p_A is its distribution function. We say that the source $[A,p_A]$ *drives* the channel $[A,p_{B|A},B]$. To continue the example concerning the elevator channel, $[b,p_{f|b},f]$, suppose it is driven by the uncertainty scheme

11-3b_3

$$[b,p_b] = \begin{pmatrix} B & 1 & 2 & 3 & 4 \\ \dfrac{1}{6} & \dfrac{1}{3} & \dfrac{1}{4} & \dfrac{1}{8} & \dfrac{1}{8} \end{pmatrix}$$

Given a channel and a source which drives it, what we would like to know is the set function $p_{A|B}$ on $A \times B$, where $p_{A|B}(a_i,b_j)$ would tell us the probability that a_i was sent if b_j was received. We would like to find on $A \times B$ the function

$$p_{A|B} = \{((a_i,b_j),p_{b_j}(a_i)) \mid \sum_{a_i \in A} p_{b_j}(a_i) = 1\}$$ where the condition imposed by the

summation implies that the channel yields output only when it is activated by the source. (If b_j is received, one of the a_i's was sent.) If there is an $a_v \in A$ such that $p_{A|B}(a_v,b_j)$ is "large" while all $p_{A|B}(a_i,b_j)$, $i \neq v$, are "small" then we can be fairly sure that if b_j is received, a_v was sent. To find the set function $p_{A|B}$ we observe that since we know both p_A and $p_{B|A}$ we can find a probability on $A \times B$. The set function p on $A \times B$ defined by $p(a_i,b_j) = p_A(a_i) p_{B|A}(a_i,b_j) = p_A(a_i)p_{a_i}(b_j)$ is a probability, since

$$\sum_{a_i \in A} \sum_{b_j \in B} p(a_i,b_j) = \sum_{a_i \in A} \sum_{b_j \in B} p_A(a_i) p_{a_i}(b_j)$$

$$= \sum_{a_i \in A} p_A(a_i) \sum_{b_j \in B} p_{a_i}(b_j)$$

$$= \sum_{a_i \in A} p_A(a_i)$$

$$= 1.$$

The function p on $A \times B$ is a joint distribution of A and B since it gives the probability of the simultaneous occurrence of $a_i \in A$ and $b_j \in B$ for every pair $(a_i,b_j) \in A \times B$. If we fix j, the sum over i of the $p(a_i,b_j)$ gives a function of b_j, that is

11-3b_4
$$\sum_{a_i \in A} p(a_i,b_j) = q_B(b_j),$$

and since further

$$\sum_{b_j \in B} q_B(b_j) = \sum_{b_j \in B} \sum_{a_i \in A} p(a_i,b_j) = 1$$

the function q_B is a probability distribution on the output alphabet B. The received signals generate the uncertainty scheme

$$\begin{pmatrix} b_1 & b_2 & \cdots & b_m \\ q_B(b_1) & q_B(b_2) & & q_B(b_m) \end{pmatrix}$$

Having found the distribution p and from it the distribution q_B, we can now compute the range values of the function $p_{A|B}$ on $A \times B$. From Bayes' rule (7-1b_3) we have

$$p_{A|B}(a_i,b_j) = p_{b_j}(a_i) = \frac{p_{a_i}(b_j) p_A(a_i)}{\sum_{a_k \in A} p_{a_k}(b_j) p_A(a_k)} = \frac{p(a_i,b_j)}{q_B(b_j)}.$$

We see that for fixed j,

$$\sum_{a_i \in A} p_{b_j}(a_i) = \frac{\sum_{a_i \in A} p(a_i b_j)}{q_B(b_j)} = \frac{q_B(b_j)}{q_B(b_j)} = 1 \qquad \text{(see 11-3b_4)}$$

so that if b_j is received one of the a_i's was sent. We now have the function $p_{A|B}$ which we sought:

11-3b₅
$$p_{A|B} = \left\{ \left((a_i,b_j), \frac{p(a_i,b_j)}{q_B(b_j)} \right) \Big| (a_i,b_j) \in A \times B \right\}.$$

In Figure 11-3b_2 we displayed the range values of the distribution function $p_{f|b}$ on $b \times f$, and in 11-3b_3 we have the source $[b,p_b]$. Let us compute the distribution $p_{b|f}$. Using the formulas

$$p(b_v,f_\mu) = p_b(b_v)\, p_{f|b}(b_v,f_\mu) \quad (b_v \in b,\, f_\mu \in f)$$

we construct Figure 11-3b_6 which is the joint distribution p on $b \times f$. Here the row sums give the distribution q_f. With p_b and q_f known, Bayes' theorem enables us to compute the function $p_{b|f}$.

11-3b_6 Figure

The joint distribution of b and f

f \ b B	1	2	3	4	q_f
β	.163	.003	0	0	0	.166
I	.003	.327	.002	0	0	.332
II	0	.003	.245	.001	0	.249
III	0	0	.003	.122	.002	.127
IV	0	0	0	.002	.123	.125
p_b	.166	.333	.250	.125	.125	1

$$p_{b|f}(b_v,f_\mu) = p_{f_\mu}(b_v) = \frac{p_{b_v}(f_\mu)p_b(b_v)}{\displaystyle\sum_{b_\lambda \in b} p_{b_\lambda}(f_\mu)p_b(b_\lambda)} = \frac{p(b_v,f_\mu)}{q_f(f_\mu)}$$

$$p_{b|f}(2,\text{II}) = \frac{p_2(\text{II})p_b(2)}{p_B(\text{II})p_b(B) + p_1(\text{II})p_b(1) + p_2(\text{II})p_b(2) + p_3(\text{II})p_b(3) + p_4(\text{II})p_b(4)}$$

$$= \frac{p(2,\text{II})}{q_f(\text{II})} = \frac{.245}{.249} = .984.$$

The other entries in Figure 11-3b_7 are similarly computed.

11-3b₇   Figure

The range values of the set function $p_{b|f}$ on $b \times f$

f＼^{b}	B	1	2	3	4
β	.982	.018	0	0	0
I	.009	.982	.006	0	0
II	0	.012	.984	.004	0
III	0	0	.023	.961	.015
IV	0	0	0	.016	.984

EXERCISES 11-3b

Let the source 11-3b₃ drive the channel $|b, p_{f|b}, f|$ where the range values of $p_{f|b}$ are given by

f＼^{b}	B	1	2	3	4
β	.500	.250	.050	0	0
I	.300	.500	.200	.050	0
II	.200	.200	.500	.200	.200
III	0	.050	.200	.500	.300
IV	0	0	.050	.250	.500

   1.   Find the joint distribution of b and f and the distribution of f.
   2.   Compute the distribution function $p_{b|f}$. (Compared to the elevator channel in the text we can consider this distribution a model of the operation of the apparatus when it is not in good working order—when there is much noise in the system.)
   3.   A telegraph line accepts at one end and emits at the other the vowels of the English language. The input and output alphabets are both {a,e,i,o,u}. Let the source be

$$|A, p_A| = \begin{pmatrix} a & e & i & o & u \\ \dfrac{3}{16} & \dfrac{5}{16} & \dfrac{3}{16} & \dfrac{3}{16} & \dfrac{2}{16} \end{pmatrix}$$

and the range of $p_{B|A}$ be given in

B \ A	a	e	i	o	u
a	$\dfrac{3}{4}$	0	$\dfrac{1}{16}$	$\dfrac{1}{24}$	0
e	$\dfrac{1}{8}$	$\dfrac{7}{8}$	$\dfrac{3}{16}$	0	$\dfrac{1}{16}$
i	0	$\dfrac{1}{12}$	$\dfrac{3}{4}$	0	$\dfrac{1}{16}$
o	$\dfrac{1}{8}$	0	0	$\dfrac{5}{6}$	$\dfrac{1}{8}$
u	0	$\dfrac{1}{24}$	0	$\dfrac{1}{8}$	$\dfrac{3}{4}$

Find the joint distribution of A and B, the distribution of B and the function $p_{A|B}$.

c.   Entropies Associated With Channels

Every source $\lfloor A, p_A \rfloor$ has an entropy

$$H(A) = - \sum_{a_i \in A} p_A(a_i) \log p_A(a_i)$$

which is the average amount of information contained in a signal of A before it has passed through the channel. Since $H_B(A) \leqslant H(A)$, (see Prob. 1, Exercises 11-3a), we can form the difference

11-3c₁ $R(A,B) = H(A) - H_B(A)$

which is called the *rate of transmission* of the channel and is the average amount of information in a signal carried through the channel.

We see immediately that $0 \leqslant R(A,B) \leqslant H(A)$. The two extreme cases $R(A,B) = H(A)$ and $R(A,B) = 0$ are worth examining. If $R(A,B) = H(A)$, $H_B(A) = 0$ and there is no loss of information in the channel; we have

$$0 = H_B(A) = - \sum_{a_i \in A} \sum_{b_j \in B} q_B(b_j) p_{b_j}(a_i) \log p_{b_j}(a_i)$$

$$= - \sum_{a_i \in A} \sum_{b_j \in B} p(a_i, b_j) \log p_{b_j}(a_i).$$

For all i and j, $p(a_i b_j) \log p_{b_j}(a_i) \leqslant 0$; and since

$$\sum_{a_i \in A} \sum_{b_j \in B} p(a_i, b_j) \log p_{b_j}(a_i) = 0$$.

every $p(a_i,b_j) \log p_{b_j}(a_i) = 0$. Hence for any pair (a_i,b_j) such that $p(a_i,b_j) \neq 0$, $\log p_{b_j}(a_i) = 0$ or $p_{b_j}(a_i) = 1$. Also given a_i, we know there must be at least one pair (a_i,b_j) with $p(a_i,b_j) \neq 0$ since we assume that every signal emitted by the source will reappear at the end of the channel as some member of the output alphabet. Furthermore,

$$1 = p_{b_j}(a_i) = \frac{p(a_i,b_j)}{q_B(b_j)} = \frac{p(a_i,b_j)}{\sum_{a_k \in A} p(a_k,b_j)}$$

implies $p(a_k,b_j) \neq 0$ only for $k = i$. This says that if $H_B(A) = 0$, the signal b_j which emerges from the channel reveals exactly which signal the source emitted. Thus if $R(A,B) = H(A)$, the output alphabet B can be partitioned (see 7-1b) into $\tau = n(A)$ parts $B^1, B^2, \dots, B^\tau$ such that if $a_i \in A$ is transmitted, $p_{b_j}(a_i) \neq 0$ if $b_j \in B^i$ and $p_{b_j}(a_i) = 0$ if $b_j \in B^k$ where $k \neq i$. (This does not mean the channel is noiseless. The channel $[X, p_{Y|X}, Y]$ with source

$$X = \begin{pmatrix} x_1 & x_2 \\ \dfrac{1}{2} & \dfrac{1}{2} \end{pmatrix}$$

output alphabet $Y = (y_1, y_2, y_3)$, and $p_{Y|X}(x_1, y_1) = 1$, $p_{Y|X}(x_2, y_2) = \dfrac{3}{4}$, $p_{Y|X}(x_2, y_3) = \dfrac{1}{4}$ is noisy, but $H_Y(X) = 0$.)

If $R(A,B) = 0$, $H(A) = H_B(A)$ and this will occur if A and B are independent (see Prob. 2 of the Exercises 11-3a). That $H(A) = H_B(A)$ only if A and B are independent is also true, although we are unable to prove it here. Thus if $R(A,B) = 0$, $p_{b_j}(a_i) = p_A(a_i)$ for all $a_i \in A$ and $b_j \in B$, and so no matter what b_j is received the probability a_i was sent remains unchanged, that is no information is transmitted.

Since $H(A,B) = H(B) + H_B(A)$ we can write

11-3c$_2$ $\qquad R(A,B) = H(A) + H(B) - H(A,B)$

a representation of R which is sometimes useful in computation.

Consider the channel $[A, p_{B|A}, B]$ where $p_{B|A}$ is given in the table:

	a_1	a_2
b_1	$\dfrac{2}{3}$	$\dfrac{1}{3}$
b_2	$\dfrac{1}{3}$	$\dfrac{2}{3}$

Suppose $p_1 = \{(a_1, \frac{1}{3}), (a_2, \frac{2}{3})\}$; then the joint distribution of A at d B is given by;

B \ A	a_1	a_2	q_B
b_1	$\frac{2}{9}$	$\frac{2}{9}$	$\frac{4}{9}$
b_2	$\frac{1}{9}$	$\frac{4}{9}$	$\frac{5}{9}$
p_A	$\frac{3}{9}$	$\frac{6}{9}$	1

Then

$$H(A) = -\left(\frac{1}{3} \log \frac{1}{3} + \frac{2}{3} \log \frac{2}{3}\right) = .9183$$

$$H(B) = -\left(\frac{4}{9} \log \frac{4}{9} + \frac{5}{9} \log \frac{5}{9}\right) = 1.4916$$

$$H(A,B) = -\left(\frac{2}{2} \log \frac{2}{9} + \frac{2}{9} \log \frac{2}{9} + \frac{1}{9} \log \frac{1}{9} + \frac{4}{9} \log \frac{4}{9}\right) = 1.8638$$

and

$$R(A,B) = H(A) + H(B) - H(AB) = .5732.$$

If the channel is unchanged but instead $p_A = \{(a_1, \frac{1}{2}), (a_2, \frac{1}{2})\}$, the joint

distribution of A and B becomes

B \ A	a_1	a_2	q_B
b_1	$\frac{1}{3}$	$\frac{1}{6}$	$\frac{1}{2}$
b_2	$\frac{1}{6}$	$\frac{1}{3}$	$\frac{1}{2}$
p_1	$\frac{1}{2}$	$\frac{1}{2}$	1

In this case

$$H(A) = H(B) = -\left(\frac{1}{2} \log \frac{1}{2} + \frac{1}{2} \log \frac{1}{2}\right) = 1$$

$$H(A,B) = -\left(\frac{2}{3} \log \frac{1}{3} + \frac{2}{3} \log \frac{1}{6}\right) = 1.8874$$

and

$$R(A,B) = H(A) + H(B) - H(A,B) = 2 - 1.8874 = .1126.$$

These simple examples show that $R(A,B)$ varies with the probability distribution of the source. The maximum value of $R(A,B)$ for all possible input distributions is called the *capacity* of the channel. That there is such a maximum follows from a standard, but not elementary, theorem of mathematical analysis. It is interesting to observe that there may be more than one input distribution that enables a channel to work at its capacity.

EXERCISES 11-3c

1. Prove that the channel $|X,p_{Y|X}, Y|$ with $p_{Y|X}$ given in

	x_1	x_2
y_1	1	0
y_2	0	$\dfrac{3}{4}$
y_3	0	$\dfrac{1}{4}$

and input distribution $p_X = \{(x_1, \dfrac{1}{2}), (x_2, \dfrac{1}{2})\}$ has $H_Y(X) = 0$.

2. Prove that for every noiseless channel $R(A,B) = H(A)$.
3. Let the source

$$A = \begin{pmatrix} a_1 & a_2 & a_3 \\ \dfrac{1}{2} & \dfrac{1}{4} & \dfrac{1}{4} \end{pmatrix}$$

drive the channel $|A, p_{B|A}, B|$ with $p_{B|A}$ given in

	a_1	a_2	a_3
b_1	$\dfrac{1}{2}$	$\dfrac{1}{4}$	$\dfrac{3}{4}$
b_2	$\dfrac{1}{2}$	$\dfrac{3}{4}$	$\dfrac{1}{4}$

Find $R(A,B)$.

4. The three sources

$$A_1 = \begin{pmatrix} a_1 & a_2 & a_3 \\ \dfrac{1}{2} & \dfrac{1}{4} & \dfrac{1}{4} \end{pmatrix}, \quad A_2 = \begin{pmatrix} a_1 & a_2 & a_3 \\ \dfrac{1}{3} & \dfrac{1}{3} & \dfrac{1}{3} \end{pmatrix},$$

and

$$A_3 = \begin{pmatrix} a_1 & a_2 & a_3 \\ \dfrac{1}{7} & \dfrac{2}{7} & \dfrac{4}{7} \end{pmatrix}$$

drive the channel $[A,p_{B|A},B]$ with $p_{B|A}$ given in

	a_1	a_2	a_3
b_1	$\dfrac{1}{2}$	$\dfrac{1}{3}$	$\dfrac{1}{4}$
b_2	$\dfrac{1}{2}$	$\dfrac{2}{3}$	$\dfrac{3}{4}$

Find $R(A_1,B)$, $R(A_2,B)$ and $R(A_3,B)$.

d. Some Generalizations

The channel described in the preceding paragraphs is useful for introducing the various concepts associated with the role of the channel in information theory; but it is a very simple channel. We proceed to suggest several generalizations that will make the theory more closely approximate the actual situation. Unfortunately, our mathematical backgrounds do not enable us to work with these generalizations.

Most channels have a memory; that is, the signal that emerges when a_i is transmitted depends not only on a_i but also on $a_{i-1}, a_{i-2}, \ldots, a_{i-\tau}$ where τ varies with the channel. Then the probability distributions that describe the channel with input alphabet A and output alphabet B is a function of the ordered pairs $(a_{i-\tau} \ldots a_{i-1} a_i, b_j)$ in which the first member of the pair is the sequence $a_{i-\tau} \ldots a_{i-1} a_i$. When the channel has a memory of length τ, that is when the output signal is conditioned not only by the input signal but also upon the τ signals which immediately precede it, we designate the distribution by $p_{B|A(\tau)}$ and we describe the channel by the triplet $[A,p_{B|A(\tau)},B]$. We have

$$p_{B|A(\tau)}(a_{i-\tau} \ldots a_{i-1}, a_i, b_j) = p(b_j | a_{i-\tau} \ldots a_{i-1} a_i).$$

As a very simple example let A be the Morse code and B the English alphabet and let the transmitting channel have a memory of one. If we designate the alphabet A by lowercase letters, $a,b,\ldots,z$ (instead of dots and dashes) and that of B by uppercase letters, $A,B,\ldots,Z$, then $p_{qu}(U)$ and $p_{xu}(U)$ are not necessarily the same. In a memoryless channel $p_{qu}(U) = p_{xu}(U) = p_u(U)$. If we are working with $[A,p_{B|A(2)},B]$ then $p_{que}(E)$ and $p_{ine}(E)$ may be different. If A has n letters and B m letters the distribution function $p_{B|A(\tau)}$ must be defined for

$m \times n_{\tau+1}$ pairs $(a_{i-\tau}\ldots a_{i-1}a_i,b_j)$. Then $p(a_i,b_j) = \sum p_{a_{i-\tau}\ldots a_{i-1}a_i}(b_j)$ where the summation is over n^τ addends (a_i is fixed, but there are n^τ sequences $a_{i-\tau}\ldots a_{i-1}$ which can precede it). Once $p(a_i,b_j)$ is found, q_B and $R(A,B)$ are readily computed.

If a source $[A,p_1]$ feeds a channel it will produce a string of signals

$$\ldots\ a_{-2},\ a_{-1},\ a_0,\ a_1,\ a_2,\ a_3,\ \ldots$$

which will be carried by the channel and received at its end as

$$\ldots\ b_{-2},\ b_{-1},\ b_0,\ b_1,\ b_2,\ b_3,\ \ldots$$

Let A_v be the finite space of all possible messages of length v which can be emitted by the source (since the alphabet A is finite with n letters, $n(A_v) = n^v = M$), and let $H(A_v)$ be the entropy of the scheme

$$\begin{pmatrix} \alpha_1 & \alpha_2 & \ldots & \alpha_M \\ \mu_1 & \mu_2 & \ldots & \mu_M \end{pmatrix}$$

where $\alpha_i \in A_v$ and $\sum_{i=1}^{M} \alpha_i = A_v$. Each v length sequence of signals from the source $[A,p_A]$ (each $\alpha_i \in A_v$) will produce a v length sequence of received signals. If we know the distribution $p_{B|A(\tau)}$, $0 \leqslant \tau \leqslant v$, we can calculate $H(B_v)$ and $H(A_v,B_v)$.

If we let v increase, we find that $\frac{1}{v} H(A_v)$, $\frac{1}{v} H(B_v)$ and $\frac{1}{v} H(A_v,B_v)$ become

stationary in the sense that given any positive number ϵ, there exists an N such that if $v > N$ and $k \in I^+$

$$\left| \frac{H(A_v)}{v} - \frac{H(A_{v+k})}{v} \right| < \epsilon$$

Once this point has been attained we write these stationary values as $\overline{H(A)}$, $\overline{H(B)}$ and $\overline{H(A,B)}$. We then define the rate of transmission by

$$R(A,B) = \overline{H(A)} + \overline{H(B)} - \overline{H(A,B)} = \overline{H(A)} - \overline{H_B(A)} \ .$$

This is truly a rate in the sense that it is the average amount of information per signal. As we did earlier, we define the capacity of the channel as the maximum value of the rate for all possible input distributions.

In this introduction to Information Theory we have looked at the problems of quantifying information and of the effectiveness of the channel in carrying to the receiver the transmitted message. The important problems of coding and of the elimination of noise have not even been looked at because of the assumed limitations of mathematical background. Our aim in giving this limited presentation of the theory was twofold: first, to acquaint the reader with still another developing field heavily dependent on probability theory and second, to perhaps whet his interest sufficiently that he will continue his explorations elsewhere.

appendix A

Explanatory Notes

This Appendix contains notes on the text which, if included in the text, would have distracted the reader from the main stream of development by causing him to mistake the emphasis or to stumble over more difficult material. They are presented here to offer completeness to the teacher and the mathematically more mature student.

[1] (page 56) Given $0 < r < 1$ and ϵ an arbitrary positive number, no matter how small. There exists an n so large that

$$0 < \frac{a}{1-r} - s_n = a\frac{r^n}{1-r} < \epsilon.$$

Now

$$a\frac{r^n}{1-r} < \epsilon$$

if

$$r^n < \frac{1-r}{a}\epsilon, \quad \text{which is true}$$

if

$$n \log r < \log\left(\frac{1-r}{a}\epsilon\right), \quad \text{which is true}$$

if

$$n > \frac{\log\left(\frac{1-r}{a}\epsilon\right)}{\log r} \quad (\text{since } r < 1, \log r < 0)$$

Hence if $N = \left[\dfrac{\log\left(\dfrac{1-r}{a}\epsilon\right)}{\log r}\right]$ and $n > N$, reversing the order of the above

steps, we obtain

$$a\frac{r^n}{1-r} < \epsilon.$$

We recall that $[x]$ is the greatest integer in x; for example, $[\pi] = 3$, $[-2,16] = -3$, $\left[\dfrac{4}{3}\right] = 1$

[2] (page 65) If Ω is a countable set, A_Ω is noncountable. $\Omega = \{w_1, w_2, w_3, \ldots\}$ can be put into one-to-one correspondence with I^+ by setting $w_i \longleftrightarrow i$. The subsets of Ω can be put into one-to-one correspondence with the subsets of I^+ by coupling the subset $\{w_{a_1}, w_{a_2}, w_{a_3}, \ldots\} \subset \Omega$ with $\{a_1, a_2, a_3, \ldots\} \subset I^+$. That is, we set $\{w_{a_1}, w_{a_2}, w_{a_3}, \ldots\} \longleftrightarrow \{a_1, a_2, a_3, \ldots\}$. Thus the subsets of Ω are countable if, and only if, the subsets of I^+ are countable. We will show there are more than a countable number of nonfinite subsets in I^+. If there were at most a countable number of such subsets, they could in turn be put into one-to-one correspondence with I^+ and hence be listed as $A_1, A_2, A_3, \ldots$ where $A_i = \{a_{i1}, a_{i2}, a_{i3}, \ldots\}$ with $a_{ij} = k \in I^+$ and $k \geqslant j$ (thus if any $a_{ij} = k$, $a_{ij+1} = k' \geqslant k + 1$). Form $B \in I^+$, where

$$B = \{b_1, b_2, b_3, \ldots\}$$

$$b_1 = a_{11} + 1$$

$$b_2 = (a_{11} + a_{22}) + 2$$

$$b_3 = (a_{11} + a_{22} + a_{33}) + 3$$

and in general,

$$b_i = (a_{11} + a_{22} + \cdots + a_{ii}) + i.$$

B is not one of the A_i since $b_i \neq a_{ii}$ for all i; also for $j < i$, $b_j < b_i$, and for $k > i$, $b_k > b_i$, i.e., we maintain the order present in the A_i so that B is not a rearrangement of some A_i. Hence the infinite subsets of I^+ are not countable, so the entire set of subsets is uncountable. Thus we have proved the set of subsets of a countable set is uncountable.

[3] (page 72) If A is finite, so is the number of terms in $\sum_{w \in A} P(\omega_i)$ and ordinary arithmetic operations justify the various steps; if however, A is countable, the absolute convergence of $\sum_{w_i \in A} P(\omega_i)$ is needed to justify the calculations. See, for example: Konrad Knopp, *Introduction to Sequences and Series* (New York: Dover Publications, 1955).

[4] (page 109) This is only one of two common definitions; the other is

$$F_X = \{(x, P(X \leqslant x))\mid x \text{ is a real number}\}.$$

To illustrate the difference let $F_X^{(1)}(x) = P(X < x)$ and $F_X^{(2)}(x) = P(X \leqslant x)$, then $F_X^{(2)}(x) = P(X < x) + P(X = x) = F_X^{(1)}(x) + P(X = x) \geqslant F_X^{(1)}(x)$. Equality will hold whenever $x \not\in R_X$. If B is a random variable distributed according to the Bernoulli law,

$$F_\beta^{(1)}(0) = 0 \qquad \text{but} \quad F_\beta^{(2)}(0) = 1 - p$$

$$F_\beta^{(1)}(1) = 1 - p \quad \text{but} \quad F_\beta^{(2)}(0) = 1$$

If $x \neq 0$ or $x \neq 1$, $F_\beta^{(1)}(x) = F_\beta^{(2)}(x)$.

[5] (page 116) If $p + q = 1$ and $\epsilon > 0$ is arbitrary, there is a $n \in I^+$ such that

$$nq^n + \frac{q^n}{p} < \epsilon.$$

If $n > \dfrac{\log \dfrac{\epsilon p}{2}}{\log q}$,   then

$$(1) \quad \frac{q^n}{p} < \frac{\epsilon}{2}.$$

Since $1 > \sqrt{q} = \dfrac{1}{1 + h}$ for some $h > 0$, $\sqrt{q^n} = \dfrac{1}{(1 + h)^n} < \dfrac{1}{nh}$ and $nq^n < \dfrac{1}{nh^2}$, hence, if $n > \dfrac{2}{\epsilon h^2}$, $nq^n < \dfrac{1}{\dfrac{2}{(\epsilon h^2)}h^2}$, or

$$(2) \quad nq^n < \frac{\epsilon}{2}.$$

From (1) and (2), if

$$n > \max \left(\frac{\log \dfrac{\epsilon p}{2}}{\log q}, \frac{2}{\epsilon h^2} \right),$$

$$nq^n + \frac{q^n}{p} < \epsilon.$$

[6] (page 163) A *matrix* is an array of numbers consisting of m rows and n columns ($m \geq 1, n \geq 1$). In general the entries of a matrix A are designated by a_{ij}, where i indicates the row and j the column in which the number appears. For example:

$$A = \begin{pmatrix} a_{11} & a_{12} & a_{13} & a_{14} \\ a_{21} & a_{22} & a_{23} & a_{24} \\ a_{31} & a_{32} & a_{33} & a_{34} \end{pmatrix}$$

is a 3 × 4 (three by four) matrix—it has three rows and four columns. The curved brackets are universally employed to enclose the array. Matrices appear in many contexts; a vector is often written as a matrix having one row (an $m \times 1$ matrix) or one column (a $1 \times n$ matrix) and a determinant is a number computed from a square matrix (one in which $m = n$). Two matrices are equal if their entries are identical, i.e., $A = B$ if $a_{ij} = b_{ij}$ for all i and j. For a fuller exposition of matrices and some of their applications see J. G. Kemeny, H. Merkel, J. S. Snell, and G. L. Thompson, *Finite Mathematical Structures* (Englewood Cliffs, N. J.: Prentice-Hall, Inc. 1959), Chap. 4, pp. 205ff.

[7] (page 248) To find $V(X)$ where $P_X = \{(n, r^n(1-r)) | n = 0, 1, 2,...\}$, we need

$$\sum_{n=0} n^2 r^n (1-r) = \sum_{n=0} (n(n-1) + n) r^n (1-r)$$

$$= (1-r) r^2 \sum_{n=2} n(n-1) r^{n-2} + \sum_{n=0} n r^n (1-r).$$

From 9-2b. $\sum_{n=0} n r^n (1-r) = E(X) = \dfrac{r}{1-r}$. We can prove by Induction that

$$\frac{2}{(1-r)^3} - \sum_{n=2}^{N} n(n-1) r^{n-2} = \frac{(N^2-1) r^{N-1}}{(1-r)^3} \left\{ \frac{N}{N-1} - 2r + \frac{N}{N+1} r^2 \right\} = R_N,$$

and we can show that by choosing N large enough we can make R_N as small as we please. Hence we define $\sum_{n=2} n(n-1) r^{N-2} = \dfrac{2}{(1-r)^3}$ and proceed from this point. Review 4-2b and note [1] for a fuller exposition of the type of reasoning used.

[8] (page 303) If $f(x) = -x \log x - (1-x) \log (1-x)$, $f'(x) = \log \dfrac{1-x}{x} > 0$ if $x \leq \dfrac{1}{2}$, and hence f is an increasing function on the interval $\left(0, \dfrac{1}{2}\right)$.

[9] (page 303) This follows from the facts (1) that $x \log x$ is a convex function and (2) that for any convex function φ and any set of positive numbers

$$x_1, \ x_2, \ldots, x_n, \frac{1}{n} \sum_{l=1}^{n} \varphi(x_l) \geqslant \varphi \left(\frac{1}{n} \sum_{l=1}^{n} x_l \right). \quad \text{Given} \ \varphi(x) = x \log x \ \text{and the set}$$

of positive numbers $p_1, \ p_2, \ldots, p_n$ with $\sum_{l=1}^{n} p_l = 1$,

$$\frac{1}{n} \sum_{l=1}^{n} \varphi(p_l) = \frac{1}{n} \sum_{l=1}^{n} p_l \log p_l \geqslant \left(\frac{1}{n} \sum_{l=1}^{n} p_l \right) \log \left(\frac{1}{n} \sum_{l=1}^{n} p_l \right) = \varphi \left(\frac{1}{n} \right).$$

Translating this into entropies,

$$- \frac{1}{n} H(p_1, \ p_2, \ldots, p_n) \geqslant \frac{1}{n} \log \frac{1}{n} = - \frac{1}{n} \log n = - \frac{1}{n} H \left(\frac{1}{n} \frac{1}{n}, \ldots, \frac{1}{n} \right).$$

Hence

$$H(p_1, \ p_2, \ldots, p_n) \leqslant H \left(\frac{1}{n}, \frac{1}{n}, \ldots, \frac{1}{n} \right).$$

appendix **B**

Table of –p log₂ p

p	0	1	2	3	4	5	6	7	8	9
0.00	0.00000	.00996	.01793	.02514	.03186	.03821	.04428	.05010	.05572	.06116
.01	.06643	.07156	.07656	.08144	.08621	.09088	.09545	.09993	.10432	.10863
.02	.11287	.11704	.12113	.12517	.12913	.13304	.13689	.14069	.14443	.14812
.03	.15176	.15535	.15890	.16240	.16586	.16927	.17265	.17598	.17927	.18253
.04	.18575	.18893	.19208	.19519	.19827	.20132	.20434	.20732	.21027	.21320
.05	.21609	.21896	.22179	.22460	.22738	.23014	.23287	.23557	.23825	.24090
.06	.24353	.24613	.24871	.25127	.25381	.25632	.25881	.26127	.26372	.26615
.07	.26855	.27093	.27330	.27564	.27796	.28027	.28255	.28482	.28706	.28929
.08	.29150	.29370	.29587	.29803	.30017	.30229	.30439	.30648	.30855	.31061
.09	.31265	.31467	.31668	.31867	.32065	.32261	.32455	.32648	.32840	.33030
.10	.33219	.33406	.33592	.33776	.33959	.34141	.34321	.34500	.34677	.34853
.11	.35028	.35202	.35374	.35545	.35714	.35883	.36050	.36216	.36381	.36544
.12	.36706	.36867	.37027	.37186	.37343	.37499	.37655	.37809	.37962	.38113
.13	.38264	.38413	.38562	.38709	.38855	.39001	.39145	.39288	.39430	.39571
.14	.39711	.39849	.39987	.40124	.40260	.40395	.40529	.40661	.40793	.40924
.15	.41054	.41183	.41311	.41438	.41564	.41689	.41813	.41937	.42059	.42181
.16	.42301	.42421	.42540	.42658	.42775	.42891	.43006	.43120	.43234	.43346
.17	.43458	.43569	.43679	.43788	.43897	.44005	.44111	.44217	.44322	.44427
.18	.44530	.44633	.44735	.44836	.44936	.45036	.45135	.45233	.45330	.45426
.19	.45522	.45617	.45711	.45805	.45897	.45989	.46081	.46171	.46261	.46350
.20	.46438	.46526	.46612	.46699	.46784	.46869	.46953	.47036	.47119	.47201
.21	.47282	.47362	.47442	.47521	.47600	.47678	.47755	.47831	.47907	.47982
.22	.48057	.48131	.48204	.48276	.48348	.48420	.48490	.48560	.48629	.48698
.23	.48766	.48834	.48901	.48967	.49032	.49097	.49162	.49225	.49289	.49351
.24	.49413	.49474	.49535	.49595	.49655	.49714	.49772	.49830	.49887	.49943
.25	.49999	.50055	.50110	.50164	.50218	.50271	.50324	.50376	.50427	.50478
.26	.50528	.50578	.50627	.50676	.50724	.50772	.50819	.50865	.50911	.50957
.27	.51002	.51046	.51090	.51133	.51176	.51218	.51260	.51301	.51342	.51382
.28	.51422	.51461	.51499	.51537	.51575	.51612	.51649	.51685	.51720	.51755
.29	.51790	.51824	.51858	.51891	.51923	.51955	.51987	.52018	.52049	.52079
.30	.52108	.52138	.52166	.52195	.52222	.52250	.52276	.52303	.52329	.52354
.31	.52379	.52403	.52427	.52451	.52474	.52497	.52519	.52541	.52562	.52583
.32	.52603	.52623	.52642	.52661	.52680	.52698	.52716	.52733	.52750	.52766
.33	.52782	.52797	.52812	.52827	.52841	.52855	.52868	.52881	.52893	.52905
.34	.52917	.52928	.52939	.52949	.52959	.52968	.52977	.52986	.52994	.53002
.35	.53010	.53017	.53023	.53029	.53035	.53040	.53045	.53050	.53054	.53058
.36	.53061	.53064	.53066	.53069	.53070	.53072	.53073	.53073	.53073	.53073
.37	.53072	.53071	.53070	.53068	.53066	.53063	.53060	.53057	.53053	.53049
.38	.53045	.53040	.53035	.53029	.53023	.53017	.53010	.53003	.52995	.52987
.39	.52979	.52971	.52962	.52952	.52943	.52932	.52922	.52911	.52900	.52889
.40	.52877	.52864	.52852	.52839	.52825	.52812	.52798	.52783	.52769	.52753
.41	.52738	.52722	.52706	.52689	.52673	.52655	.52638	.52620	.52602	.52583
.42	.52564	.52545	.52525	.52505	.52485	.52464	.52443	.52422	.52400	.52378
.43	.52356	.52333	.52310	.52287	.52263	.52239	.52215	.52190	.52165	.52140
.44	.52114	.52088	.52062	.52035	.52008	.51981	.51953	51925	.51897	.51869

Table of $-p \log_2 p$

p	0	1	2	3	4	5	6	7	8	9
.45	.51840	.51810	.51781	.51751	.51721	.51690	.51659	.51628	.51597	.51565
.46	.51533	.51501	.51468	.51435	.51402	.51368	.51334	.51300	.51265	.51230
.47	.51195	.51160	.51124	.51088	.51051	.51015	.50978	.50940	.50903	.50865
.48	.50826	.50788	.50749	.50710	.50670	.50631	.50591	.50550	.50510	.50469
.49	.50428	.50386	.50344	.50302	.50260	.50217	.50174	.50131	.50087	.50044
.50	.49999	.49955	.49910	.49865	.49820	.49775	.49729	.49683	.49636	.49589
.51	.49542	.49495	.49448	.49400	.49352	.49303	.49255	.49206	.49156	.49107
.52	.49057	.49007	.48957	.48906	.48855	.48804	.48753	.48701	.48649	.48597
.53	.48544	.48491	.48438	.48385	.48331	.48277	.48223	.48169	.48114	.48059
.54	.48004	.47948	.47893	.47836	.47780	.47724	.47667	.47610	.47552	.47495
.55	.47437	.47379	.47320	.47262	.47203	.47143	.47084	.47024	.46964	.46904
.56	.46844	.46783	.46722	.46661	.46599	.46537	.46475	.46413	.46350	.46288
.57	.46225	.46161	.46098	.46034	.45970	.45906	.45841	.45776	.45711	.45646
.58	.45580	.45514	.45448	.45382	.45316	.45249	.45182	.45114	.45047	.44979
.59	.44911	.44843	.44774	.44706	.44637	.44567	.44498	.44428	.44358	.44288
.60	.44217	.44147	.44076	.44005	.43933	.43862	.43790	.43718	.43645	.43573
.61	.43500	.43427	.43353	.43280	.43206	.43132	.43058	.42983	.42909	.42834
.62	.42758	.42683	.42607	.42531	.42455	.42379	.42302	.42226	.42149	.42071
.63	.41994	.41916	.41838	.41760	.41682	.41603	.41524	.41445	.41366	.41286
.64	.41206	.41126	.41046	.40966	.40885	.40804	.40723	.40642	.40560	.40478
.65	.40396	.40314	.40232	.40149	.40066	.39983	.39900	.39816	.39732	.39648
.66	.39564	.39480	39395	.39310	.39225	.39140	.39054	.38968	.38882	.38796
.67	.38710	.38623	.38536	.38449	.38362	.38275	.38187	.38099	.38011	.37923
.68	.37834	.37746	.37657	.37567	.37478	.37388	.37299	.37209	.37118	.37028
.69	.36937	.36847	.36755	.36664	.36573	.36481	.36389	.36297	.36205	.36112
.70	.36020	.35927	.35834	.35740	.35647	.35553	.35459	.35365	.35271	.35176
.71	.35081	.34986	.34891	.34796	.34700	.34604	.34508	.34412	.34316	.34219
.72	.34123	.34026	.33928	.33831	.33733	.33636	.33538	.33440	.33341	.33243
.73	.33144	.33045	.32946	.32846	.32747	.32647	.32547	.32447	.32347	.32246
.74	.32145	.32044	.31943	31842	.31740	.31639	.31537	.31435	.31332	.31230
.75	.31127	.31024	.30921	.30818	.30715	.30611	.30507	.30403	.30299	.30195
.76	.30090	.29985	.29880	.29775	.29670	.29564	.29459	.29353	.29247	.29140
.77	.29034	.28927	.28820	.28713	.28606	.28499	.28391	.28283	.28175	.28067
.78	.27959	.27850	.27742	.27633	.27524	.27414	.27305	.27195	.27086	.26976
.79	.26865	.26755	.26645	.26534	.26423	.26312	.26201	.26089	.25978	.25866
.80	.25754	.25642	.25529	.25417	.25304	.25191	.25078	.24965	.24851	.24738
.81	.24624	.24510	.24396	.24282	.24167	.24052	.23938	.23823	.23707	.23592
.82	.23476	.23361	.23245	.23129	.23012	.22896	.22779	.22663	.22546	.22429
.83	.22311	.22194	.22076	.21958	.21840	.21722	.21604	.21485	.21367	.21248
.84	.21129	.21010	.20890	.20771	.20651	.20531	.20411	.20291	.20170	.20050
.85	.19929	.19808	.19687	.19566	.19444	.19323	.19201	.19079	.18957	.18835
.86	.18712	.18590	.18467	.18344	.18221	.18098	.17974	.17851	.17727	.17603
.87	.17479	.17355	.17230	.17106	.16981	.16856	.16731	.16606	.16480	.16355
.88	.16229	.16103	.15977	.15851	.15724	.15598	.15471	.15344	.15217	.15090
.89	.14962	.14835	.14707	.14579	.14451	.14323	.14195	.14066	.13938	.13809
.90	.13680	.13551	.13421	.13292	.13162	.13032	.12902	.12772	.12642	.12512
.91	.12381	.12250	.12119	.11988	.11857	.11726	.11594	.11463	.11331	.11199
.92	.11067	.10934	.10802	.10669	.10536	.10403	.10270	.10137	.10004	.09870
.93	.09736	.09602	.09468	.09334	.09200	.09065	.08931	.08796	.08661	.08526
.94	.08391	.08255	.08120	.07984	.07848	.07712	.07576	.07439	.07303	.07166
.95	.07030	.06893	.06756	.06618	.06481	.06343	.06206	.06068	.05930	.05792
.96	.05653	.05515	.05376	.05237	.05099	.04960	.04820	.04681	.04541	.04402
.97	.04262	.04122	.03982	.03842	.03701	.03561	.03420	.03279	.03138	.02997
.98	.02856	.02714	.02573	.02431	.02289	.02147	.02005	.01863	.01720	.01578
.99	.01435	.01292	.01149	.01006	.00863	.00719	.00575	.00432	.00288	.00144

appendix C

Answers to Exercises

CHAPTER 2

Exercises 2-1b page 12

3. $\{2,3,4,\ldots,12\}$

5. $3,9,27,81,243 \in \{3^n | n \in I^+\}$ $\frac{1}{4},\frac{1}{8},\frac{1}{32},\frac{1}{48},\frac{1}{56} \in \left\{\frac{1}{4j} \,\middle|\, j \in I^+\right\}$

6. (a) $\{8\}$, (b) $\{6\}$, (c) $\varnothing$, (d) $\{a,b,c,d\}$

7. (b) $\{-6,-2,6\}$, (c) Fords and Chevrolets, (d) $\{a,b,c,\ldots,z\}$

8. $B = A, C \subset A, A = D, C \subset B, B \subset D, C \subset D$

9. $\varnothing,\{a\}, \{b\}, \{c\}, \{a,b\}, \{a,c\}, \{b,c\}, \{a,b,c\}$

10. 2^6

11. (a) $A \subset B$, (b) $A \not\subset B$, (c) $A = B$, (d) $A = B$

Exercises 2-2b page 19

1. $A = \{a | \frac{1}{2} < a \leqslant 1\}$, $B' = \{i | i \text{ an irrational number}\}$,

$C = \left\{c | 0 \leqslant c < \frac{1}{3}, \frac{1}{3} < c \leqslant 1\right\} = U - \left\{\frac{1}{3}\right\}$, $D'\left\{x | \frac{1}{4} \leqslant x \leqslant \frac{3}{4}\right\}$,

$E = \{0,1\}$

2. $A = \{(r,g) | r + g < 9\}$, $A \cap B = \{(3,6)\}$, $A \cup B = \{(r,g) | r = 3 \text{ or } r + g \geqslant 9\}$,

$A - B = \{(r,g) | r \geqslant 4, r + g \geqslant 9\}$, $B' = \{(r,g) | r \neq 3\}$,

$B \cup A = A \cup B, B \cap C = \{(3,g) | g \text{ is odd}\}$

$B \cap D = \{(3,6)\}$, $B \cup D = \{(r,g) | r = 3 \text{ or } g = 6\}$

$C \cup D = \{(r,g) | r + g \text{ even or } g = 6\}$

$C \cap D = \{(2,6),(4,6),(6,6)\}$, $B - A = \{(r,g) | r = 3, r + g < 9\}$

$D - C = \{(r,6) | r \text{ is odd}\}$

3. $n(A) = 2$ $n(B) = 3$, $A' = \{HH,TT\}$ $B' = \{TT\}$ $A - B = \varnothing$
   $B - A = \{HH\}$ $A \cup B = B$ $A \cap B = A$

4. (a) D (b) C (c) $\varnothing$ (d) DC' (e) C (f) D (g) C (h) $\varnothing$

5. (a) yes   (b) no   (c) $Q \cap P = Q$, $V \cup P = \{p, a, r, e, l, i, o, u\}$
   $V - Q = \{i,o,u\}$

Exercises 2-2c page 21

1. $C \cup D = \{-2,1,3\}$ $CD = \{3\}$ $C'D = \{-2\}$ $D - C = \{-2\}$
   $C \times D = \{(1,-2), (1,3), (3,-2), (3,3)\}$

2. $n(A) = 3$, $n(B) = 7$, $A' = \{(H,H,H), (H,H,T), (H,T,H), (T,H,H),$
   $(T,T,T)\}$,
   $B' = \{(T,T,T)\}$ $A - B = \varnothing$, $B - A = \{(H,H,H), (H,H,T), (H,T,H),$
   $(T,H,H)\}$ $A \cup B = B$, $AB = A$

3. $n(S) = 6$, $n(O) = 18$, $T_1 = \{(3,d)|d \in D\}$, $T_2 = \{(d,3)|d \in D\}$, $T_1T_2 =$
   $\{(3,3)\}$, $T_1 \cup T_2 = \{(d_1,d_2)|d_1 = 3 \text{ or } d_2 = 3\}$, $n(T_1 \cup T_2) = 11$,
   $n(T_1T_2) = 1$, $T_1 \times T_2 = \{(\tau_1,\tau_2)|\tau_1 \in T, \ \tau_2 \in T_2\}$, i.e., $T_1 \times T_2$ is the
   set of all ordered pairs.

CHAPTER 3

Exercises 3-1 page 25

1. (a) $\displaystyle\sum_{j=1}^{12} 2j$ (g) $\displaystyle\sum_{k=1}^{N} \frac{1}{k^k}$

   (b) $\displaystyle\sum_{k=1}^{19} 3k$ (h) $\displaystyle\sum_{n=1}^{v} \frac{n}{(j+n)^3}$

   (c) $\displaystyle\sum_{k=1}^{19} (3k+1)$ (i) $\displaystyle\sum_{k=50}^{99} ky^k$

   (d) $\displaystyle\sum_{i=1}^{7} 2^i$ (j) $\displaystyle\sum_{i=0}^{} r^i$

   (e) $\displaystyle\sum_{n=0}^{4} 2(3^n)$ (k) $\displaystyle\sum_{n=0}^{} qp^n$

   (f) $\displaystyle\sum_{j=1}^{n} \frac{1}{j(j+1)}$ (l) $\displaystyle\sum_{j=0}^{n} a^{n-j} b^j$

2. (a) $1 + 5 + 10 + \cdots + 197$

(e) $\dfrac{1}{2} + \dfrac{1}{6} + \dfrac{1}{12} + \cdots + \dfrac{1}{420}$

(b) $x + \dfrac{x^2}{4} + \dfrac{x^3}{9} + \cdots + \dfrac{x^{30}}{900}$

(f) $a + ar + ar^2 + \cdots + ar^{10}$

(c) $\dfrac{36}{7} + \dfrac{49}{8} + \dfrac{64}{9} + \cdots + \dfrac{1600}{41}$

(g) $p + pq + pq^2 + \cdots + pq^{29}$

(d) $1 - \dfrac{1}{2} + \dfrac{1}{3} - + \cdots + - \dfrac{1}{16}$

(h) $0 + \dfrac{1}{2} a^{N-1} b + a^{N-2} b^2 + \cdots + \dfrac{N}{2} b^N$

3. (a) $\displaystyle\sum_{j=1}^{N} j^3$

5. (a) $\displaystyle\sum_{j=0} \left(\dfrac{1}{3}\right)^j$

(b) $\displaystyle\sum_{k=1}^{n} \dfrac{1}{k}$

(b) $\displaystyle\sum_{k>N} \dfrac{1}{k}$

CHAPTER 4

Exercises 4-1b page 37

1. 45

2. $(197)\,(213)\,(160)$

3. 6^3

4. (a) $8(10^6)$, (b) $9 \cdot 8 \cdot 10^8$

5. $26^2 + 26^3 + 26^4$

6. $5!$

7. (a) $21!$ (b) $21!$

8. (a) $9!$ (b) $(5!)\,(4!)$ (c) no

9. (a) $9(9!)$ (b) $9(10^9)$

   (c) $2(9!)$ (d) $9(10^6)$

10. (a) $(50)_{47}$ (b) $2(25)_{20}(25)_{23}$

11. (a) $5(7)_3$ (b) $5(7^3)$

12. (a) $6! = 720$ (b) *cadefb* (c) *dcefab*

13. $(14)_9$

14. 10!

15. $n(A_5) = 4!, \quad n(A_1 \cap A_2) = 3!$

$n(A_2 - A_3) = 3 \cdot 3!$

$n(A_3 \cup A_4) = 42, \quad n(A_3 \cap A_4 \cap A_5) = 2$

$n(A_1 \cup A_3 \cup A_5) = 56, \quad n(A_2') = 96$

$n(A_4 A'_3) = 3 \cdot 3!$

Exercises 4-1d page 45

1. $\binom{15}{4}$ 2. $\binom{10}{3}$ 3. $\binom{8}{3}$

4. $\binom{15}{2}$ 5. $(6)_4$ 6. $\binom{5}{3}$

7. $\binom{13}{3}\binom{10}{2}$ 8. $\binom{200}{75,100,25}$ 9. (a) $\binom{18}{6}$ (b) $\binom{21}{9}$

10. (a) $\binom{80}{15}$ (b) $\binom{80}{5}\binom{20}{10}$ (c) $\binom{80}{10}\binom{20}{5}$

11. (a) $\binom{13}{1}\binom{4}{3}\binom{12}{2}4^2$ (b) $\binom{4}{1}\left[\binom{13}{5} - 10\right]$ (c) $10(5^4)$

     (d) $\binom{13}{2}\binom{4}{2}^2 44$ (e) $\binom{13}{1}\binom{4}{2}\binom{12}{1}\binom{4}{3}$

12. (a) $\binom{4}{3}\binom{48}{10}$ (b) $\sum_{j=0}^{3}\binom{4}{j}\binom{48}{13-j}$ (c) $\binom{13}{7}\binom{39}{6}$

     (d) $\binom{36}{13}$ (e) $\binom{4}{2}\binom{4}{3}\binom{44}{8}$

13. (a) $\binom{8}{2,2,2,1,1}$ (b) $\binom{10}{4,2,2,1,1}$

14. $\binom{4}{1}\binom{48}{12}\binom{3}{1}\binom{36}{12}\binom{2}{1}\binom{24}{12}\binom{1}{1}\binom{12}{12}$

15. $\binom{13}{4}\binom{13}{4}\binom{13}{5}$

16. $(6)(8)\binom{5}{2}(3)(10)$

17. $n = 4$

18. $\binom{6}{1} + 3\binom{6}{2} + 3\binom{6}{3} + \binom{6}{4}$

19. (a) $\binom{16}{4}$ (b) $\sum_{j=0}^{2}\binom{16}{j}$

20. $n(C) = 2^4$, $n(D) = \sum_{j=3}^{5}\binom{5}{j}$, $n(CD) = \sum_{j=2}^{4}\binom{4}{j}$,

$n(CD') = \binom{4}{0} + \binom{4}{1} = n(C - D)$

21. $n(A) = \binom{n}{k}$ $n(B) = \sum_{j=k}^{n}\binom{n}{j}$

22. $\binom{13}{N,S,E,W}\binom{39}{13-N,13-S,13-E,13-W}$

23. $\binom{5}{4} = 5$ $\binom{14}{7} = 3432$, $\binom{100}{3} = 161{,}700$,

$\binom{200}{198} = 19900$

Exercises 4-2a page 51

1. (a) $3^4 = 81$ (b) $\left(\frac{1}{3} + \frac{2}{3}\right)^6 = 1$ (c) $5^5 = 3125$ (d) 5^5

$(1-1)^{14} = 0$ (f) $\left(\frac{1}{2} + \frac{1}{2}\right)^{30} = 1$ (g) $(2-1)^7 = 1$ (h) 1

5. (a) 1.041 (b) .942 (c) 1.014

Exercises 4-2b page 57

1. (a) 124 (b) 86 (c) $\dfrac{5465}{729}$ (d) 1.666

2. (a) 1 (b) $\dfrac{3}{2}$ (c) $\dfrac{5}{2}$ (d) $\dfrac{1}{3}$ (e) $\dfrac{1}{4}$ (f) $\dfrac{49}{198}$

CHAPTER 5

Exercises 5-1a page 64

1. $\{T,F\}$

2. $\{r,w,b\}$

3. days in a leap year

4. $\{r,y,g\}$

5. $\{Su, M, T, W\ TH, F, S\}$

6. $\{rr,rw,rb,ww,wb,bb\}$

7. $\{(b_1, b_2)\,|\,b_j \in \{r,w,b\},\, j = 1,2\}$

8. $\{(d,c)\,|\,d \in \{1,2,3,4,5,6\},\, c \in \{H,T\}\}$

9. (a) $\{0,1\}$ (b) $\{(a_1,a_2,\ldots,a_{10})\,|\,a_i \in \{0,1\},\, i = 1,2,\ldots,10\}$.

10. $(X\ D)_3$ when $D = \{1,2,3,4,5,6\}$

11. Unordered triplets in which each element belongs to $D = \{1,2,3,4,5,6\}$

12. $\{3,4,5,\ldots,18\}$

13. $(X\ C)_5$ where $C = \{M,F\}$

14. (a) See Figure 5-4a_1 (b) See Figure 5-4a_2

15. $\Omega = \{S,NS,NNS,NNNS,\ldots\}$ where S stands for six and N for nonsix.

16. $\{HH, HTH, THH, HTTH, THTH, TTHH,\ldots\}$

18. (a) and (b) are comparable to 14a and 14b.

Exercises 5-1b page 66

1. (a) $\varnothing,\{T\},\{T\},\{H,T\}$ (b) $\varnothing,\{r\},\{b\},\{g\},\{r,b\},\{r,g\},\{b,g\},\Omega$

2. $A = \{(3,H),\, (3,T)\}$

3. $B = \{rg,\, gb,\, gg\}$

4. $C = \{17,18\}$

5. $C = \{(5,6,6),(6,5,6),(6,6,5),(6,6,6)\}$

6. (a) $\dbinom{10}{3}$ (b) $\displaystyle\sum_{k=3}^{10} \dbinom{10}{k}$

7. $C = \{(b,c_2,c_3,c_4,c_5) | c_i \in \{b,g\}, i = 2,3,4,5\}$

8. $S = \{HH, THH, HTHH, TTHH, HTTHH, THTHH, TTTHH\}, T = \varnothing$

9. $A_3 = \{1234, 1432, 2134, 2431, 4132, 4231\}$
   $A_1 \cup A_3 = \{1234, 1243, 1324, 1342, 1423, 1432, 2134, 2431, 4132, 4231\}$
   $A_2 \cap A_4 = \{1234, 3214\}$

10. $n(A) = \begin{pmatrix} 16 \\ 4 \end{pmatrix}, \quad n(B) = \begin{pmatrix} 16 \\ 0 \end{pmatrix} + \begin{pmatrix} 16 \\ 1 \end{pmatrix} + \begin{pmatrix} 16 \\ 2 \end{pmatrix}$

11. $n(C) = \begin{pmatrix} 10 \\ 3 \end{pmatrix}\begin{pmatrix} 7 \\ 4 \end{pmatrix}4^3, \quad n(D) = \begin{pmatrix} 10 \\ 2 \end{pmatrix}\begin{pmatrix} 8 \\ 2 \end{pmatrix}\begin{pmatrix} 6 \\ 3 \end{pmatrix}3^3$

12. $\begin{pmatrix} 4 \\ 1,2,1 \end{pmatrix}\begin{pmatrix} 4 \\ 2,1\ 1 \end{pmatrix}$

13. (a) $AB'C'$ (b) $AB'C$ (c) ABC (d) $A \cup B \cup C$,
   (e) $A'B'C'$ (f) $AB \cup AC \cup BC - (ABC)$ (g) $(ABC)'$

Exercises 5-2b page 74

1. a,c,d,f and g are functions.

2. $D_a =$ citizens in Wash., D. C. $R_a =$ days in leap year; $D_c = \{t | t$ is a real number$\}$ $R_c = \{s | s$ is a nonnegative number$\}$; $D_d = R_d =$ real numbers; $D_f = \{1,2,3,4,5,6\}$, $R_f = \{+ 1, - 1\}$; $D_g =$ ticket holders, $R_g =$ seat numbers.

4. (a)$\dfrac{1}{2}$, (b)$\dfrac{3}{4}$, (c)$\dfrac{1}{2}$, (d)$\dfrac{1}{4}$; 5, yes; 6, no; 7, yes;

   8, yes – $\sum\limits_{j=0}^{3} \begin{pmatrix} 8 \\ j \end{pmatrix}\begin{pmatrix} 7 \\ 3-j \end{pmatrix} = \begin{pmatrix} 15 \\ 3 \end{pmatrix}$ (see Prob. 4 Exercises 4.2a.)

Exercises 5-3b page 82

1. $P(AB') = P(A) - P(AB)$, $P(A'B') = 1 - P(A) - P(B) + P(AB)$
   $P((AB)') = 1 - P(AB)$, $P(A \cup B') = 1 - P(B) + P(AB)$

Exercises 5-3c page 86

NOTE: In the following Ω, A_Ω and P are defined, then $\{\Omega, A_\Omega, P\}$ is the probability space.

1. $\Omega = \{r,w,b\}$, A_Ω: set of events. Set $p(r) = \dfrac{1}{3}$, $p(w) = \dfrac{2}{9}$, $p(b) = \dfrac{4}{9}$

 $A \in A_\Omega$, $P(A) = \displaystyle\sum_{\omega \in A} p(\omega)$, (a) $\dfrac{2}{9}$, (b) $\dfrac{7}{9}$, (c) $\dfrac{5}{9}$

2. Ω: set of cards, A_Ω set of events, if $\omega \in \Omega$, set $p(\omega) = \dfrac{1}{52}$ then for

 $A \in A_\Omega$ set $P(A) = \dfrac{n(A)}{52}$. (a) $\dfrac{1}{4}$, (b) $\dfrac{1}{13}$, (c) $\dfrac{1}{2}$, (d) $\dfrac{6}{26}$

3. $\Omega = \{2,3,4,\ldots,12\}$, A_Ω set of events. Set $p(2) = p(12) = \dfrac{1}{36}$, $p(3) =$

 $p(11) = \dfrac{2}{36}$, $p(4) = p(10) = \dfrac{3}{36}$, $p(5) = p(9) = \dfrac{4}{36}$, $p(6) = p(8) =$

 $\dfrac{5}{36}$, $p(7) = \dfrac{6}{36}$ and for $A \in A_\Omega$ $P(A) = \displaystyle\sum_{\omega \in A} p(\omega)$ (a) $\dfrac{1}{18}$ (b) 1

 (c) 7   (d) $\dfrac{11}{12}$

4. $\Omega = (X\,D)_5$ where $D = \{0,1,2,\ldots,9\}$ A_Ω equals set of events. If $\omega \in \Omega$
 set $p(\omega) = 10^{-5}$ and if $A \in A_\Omega$, $P(A) = n(A)10^{-5}$, $P(A) = (10)_5\,10^{-5}$;
 $P(B) = \left(\dfrac{9}{10}\right)^5$, $P(C) = \dfrac{9^4}{10^5}$, $P(D) = \dfrac{4(9^4-1)}{10^5}$, $P(AB) = (9)_5\,10^{-5}$
 $P(A \cup B) = [(10)_5 - 9^5 + (9)_5]10^{-5}$, $P(CD) = 0$

5. Ω: all groups of 5; A_Ω set of events. For $\omega \in \Omega$, $p(\omega) = \dbinom{23}{5}^{-1}$ and for

 $A \in A_\Omega$ $P(A) = n(A) \dbinom{23}{5}^{-1}$, $P(A) = \dbinom{13}{2}\dbinom{10}{3}\dbinom{23}{5}^{-1}$

 $P(B) = \displaystyle\sum_{k=1}^{5} \dbinom{13}{k}\dbinom{10}{5-k}\dbinom{23}{5}^{-1}$,

 $P(C) = \left[\dbinom{13}{3}\dbinom{10}{2} + \dbinom{13}{1}\dbinom{10}{4}\right]\dbinom{23}{5}^{-1}$

6. Ω is set of all bridge hands $n(\Omega) = \dbinom{52}{13}^{-1}$ A_Ω set of events. For

 $\omega \in \Omega$, set $p(\omega) = \dbinom{52}{13}^{-1}$ and for $A \in A_\Omega$, set $P(A) = n(A)\dbinom{52}{13}^{-1}$

 (a) $\dbinom{4}{3}\dbinom{48}{10}\dbinom{53}{13}^{-1}$

 (b) $\left[\dbinom{4}{3}\dbinom{48}{10} + \dbinom{4}{4}\dbinom{48}{9}\right]\dbinom{52}{13}^{-1}$

(c) $\left[\binom{4}{0}\binom{48}{13} + \binom{4}{1}\binom{48}{12}\right]\binom{52}{13}^{-1}$

(d) $\binom{4}{2}^3\binom{40}{7}\binom{52}{13}^{-1}$ (e) $\binom{4}{2}\binom{40}{7}\binom{52}{13}^{-1}$

(f) $\binom{26}{2}\binom{26}{11}\binom{52}{13}^{-1}$

7. $Y = \{Jan, Feb,\ldots,Dec.\}$, $\Omega = (\text{X } Y)_n$ where in a, $n = 12$, in b, $n = 8$, in c, $n = 15$. A_Ω set of events. For $\omega \in \Omega$, set $p(\omega) = 12^{-n}$ and for $\in A_\Omega$, set $P(A) = n(A)12^{-n}$

(a) $\dfrac{12!}{12^{-12}}$ (b) $\binom{12}{3}(3^8 - 765)12^{-8}$ where $765 = 3 +\binom{3}{2}(2^8 - 2)$

(c) $\binom{12}{2,3,1,6}\binom{15}{4,4,2,2,2,1}12^{-15}$

8. Let $n \in \{1,2,3,4,5\}$, $\Omega = \{6,n6, nn6, nnn6,\ldots\}$, A_Ω set of events. For $\omega \in \Omega$ set $p(\omega) = \left(\dfrac{5}{6}\right)^i \dfrac{1}{6}$ where i counts the number of n's in ω

$\left(\sum\limits_{\omega \in \Omega} p(\omega) = 1\right)$. For $A \in A_\Omega$, $P(A) = \sum\limits_{\omega \in A} p(\omega)$ (a) $\sum\limits_{i=5} \dfrac{1}{6}\left(\dfrac{5}{6}\right)^i$

(b) $P(odd) = \dfrac{6}{11} > \dfrac{1}{2}$ (c) $\dfrac{91}{216}$

9. Ω: all groups of 20, A_Ω set of events. For $\Omega \in \omega$, $p(\Omega) = \left(\dfrac{300}{20}\right)^{-1}$

and for $A \in A_\Omega$, $P(A) = n(A)\left(\dfrac{300}{20}\right)^{-1}$.   (a) $\binom{200}{20}\left(\dfrac{300}{20}\right)^{-1}$,

(b) $\binom{100}{10}\binom{200}{10}\left(\dfrac{300}{10}\right)^{-1}$

(c) $\sum\limits_{i=0}^{2}\binom{100}{i}\binom{200}{20\text{-}i}\left(\dfrac{300}{20}\right)^{-1}$

(d) $1 - \binom{200}{20}\left(\dfrac{300}{20}\right)^{-1}$

10. $\Omega = \{HH,HTH,THH,HTTH,THTH,TTHH,\ldots\}$, A_Ω set of events. For $\omega \in \Omega$ $p(\omega) = \left(\dfrac{1}{2}\right)^i$ where i counts the number of H's and T's in ω and

$A \in A_\Omega$, $P(A) = \sum\limits_{\omega \in A} p(\omega)$ (a) $\dfrac{1}{2}$, (b) $\dfrac{1}{4}$ (c) $(k-1)\left(\dfrac{1}{2}\right)^k$

Exercises 5-4b page 93

1. $C = \{H,T\}$, $\Omega = (X\ C)_{10}$ $\omega \in \Omega$, $p(\omega) = 2^{-10}$ $A \in A_\Omega$

   $P(A) = n(A)\,(2^{-10})$

   $$P(A) = \binom{10}{5} 2^{-10} \qquad P(B) = \sum_{j=1}^{5} \binom{10}{2j} 2^{-10}, \quad P(C) = \sum_{k=0}^{3} \binom{10}{k} 2^{-10}$$

   $$P(BC) = \binom{10}{2} 2^{-10}, \quad P(AB) = 0, \quad P(B \cup C) = P(B) + P(C) - P(BC)$$

2. $L = \{a,n\}$, $\Omega = (X\ L)_{100}$; $\omega \in \Omega$, $p(\omega) = (.97)^k(.03)^{100-k}$ where k is the number of a's in ω; $A \in A_\Omega$, $P(A) = \sum_{\omega \in A} p(\omega)$ (a) $(.97)^{100}$

   (b) $\binom{100}{5} (.97)^{95}\,(.03)^5$

3. $D = \{S,F\}$, $\Omega = (X\ D)_7$; $\omega \in \Omega$, $p(\omega) = \left(\dfrac{4}{9}\right)^k \left(\dfrac{5}{9}\right)^{7-k}$, where k counts the number of S's in ω; $A \in A_\Omega$, $P(A) = \sum_{\omega \in A} p(\omega)$

   (a) $\dbinom{7}{2}\left(\dfrac{4}{9}\right)^2\left(\dfrac{5}{9}\right)^5$ (b) $\displaystyle\sum_{i=4}^{7} \binom{7}{i}\left(\dfrac{4}{9}\right)^i\left(\dfrac{5}{9}\right)^{7-i}$

   (c) $\displaystyle\sum_{j=0}^{3} \binom{7}{2j+1}\left(\dfrac{5}{9}\right)^{2j+1}\left(\dfrac{4}{9}\right)^{6-2j}$

4. (a) $\dbinom{5}{3}\left(\dfrac{1}{2}\right)^5$ (b) $1 - \left(\dfrac{1}{2}\right)^5$ (c) $\dfrac{1}{4}$

5. $D = \{r,w,b\}$, $\Omega = (X\ D)_{10}$; $\omega \in \Omega$, $p(\omega) = \left(\dfrac{1}{6}\right)^i\left(\dfrac{1}{3}\right)^j\left(\dfrac{1}{2}\right)^{10-i-j}$, where i counts the number of r's and j the number of w's in ω; $A \in A_\Omega$,

   $P(A) = \displaystyle\sum_{\omega \in A} p(\omega)$ (a) $\dbinom{10}{4,2,4}\left(\dfrac{1}{6}\right)^4\left(\dfrac{1}{3}\right)^2\left(\dfrac{1}{2}\right)^4$ (b) $\left(\dfrac{1}{2}\right)^{10}$

   (c) $\dbinom{10}{5,3,2}\left(\dfrac{1}{6}\right)^5\left(\dfrac{1}{3}\right)^3\left(\dfrac{1}{2}\right)^2$ (d) $\dbinom{10}{3}\left(\dfrac{1}{3}\right)^3\left(\dfrac{2}{3}\right)^7$ (e) $\dfrac{11}{2^{10}}$

6. (a) $\dbinom{5}{1,1,1,1,1}(.2)\,(.3)\,(.25)\,(.15)\,(.10)$ (b) $\dbinom{5}{2,2,1}(.2)^5(.15)^2(.1)$

   (c) $\displaystyle\sum_{\substack{i+j=5\\ i\geqslant 1}}^{4} \binom{5}{i}(.3)^i(.25)^j$

CHAPTER 6

Exercises 6-1b page 98

1. $R_X = \{-1, 0, +1\}$, $P_X = \left\{\left(-1, \frac{2}{9}\right), \left(0, \frac{3}{9}\right), \left(1, \frac{4}{9}\right)\right\}$

2. $R_Y = \{1,2,3,4\}$, $P_Y = \left\{\left(i, \frac{1}{4}\right) \middle| i \in R_Y\right\}$, $R_Z = \{1,2,\ldots,13\}$

   $P_Z = \left\{\left(j, \frac{1}{13}\right) \middle| j \in R_Z\right\}$

3. $R_X = \{1,2,\ldots,9\}$, S_X: subsets of R_X, $P_X = \left\{\left(k, \frac{1}{9}\right) \middle| k \in R_X\right\}$,

   $\{R_X, S_X, P_X\}$ is the induced probability space, (a) $\frac{1}{9}$ (b) $\frac{2}{9}$ (c) $\frac{2}{9}$

4. $R_Y = \{0,1,2,\ldots,7\}$, S_Y: subsets of

   $$R_Y, \quad P_Y = \left\{\left(i, \binom{15}{i}\binom{25}{7-i}\binom{40}{7}^{-1}\right) \middle| i \in R_Y\right\},$$

   $\{R_Y, S_Y, P_Y\}$ induced probability space, (a) $\binom{25}{7}\binom{40}{7}^{-1}$

   (b) $\binom{15}{2}\binom{25}{5}\binom{40}{7}^{-1}$ (c) $\left[\binom{15}{6}\binom{25}{1} + \binom{15}{7}\right]\binom{40}{7}^{-1}$

   (d) $\binom{15}{7}\binom{40}{7}^{-1}$. $P_Z(i) = P_Y(7-i)$ where $i \in R_Y$

5. $R_X = \{3,4,5,\ldots,18\}$ $R_Y = R_Z = \{1,2,3,4,5,6\}$.

   $P_X(3) = P_X(4) = P_X(17) = P_X(18) = \dfrac{1}{56}$

   $P_X(5) = P_X(16) = \dfrac{2}{56}$, $P_X(6) = P_X(15) = \dfrac{3}{56}$, $P_X(7) = P_X(14) = \dfrac{4}{56}$

   $P_X(8) = P_X(13) = \dfrac{5}{56}$, $P_X(9) = P_X(10) = P_X(11) = P_X(12) = \dfrac{6}{56}$

   $P_Y(1) = P_Z(6) = \dfrac{1}{56}$, $P_Y(2) = P_Z(5) = \dfrac{3}{56}$, $P_Y(3) = P_Z(4) = \dfrac{6}{56}$

   $P_Y(4) = P_Z(3) = \dfrac{10}{56}$, $P_Y(5) = P_Z(2) = \dfrac{15}{56}$, $P_Y(6) = P_Z(1) = \dfrac{21}{56}$

(a) $P_Y(6) = \dfrac{21}{56}$ (b) $\{Z \geqslant 4\}$ (c) $\{X \geqslant 5\}$ (d) $\displaystyle\sum_{j=1}^{4} P_Y(j) = \dfrac{20}{56}$

(e) $P_Z(Z > 4) = \dfrac{4}{56}$ (f) $P(X < 5)$ (g) $\{Y < 3\}$

(h) $\{Z = 3\} \cap \{X > 9\}$

(i) $\{Y = Z\}$ (j) $\{Y = 6\} \cup \{Z = 1\}$

6. $R_X = \{0,1,2,3,4\}, \; P_X = \left\{ \left(i, \dbinom{4}{i} \dbinom{48}{13-i} \dbinom{52}{13}^{-1} \right) \middle| i \in R_X \right\},$

$R_Y = \{0,1,2,\ldots,13\}$

$P_Y = \left\{ \left(j, \dbinom{13}{j} \dbinom{39}{13-j} \dbinom{52}{13}^{-1} \right) \middle| j \in R_Y \right\}, \; R_Z = \{0,1,2,\ldots,13\}$

$P_Z = \left\{ \left(k, \dbinom{26}{k} \dbinom{26}{13-k} \dbinom{52}{13}^{-1} \right) \middle| k \in R_Z \right\}$

(a) $\dbinom{4}{3} \dbinom{48}{10} \dbinom{52}{13}^{-1}$ (b) $\{X > 1\}$ (c) $\dbinom{13}{5} \dbinom{39}{8} \dbinom{52}{13}^{-1}$

(d) $\{X \leqslant 3\}$ (e) $\{Z = 13\}$ (f) $\{X = 0\} \cap \{Y = 0\}$

7. $R_1 = \{1,2,3,\ldots\}$ $P_1 = \left\{ \left(i, \dfrac{1}{6} \left(\dfrac{5}{6}\right)^{i-1} \right) \middle| i \in R_1 \right\}$ (a) $\dfrac{5}{11}$

(b) $\dfrac{25}{91}$ (c) $1 - \left(\dfrac{5}{6}\right)^{5}$

8. $R_Y = \{0,1,2,\ldots,100\}, \; P_Y = \{(k, (.05)^k (.95)^{100-k}) | k \in R_Y\}$

(a) $\displaystyle\sum_{i=0}^{3} P_Y(i)$ (b) $\displaystyle\sum_{k=11}^{100} P_Y(k)$

9. $P_X = \{(i, p_i) | i \in I^+\}$ (a) $\displaystyle\sum_{k=1} p_{2k}$ (b) $\displaystyle\sum_{k=1} p_{8k}$ (c) $\displaystyle\sum_{j=0}^{5} p_j$

11. $P_Y = \left\{ \left(k, (k-1) \left(\dfrac{1}{2}\right)^k \right) \middle| k \in I^+ - \{1\} \right\}$

12. $R_Z = \{1,2,3,\ldots,40\} \; P_Z = \left\{ \left(j, \dfrac{(39)_{j-1}(13)}{(52)_j} \right) \middle| j \in R_Z \right\}$

To show $\displaystyle\sum_{j=1}^{40} P_Z(j) = 1$, use Prob. 4b, Exercises 4-2a.

Exercises 6-1d page 105

1. A 4 by 13 table in which each entry is $\dfrac{1}{52}$.

2. (All entries in table are divided by $\dbinom{40}{7}$).

$Z \backslash^{\!Y}$	0	1	2	3
0				
1				
2				
3				
4				$\dbinom{15}{3}\dbinom{25}{4}$
5			$\dbinom{15}{2}\dbinom{25}{5}$	
6		$\dbinom{15}{1}\dbinom{25}{6}$		
7	$\dbinom{15}{0}\dbinom{25}{7}$			
P_Y	$\dbinom{15}{0}\dbinom{25}{7}$	$\dbinom{15}{1}\dbinom{25}{6}$	$\dbinom{15}{2}\dbinom{25}{5}$	$\dbinom{15}{3}\dbinom{25}{4}$

(Middle region of table is 0.)

3. (a)

$Y \backslash^{\!X}$	0	1	2	3	P_Y
0	$\dfrac{1}{27}$	$\dfrac{1}{9}$	$\dfrac{1}{9}$	$\dfrac{1}{27}$	$\dfrac{8}{27}$
1	$\dfrac{1}{9}$	$\dfrac{2}{9}$	$\dfrac{1}{9}$	0	$\dfrac{4}{9}$
2	$\dfrac{1}{9}$	$\dfrac{1}{9}$	0	0	$\dfrac{2}{9}$
3	$\dfrac{1}{27}$	0	0	0	$\dfrac{1}{27}$
P_X	$\dfrac{8}{27}$	$\dfrac{4}{9}$	$\dfrac{2}{9}$	$\dfrac{1}{27}$	1

4	5	6	7	P_Z
			$\binom{15}{7}\binom{25}{0}$	$\binom{15}{7}\binom{25}{0}$
		$\binom{15}{6}\binom{25}{1}$		$\binom{15}{6}\binom{25}{1}$
	$\binom{15}{5}\binom{25}{2}$			$\binom{15}{5}\binom{25}{2}$
$\binom{15}{4}\binom{25}{3}$				$\binom{15}{4}\binom{25}{3}$
				$\binom{15}{3}\binom{25}{4}$
		0		$\binom{15}{2}\binom{25}{5}$
				$\binom{15}{1}\binom{25}{6}$
				$\binom{15}{0}\binom{25}{7}$
$\binom{15}{4}\binom{25}{3}$	$\binom{15}{5}\binom{25}{2}$	$\binom{15}{6}\binom{25}{1}$	$\binom{15}{7}\binom{25}{0}$	$\binom{40}{7}$

(b)

Z \ Y	0	1	2	3	P_Z
1	$\frac{2}{27}$	0	0	$\frac{1}{27}$	$\frac{3}{27}$
2	$\frac{6}{27}$	$\frac{6}{27}$	$\frac{6}{27}$	0	$\frac{18}{27}$
3	0	$\frac{6}{27}$	0	0	$\frac{6}{27}$
P_Y	$\frac{8}{27}$	$\frac{12}{27}$	$\frac{6}{27}$	$\frac{1}{27}$	1

4. $R_X = \{0,1,2,3,4,5\}$, S_X: subsets of R_X,

$$P_X = \left\{ \left(j, \binom{15}{j}\binom{12}{5-j}\binom{27}{5}^{-1} \right) \middle| j \in R_X \right\} \{R_X, S_X, P_X\} \text{ induced}$$

probability space. (All entries divided by $\binom{27}{5}$).

Y\X	0	1	2
0			
1		0	
2			
3			$\binom{15}{2}\binom{12}{3}$
4		$\binom{15}{1}\binom{12}{4}$	
5	$\binom{15}{0}\binom{12}{5}$		
P_X	$\binom{15}{0}\binom{12}{5}$	$\binom{15}{1}\binom{12}{4}$	$\binom{15}{2}\binom{12}{3}$

5. (All entries divided by 56.)

Z\Y	1	2	3	4	5	6	P_Z
1	1	2	3	4	5	6	21
2	0	1	2	3	4	5	15
3	0	0	1	2	3	4	10
4	0	0	0	1	2	3	6
5	0	0	0	0	1	2	3
6	0	0	0	0	0	1	1
P_Y	1	3	6	10	15	21	56

3	4	5	P_Y
		$\binom{15}{5}\binom{12}{0}$	$\binom{15}{5}\binom{12}{0}$
	$\binom{15}{4}\binom{12}{1}$		$\binom{15}{4}\binom{12}{1}$
$\binom{15}{3}\binom{12}{2}$			$\binom{15}{3}\binom{12}{2}$
			$\binom{15}{2}\binom{12}{3}$
		0	$\binom{15}{1}\binom{12}{4}$
			$\binom{15}{0}\binom{12}{5}$
$\binom{15}{3}\binom{12}{2}$	$\binom{15}{4}\binom{12}{1}$	$\binom{15}{5}\binom{12}{0}$	$\binom{27}{5}$

(All entries divided by 56.)

Y \ X	1	2	3	4	5	6	P_X
3	1						1
4		1					1
5		1	1				2
6		1	1	1			3
7			2	1	1		4
8			1	2	1	1	5
9			1	2	2	1	6
10				2	2	2	6
11				1	3	2	6
12				1	2	3	6
13					2	3	5
14					1	3	4
15					1	2	3
16						2	2
17						1	1
18						1	1
P_Y	1	3	6	10	15	21	56

6.

Y \ X	0	1	2	P_Y
0	0	$\dfrac{1}{54}$	$\dfrac{2}{54}$	$\dfrac{3}{54}$
1	$\dfrac{1}{54}$	$\dfrac{2}{54}$	$\dfrac{3}{54}$	$\dfrac{6}{54}$
2	$\dfrac{4}{54}$	$\dfrac{5}{54}$	$\dfrac{6}{54}$	$\dfrac{15}{54}$
3	$\dfrac{9}{54}$	$\dfrac{10}{54}$	$\dfrac{11}{54}$	$\dfrac{30}{54}$
P_X	$\dfrac{14}{54}$	$\dfrac{18}{54}$	$\dfrac{22}{54}$	1

$$P_X(2) = P(X = x)$$

$$= \sum_{j=0}^{3} P(X = x, Y = j)$$

$$= \sum_{j=0}^{3} \frac{1}{54}(x + j^2)$$

$$= \frac{1}{54}(x + (x + 1) + (x + 4) + (x + 9))$$

$$= \frac{1}{54}(4x + 14) = \frac{1}{27}(2x + 7)$$

Exercises 6-1e page 108

1. X has a Bernoulli distribution with $p = \dfrac{1}{4}$.

2. Y obeys the binomial law with $p = \dfrac{1}{4}$ and $n = 5$.

3. Z is binomial with $p = \dfrac{1}{4}$ and $n = 10$.

4. X, Y, Z are all binomial with $p = \dfrac{1}{2}$, $n = 6$,

$$P_X(4) = P_Y(4) = P_Z(4) = \binom{6}{4}\left(\frac{1}{2}\right)^6.$$

Exercises 6-2b page 113

2. $F_Y(-3) = 0$, $F_Y\left(\dfrac{5}{2}\right) = \displaystyle\sum_{j=0}^{2} \binom{5}{j} \left(\dfrac{1}{4}\right)^{j} \left(\dfrac{3}{4}\right)^{5-j}$,

$F_Y\left(\dfrac{15}{4}\right) = \displaystyle\sum_{j=0}^{3} \binom{5}{j} \left(\dfrac{1}{4}\right)^{j} \left(\dfrac{3}{4}\right)^{5-j}$, $F_Y(6) = 1$

$F_X\left(\dfrac{3}{2}\right) = F_Y\left(\dfrac{3}{2}\right) = F_Z\left(\dfrac{3}{2}\right) = \displaystyle\sum_{k=0}^{1} \binom{6}{k} \left(\dfrac{1}{2}\right)^{6}$;

5. $F_X\left(\dfrac{7}{2}\right) = \displaystyle\sum_{k=1}^{3} \left(\dfrac{1}{6}\right)\left(\dfrac{5}{6}\right)^{k-1}$, $F_X\left(\dfrac{41}{2}\right) = \displaystyle\sum_{k=1}^{20} \left(\dfrac{1}{6}\right) \left(\dfrac{5}{6}\right)^{k-1}$.

Exercises 6-3b page 118

1. $E(X) = \dfrac{2}{9}$

2. $E(Y) = \dfrac{5}{2}$, $E(Z) = 7$

3. $E(X) = 5$

4. $E(Y) = \displaystyle\sum_{i=1}^{7} i \binom{15}{i} \binom{25}{7-i}\binom{40}{7}^{-1}$

$= \binom{40}{7}^{-1} (15) \displaystyle\sum_{i=1}^{7} \binom{14}{i-1}\binom{25}{7-i} = 15 \binom{39}{6}\binom{40}{7}^{-1} = \dfrac{21}{8}$

(See Prob. 4b, Exercises 4-2a.)

5. $E(X) = 10.5$, $E(Y) = 4.75$, $E(Z) = 2.25$

6. $E(X) = 1$, $E(Y) = \dfrac{13}{4}$, $E(Z) = \dfrac{13}{2}$

7. $E(X) = 6$

8. $E(Y) = 5$

9. $E(X) = \dfrac{988}{275}$

10. (a) $+\$.053 \left(\dfrac{1}{19} \text{ of } \$1\right)$ (b) $+\$.053$ (c) $-\$.106$

11. 2.97 minutes

12. $\$15,750$

Exercises 6-3c page 128

1. $E(X^2) = \dfrac{6}{9}, \quad E(X^3) = \dfrac{2}{9}$

2. $E(Y + Z) = \dfrac{19}{2}, \quad E(Y^2) = \dfrac{15}{2}$

3. $E(X^2) = \dfrac{285}{9}$

4. $E(2Y - 3) = \dfrac{9}{4}$

5. $E(Z^2 - E(Z)) = 4.50 \qquad E(X - E(X))^2 = 11.25$

Exercises 6-5 page 136

1. $V(X) = \dfrac{50}{81}, \; V(Y) = \dfrac{5}{4}, \; V(Z) = 14, \; V(X) = 2.96.$

2. $\text{Cov}(Y,Z) = .56.$

3. $V(Y) = V(Z) = 1.69, \quad V(Y + Z) = 4.50.$

4. $E(X_i) = \dfrac{1}{5}, \; V(X_i) = \dfrac{4}{25}, \; \text{cov}(X_i,X_j) = \dfrac{1}{100}, \; E(S_5) = 1, \; V(S_5) = 1.$

5. $E(X_i) = \dfrac{1}{n}, \; V(X_i) = \dfrac{n-1}{n}, \; \text{cov}(X_i,X_j) = \dfrac{1}{n^2(n-1)}, E(S_n) = 1, \; V(S_n) = 1.$

6. (a) $1 - \dfrac{1}{2!} + \dfrac{1}{3!} - \dfrac{1}{4!} + \dfrac{1}{5!},$

   (b) $1 - 1 + \dfrac{1}{2!} - \dfrac{1}{3!} + \cdots \pm \dfrac{1}{n!}.$

CHAPTER 7

Exercises 7-1a page 140

1. $\dfrac{1}{3}$

2. $\dfrac{1}{2}, \dfrac{1}{13}, \dfrac{3}{13}$

3. $\dfrac{1}{8}$

4. $\dfrac{1}{2}$

5. $\left(\sum\limits_{j=2}^{5} \binom{5}{j} 5^{5-j} 6^{-5} \right) / \left(1 - \left(\frac{5}{6} \right)^{5} \right)$

6. (a) $\left(\sum_{j=4} \left(\frac{1}{6} \right) \left(\frac{5}{6} \right)^{j-1} \right) / \left(\frac{5}{6} \right),$ 　　(b) $\dfrac{11}{36}$

7. $\dfrac{1}{13}$

8. $P_{\text{def}}(X) = \dfrac{(.20)(.04)}{(.20)(.04) + (.35)(.04) + (.45)(.03)}$

9. $\left(\sum\limits_{i=2}^{4} \binom{4}{i} \binom{35}{13-i} \right) / \binom{39}{13}$

10. (a) $2 \binom{23}{10} / \binom{26}{13},$ 　　(b) $2 \binom{23}{12} / \binom{26}{13}$

11. $\left(\dfrac{b}{b+w} \right) \left(\dfrac{b+n}{b+w+m+n} \right) \left(\dfrac{b+2n}{b+w+2m+2n} \right)$

12. $\dfrac{b(b+n)}{w(b+m) + b(b+n)}$

14. $\dfrac{1}{36}, \dfrac{5}{18}, \dfrac{5}{9}$

Exercises 7-1b page 143

1. (a) .285　　(b) $\dfrac{6}{19}$

2. (a) $\dfrac{11}{75}$　　(b) $\dfrac{2}{11}$

3. $P(\text{spec}) = \sum\limits_{j=1}^{N} p_j \sigma_j,\ P_{\text{spec}}(S_N) = \dfrac{P_{S_N}(\text{spec})\, P(S_N)}{P(\text{spec})} = \dfrac{\sigma_N p_N}{\sum\limits_{j=1}^{N} p_j \sigma_j}$

4. $\sum\limits_{n \geqslant k} \left\{ \binom{n}{k} \left(\frac{1}{3} \right)^{k} \left(\frac{2}{3} \right)^{n-k} \right\}\ p'' = p^k \left(\frac{1}{3} \right)^{k} \left(1 - \frac{2}{3} p \right)^{-(k+1)}$

5. $\dfrac{p}{3-2p}$

Exercises 7-2b page 147

1. yes   2. yes   3. (a) yes (b) no
5. (a) yes (b) yes (c) no.   6 (a) no (b) no

Exercises 7-2c page 154

1. $\dfrac{1}{3}, \dfrac{1}{3}, \dfrac{1}{9}, \left(\dfrac{2}{3}\right)^{100}$

2. (a) $\dfrac{142}{315}$ (b) $\dfrac{133}{315}$ (c) $\dfrac{182}{315}$

3. (a) $\dfrac{1}{2}, \dfrac{1}{2}$ (b) $\dfrac{3}{5}$ (c) $\dfrac{2^n}{1 + 2^n}$

4. (a) $\dbinom{n}{i,j,k} p_a^i p_b^j p_c^k$ (b) p_a.   (c) $p_a\, p_b$

5. 5

6. (a) $\dbinom{5}{2,0,3} (.3)^2 (.2)^0 (.5)^3$ (b) $\dbinom{5}{2,1,2} (.3)^2 (.2) (.5)^2$

Exercises 7-4a page 158

1. (c)

X_j \ X_i	0	1	P_{X_j}
0	$\dfrac{1}{4}$	$\dfrac{1}{4}$	$\dfrac{1}{2}$
1	$\dfrac{1}{4}$	$\dfrac{1}{4}$	$\dfrac{1}{2}$
P_{X_i}	$\dfrac{1}{2}$	$\dfrac{1}{2}$	1

(d) $\text{Cov}\,(X_i, X_j) = 0$

(e) $E(X_i) = \dfrac{1}{2}$, $V(X_i) = \dfrac{1}{2}$

(f) $\displaystyle\sum_{i=1}^{n} E(X_i) = E\left(\sum_{i=1}^{n} X_i\right) = E(Y) = \dfrac{n}{2}$

(g) $\displaystyle\sum_{i=1}^{n} V(X_i) = V\left(\sum_{i=1}^{n} X_i\right) = V(Y) = \dfrac{n}{2}$

2. No pair is independent (See Prob. 5 Exercises 6-1*d* for joint distribution tables.)

3. $E(X|Y = 3) = \sum_{j=1}^{7} j\, P_{\{Y=3\}}(X = j) = \sum_{j=1}^{7} \binom{10}{j,3,7-j}\left(\frac{1}{6}\right)^{j}\left(\frac{1}{3}\right)^{3}\left(\frac{1}{2}\right)^{7-j}$

$$= \frac{140(2^{6})}{3^{9}}$$

$E(Y|Z = 3) = \dfrac{280}{2^{9}}, \quad P_{\{Z=7\}}(W = 3) = 1, \quad P_{\{Z=6\}}(W = 3) = 0$

$$P_{\{X=2\}}(W = 4) = \binom{8}{2}\left(\frac{1}{3}\right)^{2}\left(\frac{1}{2}\right)^{6}\left(\frac{5}{6}\right)^{-8} \approx .04,$$

$$E(W|X = 2) = \sum_{j=2}^{10} j\, P_{\{X=2\}}(W = j)$$

4. $P((X + Y) > 7) = \dfrac{14}{30}, P((Y - X) > 0) = \dfrac{163}{300}$

(c) $P_{Y|3} = \left\{(1,0),(2,0),\left(3,\dfrac{1}{3}\right),\left(4,\dfrac{1}{3}\right),\left(5,\dfrac{1}{3}\right)\right\}, \quad E(Y|3) = 4$

CHAPTER 8

Exercises 8-1c page 169

1. $\begin{pmatrix} p & q \\ p & q \end{pmatrix}$

2. $\begin{pmatrix} 1 & 0 & 0 & 0 & 0 \\ \frac{3}{7} & 0 & \frac{4}{7} & 0 & 0 \\ 0 & \frac{3}{7} & 0 & \frac{4}{7} & 0 \\ 0 & 0 & \frac{3}{7} & 0 & \frac{4}{7} \\ 0 & 0 & 0 & 0 & 1 \end{pmatrix}$ (b) $\sum_{n=0} \left(\frac{3}{7}\right)^{2}\left(2 \cdot \frac{3}{7} \cdot \frac{4}{7}\right)^{n} = \dfrac{9}{25}$

3.
$$\begin{pmatrix} 1 & 0 & 0 & 0 & 0 \\ p & 0 & q & 0 & 0 \\ 0 & p & 0 & q & 0 \\ 0 & 0 & p & 0 & q \\ 0 & 0 & 0 & 0 & 1 \end{pmatrix} \qquad \sum_{n=0} q^2 (2pq)^n = \frac{q^2}{1 - 2pq}$$

4.

	E_r	E_w	E_b
E_r	$\frac{3}{12}$	$\frac{3}{12}$	$\frac{6}{12}$
E_w	$\frac{4}{12}$	$\frac{2}{12}$	$\frac{6}{12}$
E_n	$\frac{4}{12}$	$\frac{3}{12}$	$\frac{5}{12}$

5.

	E_r	E_w	E_b	E_g
E_r	$\frac{3}{15}$	$\frac{3}{15}$	$\frac{6}{15}$	$\frac{3}{15}$
E_w	$\frac{4}{15}$	$\frac{2}{15}$	$\frac{6}{15}$	$\frac{3}{15}$
E_b	$\frac{4}{15}$	$\frac{3}{15}$	$\frac{5}{15}$	$\frac{3}{15}$
E_g	$\frac{4}{15}$	$\frac{3}{15}$	$\frac{6}{15}$	$\frac{2}{15}$

6.
$$\begin{pmatrix} 0 & \frac{1}{2} & 0 & \frac{1}{2} & 0 & 0 & 0 & 0 & 0 \\ \frac{1}{3} & 0 & \frac{1}{3} & 0 & \frac{1}{3} & 0 & 0 & 0 & 0 \\ 0 & \frac{1}{2} & 0 & 0 & 0 & \frac{1}{2} & 0 & 0 & 0 \\ \frac{1}{3} & 0 & 0 & 0 & \frac{1}{3} & 0 & \frac{1}{3} & 0 & 0 \\ 0 & \frac{1}{4} & 0 & \frac{1}{4} & 0 & \frac{1}{4} & 0 & \frac{1}{4} & 0 \\ 0 & 0 & \frac{1}{3} & 0 & \frac{1}{3} & 0 & 0 & 0 & \frac{1}{3} \\ 0 & 0 & 0 & \frac{1}{2} & 0 & 0 & 0 & \frac{1}{2} & 0 \\ 0 & 0 & 0 & 0 & \frac{1}{3} & 0 & \frac{1}{3} & 0 & \frac{1}{3} \\ 0 & 0 & 0 & 0 & 0 & \frac{1}{2} & 0 & \frac{1}{2} & 0 \end{pmatrix}$$

7. (a)

$$
\begin{pmatrix}
1 & 0 & 0 & 0 & 0 & 0 & 0 & 0 & 0 \\
\frac{1}{3} & 0 & \frac{1}{3} & 0 & \frac{1}{3} & 0 & 0 & 0 & 0 \\
0 & 0 & 1 & 0 & 0 & 0 & 0 & 0 & 0 \\
\frac{1}{3} & 0 & 0 & 0 & \frac{1}{3} & 0 & \frac{1}{3} & 0 & 0 \\
0 & \frac{1}{4} & 0 & \frac{1}{4} & 0 & \frac{1}{4} & 0 & \frac{1}{4} & 0 \\
0 & 0 & \frac{1}{3} & 0 & \frac{1}{3} & 0 & 0 & 0 & \frac{1}{3} \\
0 & 0 & 0 & 0 & 0 & 0 & 1 & 0 & 0 \\
0 & 0 & 0 & 0 & \frac{1}{3} & 0 & \frac{1}{3} & 0 & \frac{1}{3} \\
0 & 0 & 0 & 0 & 0 & 0 & 0 & 0 & 1
\end{pmatrix}
$$

(b)

$$
\begin{pmatrix}
0 & \frac{1}{2} & 0 & \frac{1}{2} & 0 & 0 & 0 & 0 & 0 \\
\frac{1}{3} & 0 & \frac{1}{3} & 0 & \frac{1}{3} & 0 & 0 & 0 & 0 \\
0 & \frac{1}{2} & 0 & 0 & 0 & \frac{1}{2} & 0 & 0 & 0 \\
\frac{1}{3} & 0 & 0 & 0 & \frac{1}{3} & 0 & \frac{1}{3} & 0 & 0 \\
0 & 0 & 0 & 0 & 1 & 0 & 0 & 0 & 0 \\
0 & 0 & \frac{1}{3} & 0 & \frac{1}{3} & 0 & 0 & 0 & \frac{1}{3} \\
0 & 0 & 0 & \frac{1}{2} & 0 & 0 & 0 & \frac{1}{2} & 0 \\
0 & 0 & 0 & 0 & \frac{1}{3} & 0 & \frac{1}{3} & 0 & \frac{1}{3} \\
0 & 0 & 0 & 0 & 0 & \frac{1}{2} & 0 & \frac{1}{2} & 0
\end{pmatrix}
$$

8.

$$
\begin{pmatrix}
0 & \tfrac{1}{2} & 0 & 0 & \tfrac{1}{2} & & & & & & & & & & & \\
\tfrac{1}{3} & 0 & \tfrac{1}{3} & 0 & 0 & \tfrac{1}{3} & & & & & & & & & & \\
0 & \tfrac{1}{3} & 0 & \tfrac{1}{3} & 0 & 0 & \tfrac{1}{3} & & & & & & & & & \\
0 & 0 & \tfrac{1}{2} & 0 & 0 & 0 & 0 & \tfrac{1}{2} & & & & & \text{\Large 0} & & & \\
\tfrac{1}{3} & 0 & 0 & 0 & 0 & \tfrac{1}{3} & 0 & 0 & \tfrac{1}{3} & & & & & & & \\
& \tfrac{1}{4} & 0 & 0 & \tfrac{1}{4} & 0 & \tfrac{1}{4} & 0 & 0 & \tfrac{1}{4} & & & & & & \\
& & \tfrac{1}{4} & 0 & 0 & \tfrac{1}{4} & 0 & \tfrac{1}{4} & 0 & 0 & \tfrac{1}{4} & & & & & \\
& & & \tfrac{1}{3} & 0 & 0 & \tfrac{1}{3} & 0 & 0 & 0 & 0 & \tfrac{1}{3} & & & & \\
& & & & \tfrac{1}{3} & 0 & 0 & 0 & 0 & \tfrac{1}{3} & 0 & 0 & \tfrac{1}{3} & & & \\
& & & & & \tfrac{1}{4} & 0 & 0 & \tfrac{1}{4} & 0 & \tfrac{1}{4} & 0 & 0 & \tfrac{1}{4} & & \\
& & & & & & \tfrac{1}{4} & 0 & 0 & \tfrac{1}{4} & 0 & \tfrac{1}{4} & 0 & 0 & \tfrac{1}{4} & \\
& \text{\Large 0} & & & & & & \tfrac{1}{3} & 0 & 0 & \tfrac{1}{3} & 0 & 0 & 0 & 0 & \tfrac{1}{3} \\
& & & & & & & & \tfrac{1}{2} & 0 & 0 & 0 & 0 & \tfrac{1}{2} & 0 & 0 \\
& & & & & & & & & \tfrac{1}{3} & 0 & 0 & \tfrac{1}{3} & 0 & \tfrac{1}{3} & 0 \\
& & & & & & & & & & \tfrac{1}{3} & 0 & 0 & \tfrac{1}{3} & 0 & \tfrac{1}{3} \\
& & & & & & & & & & & \tfrac{1}{2} & 0 & 0 & \tfrac{1}{2} & 0 \\
\end{pmatrix}
$$

9. (a)

$$\begin{pmatrix}
1 & 0 & 0 & 0 & 0 \\
\frac{1}{3} & 0 & \frac{1}{3} & 0 & 0 & \frac{1}{3} \\
0 & \frac{1}{3} & 0 & \frac{1}{3} & 0 & 0 & \frac{1}{3} \\
0 & 0 & 0 & 1 & 0 & 0 & 0 & 0 \\
\frac{1}{3} & 0 & 0 & 0 & 0 & \frac{1}{3} & 0 & 0 & \frac{1}{3} \\
\frac{1}{4} & 0 & 0 & \frac{1}{4} & 0 & \frac{1}{4} & 0 & 0 & \frac{1}{4} \\
& \frac{1}{4} & 0 & 0 & \frac{1}{4} & 0 & \frac{1}{4} & 0 & 0 & \frac{1}{4} \\
& & \frac{1}{3} & 0 & 0 & \frac{1}{3} & 0 & 0 & 0 & 0 & \frac{1}{3} \\
& & & \frac{1}{3} & 0 & 0 & 0 & 0 & \frac{1}{3} & 0 & 0 & \frac{1}{3} \\
& & & & \frac{1}{4} & 0 & 0 & \frac{1}{4} & 0 & \frac{1}{4} & 0 & 0 & \frac{1}{4} \\
& & & & & \frac{1}{4} & 0 & 0 & \frac{1}{4} & 0 & \frac{1}{4} & 0 & 0 & \frac{1}{4} \\
& & & & & & \frac{1}{3} & 0 & 0 & \frac{1}{3} & 0 & 0 & 0 & 0 & \frac{1}{3} \\
& & & & & & & 0 & 0 & 0 & 0 & 1 & 0 & 0 & 0 \\
& & & & & & & & \frac{1}{3} & 0 & 0 & \frac{1}{3} & 0 & \frac{1}{3} & 0 \\
& & & & & & & & & \frac{1}{3} & 0 & 0 & \frac{1}{3} & 0 & \frac{1}{3} \\
& & & & & & & & & & 0 & 0 & 0 & 0 & 1
\end{pmatrix}$$

$$\begin{pmatrix}
0 & \frac{1}{2} & 0 & 0 & \frac{1}{2} & & & & & & & & & & & & \\
\frac{1}{3} & 0 & \frac{1}{3} & 0 & 0 & \frac{1}{3} & & & & & & & & & & & \\
0 & \frac{1}{3} & 0 & \frac{1}{3} & 0 & 0 & \frac{1}{3} & & & & & & & & & & \\
0 & 0 & \frac{1}{2} & 0 & 0 & 0 & 0 & \frac{1}{2} & & & & & \mathbf{0} & & & & \\
\frac{1}{3} & 0 & 0 & 0 & 0 & \frac{1}{3} & 0 & 0 & \frac{1}{3} & & & & & & & & \\
& 0 & 0 & 0 & 0 & 1 & 0 & 0 & 0 & 0 & & & & & & & \\
& & 0 & 0 & 0 & 0 & 1 & 0 & 0 & 0 & 0 & & & & & & \\
& & & \frac{1}{3} & 0 & 0 & \frac{1}{3} & 0 & 0 & 0 & 0 & \frac{1}{3} & & & & & \\
& & & & \frac{1}{3} & 0 & 0 & 0 & 0 & \frac{1}{3} & 0 & 0 & \frac{1}{3} & & & & \\
& & & & & 0 & 0 & 0 & 0 & 1 & 0 & 0 & 0 & 0 & & & \\
& & & & & & 0 & 0 & 0 & 0 & 1 & 0 & 0 & 0 & 0 & & \\
& & & \mathbf{0} & & & & \frac{1}{3} & 0 & 0 & \frac{1}{3} & 0 & 0 & 0 & 0 & \frac{1}{3} & \\
& & & & & & & & \frac{1}{2} & 0 & 0 & 0 & 0 & \frac{1}{2} & 0 & 0 & \\
& & & & & & & & & \frac{1}{3} & 0 & 0 & \frac{1}{3} & 0 & \frac{1}{3} & 0 & \\
& & & & & & & & & & \frac{1}{3} & 0 & 0 & \frac{1}{3} & 0 & \frac{1}{3} & \\
& & & & & & & & & & & \frac{1}{2} & 0 & 0 & \frac{1}{2} & 0 &
\end{pmatrix}$$

Exercises 8-2a page 174

1. $(0 \ 0 \ 1 \ 0 \ 0)$; $(p^2 \ 2p^2q \ 0 \ 2pq^2 \ q^2)$

2. $\left(0 \ \dfrac{1}{4} \ 0 \ \dfrac{1}{4} \ 0 \ \dfrac{1}{4} \ 0 \ \dfrac{1}{4} \ 0\right)$; $\left(\dfrac{1}{6} \ 0 \ \dfrac{1}{6} \ 0 \ \dfrac{1}{3} \ 0 \ \dfrac{1}{6} \ 0 \ \dfrac{1}{6}\right)$

3. $(0 \ 1)$, $(p \ q)$, $(p \ q)$

4. $\left(0 \ \dfrac{1}{8} \ 0 \ \dfrac{1}{8} \ \dfrac{1}{2} \ \dfrac{1}{8} \ 0 \ \dfrac{1}{8} \ 0\right)$, $\left(\dfrac{1}{24} \ \dfrac{1}{8} \ \dfrac{1}{24} \ \dfrac{1}{8} \ \dfrac{1}{3} \ \dfrac{1}{8} \ \dfrac{1}{24} \ \dfrac{1}{8} \ \dfrac{1}{24}\right)$

Exercises 8-2d page 188

1. $\pi_1 = \left(0 \ \dfrac{1}{3} \ 0 \ \dfrac{2}{3} \ 0\right)$, $\pi_2 = \left(\dfrac{1}{9} \ 0 \ \dfrac{4}{9} \ 0 \ \dfrac{4}{9}\right)$,

   $\pi_3 = \left(\dfrac{1}{9} \ \dfrac{4}{27} \ 0 \ \dfrac{8}{27} \ \dfrac{4}{9}\right)$

   $\pi_4 = \left(\dfrac{13}{81} \ 0 \ \dfrac{16}{81} \ 0 \ \dfrac{52}{81}\right)$, $\pi_5 = \left(\dfrac{13}{81} \ \dfrac{16}{243} \ 0 \ \dfrac{32}{243} \ \dfrac{52}{81}\right)$

2. $\pi_1 = \left(0 \ \dfrac{3}{7} \ 0 \ \dfrac{4}{7} \ 0\right)$, $\pi_2 = \left(\dfrac{9}{49} \ 0 \ \dfrac{24}{49} \ 0 \ \dfrac{16}{49}\right)$,

   $\pi_3 = \left(\dfrac{9}{49} \ \dfrac{72}{343} \ 0 \ \dfrac{96}{343} \ \dfrac{16}{49}\right)$, $\pi_4 = \left(\dfrac{657}{2401} \ 0 \ \dfrac{576}{2401} \ 0 \ \dfrac{1168}{2401}\right)$,

   $\pi_5 = \left(\dfrac{657}{7^4} \ \dfrac{1728}{7^5} \ 0 \ \dfrac{2304}{7^5} \ \dfrac{1168}{7^4}\right)$

3. (a) $\pi_0 = \left(\dfrac{1}{3} \ \dfrac{1}{6} \ \dfrac{1}{2}\right)$, $\pi_1 = \left(\dfrac{11}{36} \ \dfrac{17}{72} \ \dfrac{11}{24}\right) = (.3056 \ .2361 \ .4583)$

   $\pi_2 = \left(\dfrac{266}{864} \ \dfrac{199}{864} \ \dfrac{399}{864}\right) = (.3079 \quad .2303 \quad .4618)$

   $\pi_3 = (.3077 \ .2308 \ .4615)$

   (b) $\pi_0 = \left(\dfrac{1}{3} \ \dfrac{1}{4} \ \dfrac{5}{12}\right)$, $\pi_1 = \left(\dfrac{11}{36} \ \dfrac{11}{48} \ \dfrac{67}{144}\right) = (.3055 \ .2292 \ .4653)$

   $\pi_2 = (.3079 \ .2309 \ .4612)$, $\pi_3 = (.3077 \ .2308 \ .4615)$

4. $\pi_1 = \left(0 \ 0 \ 0 \ \dfrac{1}{2} \ 0 \ 0 \ 0 \ \dfrac{1}{2} \ 0\right)$, $\pi_2 = \left(\dfrac{1}{6} \ 0 \ 0 \ 0 \ \dfrac{1}{3} \ 0 \ \dfrac{1}{3} \ 0 \ \dfrac{1}{6}\right)$

   $\pi_3 = \left(0 \ \dfrac{1}{6} \ 0 \ \dfrac{1}{3} \ 0 \ \dfrac{1}{6} \ 0 \ \dfrac{1}{3} \ 0\right)$

5. $\pi_1'' = \begin{pmatrix} \dfrac{1}{6} & 0 & 0 & \dfrac{5}{6} & 0 \end{pmatrix}$, $\pi_2'' = \begin{pmatrix} \dfrac{1}{6} & \dfrac{5}{36} & 0 & 0 & \dfrac{25}{36} \end{pmatrix}$,

$\pi_3'' = \begin{pmatrix} \dfrac{1}{6} & \dfrac{5}{36} & \dfrac{25}{216} & 0 & \dfrac{125}{216} \end{pmatrix}$

6. $\pi_1 = \begin{pmatrix} \dfrac{1}{6} & \dfrac{1}{12} & \dfrac{1}{6} & \dfrac{1}{3} & \dfrac{1}{4} \end{pmatrix}$, $\pi_2 = \begin{pmatrix} \dfrac{1}{6} & \dfrac{5}{36} & \dfrac{5}{72} & \dfrac{5}{36} & \dfrac{35}{72} \end{pmatrix}$,

$\pi_3 = \begin{pmatrix} \dfrac{1}{6} & \dfrac{5}{36} & \dfrac{25}{216} & \dfrac{25}{432} & \dfrac{225}{432} \end{pmatrix}$

7.

$$T = \begin{pmatrix} \dfrac{1}{6} & \dfrac{1}{6} & \dfrac{1}{6} & \dfrac{1}{6} & \dfrac{1}{6} & \dfrac{1}{6} \\[2mm] 0 & \dfrac{2}{6} & \dfrac{1}{6} & \dfrac{1}{6} & \dfrac{1}{6} & \dfrac{1}{6} \\[2mm] 0 & 0 & \dfrac{3}{6} & \dfrac{1}{6} & \dfrac{1}{6} & \dfrac{1}{6} \\[2mm] 0 & 0 & 0 & \dfrac{4}{6} & \dfrac{1}{6} & \dfrac{1}{6} \\[2mm] 0 & 0 & 0 & 0 & \dfrac{5}{6} & \dfrac{1}{6} \\[2mm] 0 & 0 & 0 & 0 & 0 & 1 \end{pmatrix}$$

$$\pi_0 = \begin{pmatrix} \dfrac{1}{6} & \dfrac{1}{6} & \dfrac{1}{6} & \dfrac{1}{6} & \dfrac{1}{6} & \dfrac{1}{6} \end{pmatrix}$$

$$\pi_1 = \begin{pmatrix} \dfrac{1}{36} & \dfrac{3}{36} & \dfrac{5}{36} & \dfrac{7}{36} & \dfrac{9}{36} & \dfrac{11}{36} \end{pmatrix}$$

$$\pi_2 = \begin{pmatrix} \dfrac{1}{216} & \dfrac{7}{216} & \dfrac{19}{216} & \dfrac{37}{216} & \dfrac{61}{216} & \dfrac{91}{216} \end{pmatrix}$$

Exercises 8-3a page 193

We arrange the answers in matrix-form so that the entry in the ith row and jth column is $p_{ij}^{(2)}$.

1.
$$\begin{pmatrix} 1 & 0 & 0 & 0 & 0 \\ \dfrac{3}{9} & \dfrac{2}{9} & 0 & \dfrac{4}{9} & 0 \\ \dfrac{1}{9} & 0 & \dfrac{4}{9} & 0 & \dfrac{4}{9} \\ 0 & \dfrac{1}{9} & 0 & \dfrac{2}{9} & \dfrac{6}{9} \\ 0 & 0 & 0 & 0 & 1 \end{pmatrix}$$

2.
$$\begin{pmatrix} \dfrac{1}{6} & \dfrac{5}{36} & \dfrac{25}{36} & 0 & 0 \\ \dfrac{1}{6} & \dfrac{5}{36} & 0 & \dfrac{25}{36} & 0 \\ \dfrac{1}{6} & \dfrac{5}{36} & 0 & 0 & \dfrac{25}{36} \\ \dfrac{1}{6} & \dfrac{5}{36} & 0 & 0 & \dfrac{25}{36} \\ \dfrac{1}{6} & \dfrac{5}{36} & 0 & 0 & \dfrac{25}{36} \end{pmatrix}$$

3.
$$\begin{pmatrix} .2599 & .4018 & .3383 \\ .2306 & .3728 & .3966 \\ .2007 & .3418 & .4575 \end{pmatrix}$$

Exercises 8-3b page 201

3.
$$T^2 = \begin{pmatrix} \dfrac{5}{16} & \dfrac{5}{24} & \dfrac{23}{48} \\ \dfrac{1}{8} & \dfrac{5}{12} & \dfrac{11}{24} \\ \dfrac{1}{6} & \dfrac{2}{9} & \dfrac{11}{18} \end{pmatrix} \qquad T^3 = \begin{pmatrix} \dfrac{35}{192} & \dfrac{91}{288} & \dfrac{289}{576} \\ \dfrac{23}{96} & \dfrac{31}{144} & \dfrac{157}{288} \\ \dfrac{11}{72} & \dfrac{31}{108} & \dfrac{121}{216} \end{pmatrix}$$

$$T^5 = \begin{pmatrix} .1801 & .2811 & .5388 \\ .1926 & .2608 & .5466 \\ .1770 & .2759 & .5471 \end{pmatrix}$$

$$T^2 = \begin{pmatrix} p & 3p & 5p & 7p & 9p & 11p \\ 0 & 4p & 5p & 7p & 9p & 11p \\ 0 & 0 & 9p & 7p & 9p & 11p \\ 0 & 0 & 0 & 16p & 9p & 11p \\ 0 & 0 & 0 & 0 & 25p & 11p \\ 0 & 0 & 0 & 0 & 0 & 36p \end{pmatrix} \quad \text{where } p = \frac{1}{36}$$

$$T^3 = \begin{pmatrix} q & 7q & 19q & 37q & 61q & 91q \\ 0 & 8q & 19q & 37q & 61q & 91q \\ 0 & 0 & 27q & 37q & 61q & 91q \\ 0 & 0 & 0 & 64q & 61q & 91q \\ 0 & 0 & 0 & 0 & 125q & 91q \\ 0 & 0 & 0 & 0 & 0 & 216q \end{pmatrix} \quad \text{where } q = \frac{1}{216}$$

In T^n, $p_{ii} = \left(\dfrac{i}{6}\right)^n$, $p_{ij} = \dfrac{j^n - (j-1)^n}{6^n}$, $j > i$, $p_{ij} = 0$ otherwise.

Exercises 8-4c page 218

3.

$$T = \begin{pmatrix} 0 & 1 & 0 & 0 & 0 & 0 & 0 \\ p & 0 & q & 0 & 0 & 0 & 0 \\ 0 & p & 0 & q & 0 & 0 & 0 \\ 0 & 0 & p & 0 & q & 0 & 0 \\ 0 & 0 & 0 & p & 0 & q & 0 \\ 0 & 0 & 0 & 0 & p & 0 & q \\ 0 & 0 & 0 & 1 & 0 & 0 & 0 \end{pmatrix}$$

4.
$$T = \begin{pmatrix} 0 & 0 & 1 & 0 & 0 \\ \frac{1}{4} & 0 & \frac{3}{4} & 0 & 0 \\ 0 & \frac{1}{2} & 0 & \frac{1}{2} & 0 \\ 0 & 0 & \frac{3}{4} & 0 & \frac{1}{4} \\ 0 & 0 & 1 & 0 & 0 \end{pmatrix}$$

5.
$$T = \begin{pmatrix} 0 & q_1 & 0 & 0 & 0 & p_1 \\ p_2 & 0 & q_2 & 0 & 0 & 0 \\ 0 & p_3 & 0 & q_3 & 0 & 0 \\ 0 & 0 & p_4 & 0 & q_4 & 0 \\ 0 & 0 & 0 & p_5 & 0 & q_5 \\ q_6 & 0 & 0 & 0 & p_6 & 0 \end{pmatrix}$$

Exercises 8-5c page 233

2. Yes, no.   3. Yes, T^4 is positive; $\Delta^{(32)} < .01$.

4. $p_0^{(k)} = 0, k > 0$. Let $\pi_0 = (1\,0\,0\,0\,0\,0)$, $T = \begin{pmatrix} \frac{1}{6} & \frac{5}{6} & 0 & 0 & 0 & 0 \\ 0 & \frac{1}{3} & \frac{2}{3} & 0 & 0 & 0 \\ 0 & 0 & \frac{1}{2} & \frac{1}{2} & 0 & 0 \\ 0 & 0 & 0 & \frac{2}{3} & \frac{1}{3} & 0 \\ 0 & 0 & 0 & 0 & \frac{5}{6} & \frac{1}{6} \\ 0 & 0 & 0 & 0 & 0 & 1 \end{pmatrix}$

then $\pi_1 = \left(\frac{1}{6} \quad \frac{5}{6} \quad 0 \quad 0 \quad 0 \right)$ and $\pi_2 = \left(\frac{1}{36} \quad \frac{5}{12} \quad \frac{5}{9} \quad 0 \quad 0 \quad 0 \right)$.

5. For the case $n = 5$ we have six states $E_0, E_1, E_2, E_3, E_4, E_5$.

$$T = \begin{pmatrix} 0 & 1 & 0 & 0 & 0 & 0 \\ \frac{1}{5} & 0 & \frac{4}{5} & 0 & 0 & 0 \\ 0 & \frac{2}{5} & 0 & \frac{3}{5} & 0 & 0 \\ 0 & 0 & \frac{3}{5} & 0 & \frac{2}{5} & 0 \\ 0 & 0 & 0 & \frac{4}{5} & 0 & \frac{1}{5} \\ 0 & 0 & 0 & 0 & 1 & 0 \end{pmatrix} \quad T^2 = \begin{pmatrix} \frac{1}{5} & 0 & \frac{4}{5} & 0 & 0 & 0 \\ 0 & \frac{13}{25} & 0 & \frac{12}{25} & 0 & 0 \\ \frac{2}{25} & 0 & \frac{17}{25} & 0 & \frac{6}{25} & 0 \\ 0 & \frac{6}{25} & 0 & \frac{17}{25} & 0 & \frac{2}{25} \\ 0 & 0 & \frac{12}{25} & 0 & \frac{13}{25} & 0 \\ 0 & 0 & 0 & \frac{4}{5} & 0 & \frac{1}{5} \end{pmatrix}$$

$$T^3 = \begin{pmatrix} 0 & \frac{13}{25} & 0 & \frac{12}{25} & 0 & 0 \\ \frac{13}{125} & 0 & \frac{88}{125} & 0 & \frac{24}{125} & 0 \\ 0 & \frac{44}{125} & 0 & \frac{75}{125} & 0 & \frac{6}{125} \\ \frac{6}{125} & 0 & \frac{75}{125} & 0 & \frac{44}{125} & 0 \\ 0 & \frac{24}{125} & 0 & \frac{88}{125} & 0 & \frac{13}{125} \\ 0 & 0 & \frac{12}{25} & 0 & \frac{13}{25} & 0 \end{pmatrix}$$

$$\pi_0 = (0\ 0\ 0\ 1\ 0\ 0)$$

$$\pi_1 = \left(0\ \ 0\ \ \frac{3}{5}\ \ 0\ \ \frac{2}{5}\ \ 0 \right)$$

$$\pi_2 = \left(0\ \ \frac{6}{25}\ \ 0\ \ \frac{17}{25}\ \ 0\ \ \frac{2}{25} \right)$$

$$\pi_3 = \left(\frac{6}{125}\ \ 0\ \ \frac{75}{125}\ \ 0\ \ \frac{44}{125}\ \ 0 \right)$$

$$\pi_5 = \left(\frac{180}{3125} \quad 0 \quad \frac{1923}{3125} \quad 0 \quad \frac{1022}{3125} \quad 0 \right)$$

$$\pi_8 = \left(0 \quad \frac{120840}{390625} \quad 0 \quad \frac{244961}{390625} \quad 0 \quad \frac{24824}{390625} \right)$$

6.

$$T = \begin{pmatrix} 0 & 1 & 0 & 0 & 0 & 0 \\ \frac{1}{25} & \frac{8}{25} & \frac{16}{25} & 0 & 0 & 0 \\ 0 & \frac{4}{25} & \frac{12}{25} & \frac{9}{25} & 0 & 0 \\ 0 & 0 & \frac{9}{25} & \frac{12}{25} & \frac{4}{25} & 0 \\ 0 & 0 & 0 & \frac{16}{25} & \frac{8}{25} & \frac{1}{25} \\ 0 & 0 & 0 & 0 & 1 & 0 \end{pmatrix}$$

$$T^2 = \begin{pmatrix} 25q & 200q & 400q & 0 & 0 & 0 \\ 8q & 153q & 320q & 144q & 0 & 0 \\ 4q & 80q & 289q & 216q & 36q & 0 \\ 0 & 36q & 216q & 289q & 80q & 4q \\ 0 & 0 & 144q & 320q & 153q & 8q \\ 0 & 0 & 0 & 400q & 200q & 25q \end{pmatrix} \quad (q = 625)$$

$$T^3 = \begin{pmatrix} 200q' & 3825q' & 8000q' & 3600q' & 0 & 0 \\ 153q' & 2704q' & 7584q' & 4608q' & 576q' & 0 \\ 80q' & 1896q' & 6692q' & 5769q' & 1152q' & 36q' \\ 36q' & 1152q' & 5769q' & 6692q' & 1896q' & 80q' \\ 0 & 576q' & 4608q' & 7584q' & 2704q' & 153q' \\ 0 & 0 & 3600q' & 8000q' & 3825q' & 200q' \end{pmatrix} \quad (q' = 15,625)$$

$$\pi_0 = (0\ 0\ 0\ 0\ 0\ 1)$$

$$\pi_1 = (0\ 0\ 0\ 0\ 1\ 0)$$

$$\pi_2 = \left(0\ 0\ 0\ \frac{16}{25}\ \frac{8}{25}\ \frac{1}{25}\right)$$

$$\pi_3 = \left(0\ 0\ \frac{144}{625}\ \frac{320}{625}\ \frac{153}{625}\ \frac{8}{625}\right)$$

$$\pi_5 = (576q''\quad 23{,}040q''\quad 132{,}768q''\quad 175{,}744q''\quad 55{,}793q''\quad 2{,}704q'')$$
$$(q'' = 390{,}625)$$

$$\pi_8 = (20{,}588{,}544\bar{q}\quad 550{,}825{,}344\bar{q}\quad 2348{,}522{,}496\bar{q}\quad 2495{,}025{,}472\bar{q}$$
$$660{,}637{,}472\bar{q}\quad 27{,}916{,}297\ \bar{q})$$

$$(\bar{q} = 6103{,}515{,}625)$$

CHAPTER 9

Exercises 9-1c page 242

1. 27.

2. $N = \left\langle \dfrac{nS - c_1}{T - S} \right\rangle + 1$, let $\tau = \left\langle \dfrac{(n+1)\ T - S - c_1}{T - S} \right\rangle S$, then $Q(t) = S$ if
$t \geqslant \tau$, $Q(t) = S + (kS - t)$ if $k \in I^+$ and $t \leqslant kS < t + S \leqslant \tau$, $Q(t) = ((n - i + 1) + j)S$ where $t = 0$, i is the order of arrival of the customer and $(n - i + j)S < c_1 + jT$.

3. Waiting line disappears at $t = 75$ when the customer who entered at $t = 74$ goes into service, thereafter a customer enters service upon joining the queue, $N = 2K = 2\left(\left\langle \dfrac{\frac{n}{2}S - c_1}{2T - S} \right\rangle + 1\right)$, $S < 2T$.

5. $w(o) = kS$, $k = 1, 2, \ldots, n$, $w(t) = (n + k - 1)S - t$,
$0 < t = c_1 + (k - 1)T < \tau = \left\langle \dfrac{(n+1)\ T - S - c_1}{T - S} \right\rangle \cdot S$, $w(t) = 0, t \geqslant \tau$.

6. For $0 < t < \tau$, $l(t) = n + k{-}v$ where (i) $(k - 1)T \leqslant t - c_1 < kT$ and $vS \leqslant t < (v + 1)S$, $l(0) = n$, $l(t) = 0, t \geqslant \tau$.

Exercises 9-2b page 249

1. $p_0 = \dfrac{1}{3}$, $p_5 = \left(\dfrac{2}{3}\right)^5 \dfrac{1}{3}$, $E(X) = 2$, $V(X) = 6$.

2. Boundless (actually the question is meaningless for $p_n = 0$ for all n; if we use the formula $p_n = r^n p$ we get $p_5 = \dfrac{243}{320}$, $p_6 = \dfrac{729}{640} > 1$ and the absurdity is apparent).

3. (a) $2\theta \leqslant T \leqslant 2\theta + 2k$ (b) $p(T = 2\theta + \mu) = \displaystyle\sum_{i+j=\mu} p_i p_j$

   (c) $\displaystyle\sum_{\mu=0}^{2k} p(T = 2\theta + \mu) = \sum_{i=0}^{k} p_i \sum_{j=0}^{k} p_j = 1$

   (d) $\displaystyle\sum_{\mu=0}^{2k} (2\theta + \mu)\, p(T = 2\theta + \mu) = 2\theta + \sum_{\mu=0}^{2k} \mu \left(\sum_{i+j=\mu}^{2k} p_i p_j \right)$

Exercises 9-2c page 251

2. (a) $e^{-\frac{1}{2}}$ (b) $E(X) = \dfrac{1}{2}$ (c) 6,   (d) $\dfrac{1}{2}$, 1, boundless

3. $\displaystyle\sum_{v=0}^{k} (\theta + v)p_v = \theta + \sum_{v=0}^{k} v p_v$,   none

Exercises 9-2d page 256

1.

n	0	1	2	3	4
channels in use	0	1	2	3	4
customers waiting	0	0	0	0	0
	$\dfrac{3}{231}$	$\dfrac{12}{231}$	$\dfrac{24}{231}$	$\dfrac{32}{231}$	$\dfrac{32}{231}$
p_n					
	.0129	.0519	.1038	.1385	.1385

$E(X) = \dfrac{1436}{231}$

2. $\min C = 6$; if $C = 7$, $E(X) = 5.7513$

3. $p_0 = \dfrac{1-r}{1-r^{N+1}}$; $p_n = r^n p_0$, $n = 1, 2, \ldots, N$

$$E(X) = \frac{r}{1-r^{N+1}}\left(\frac{1-(N+1)r^N + Nr^{N+1}}{1-r}\right)$$

4. $p_0 = \left[\displaystyle\sum_{n=0}^{c} \frac{r^n}{n!} + \frac{r^c}{c!}\sum_{n=c+1}^{N}\frac{r^{n-c}}{c^{n-c}}\right]^{-1}$

$$p_n = \frac{r^n}{n!}\,p_0,\; 0 \leqslant n \leqslant c,\; p_n = \frac{r^{n-c}}{c^{n-c}}\frac{r^c}{c!}\,p_0,\; c < n \leqslant N$$

5. $p_0 = (1-\lambda)\,p_0 + \mu p_1$, $p_1 = \dfrac{\lambda}{\mu}\,p_0$

$$p_n = \left(1 - \left(\lambda e^{-\frac{\alpha n}{\mu}} + \mu\right)\right)p_n + \lambda e^{-\frac{\alpha(n-1)}{\mu}}\,p_{n-1} + \mu\,p_{n+1} = \left(\frac{\lambda}{\mu}\right)^n e^{-\binom{n}{2}\frac{\alpha}{\mu}}\,p_0$$

5	6	7	...	n
5	5	5	...	5
0	1	2	...	$n-5$
$\dfrac{32}{231}\left(\dfrac{4}{5}\right)$	$\dfrac{32}{231}\left(\dfrac{4}{5}\right)^2$	$\dfrac{32}{231}\left(\dfrac{4}{5}\right)^3$	...	$\dfrac{32}{231}\left(\dfrac{4}{5}\right)^{n-4}$
.1108	.0886	.0709	...	...

CHAPTER 10

Exercises 10-1b page 264

1. Range of P_4 is 0 through 16, $P_4(0) = .2637$, $P_4(13) = .471$.

2.

G_1	0	3
P	$1-p$	p

G_2	0	3	6	9
P_2	$(1-p) + p(1-p)^3$	$3p^2(1-p)^2$	$3p^3(1-p)$	p^4

G_3	P_3
0	$(1-p) + p(1-p)^3 + 3p^2(1-p)^5 + 3p^3(1-p)^7 + p^4(1-p)^9$
3	$9p^3(1-p)^4 + 18p^4(1-p)^6 + 9p^5(1-p)^8$
6	$9p^4(1-p)^3 + 45p^5(1-p)^5 + 36p^6(1-p)^7$
9	$3p^5(1-p)^2 + 60p^6(1-p)^4 + 84p^7(1-p)^6$
12	$45p^7(1-p)^3 + 126p^8(1-p)^5$
15	$18p^8(1-p)^2 + 126p^9(1-p)^4$
18	$3p^9(1-p) + 84p^{10}(1-p)^3$
21	$36p^{11}(1-p)^2$
24	$9p^{12}(1-p)$
27	p^{13}

3.

G_1	0	m
P_1	$1-p$	p

G_2	0	m	...	km	...	m^2
P_2	$(1-p) + p(1-p)^m$	$mp^2(1-p)^{m-1}$	...	$\binom{m}{k}p^{k+1}(1-p)^{m-k}$	...	p^{m+1}

$$\sum_{k=0}^{m} P_2(km) = (1-p) + p \sum_{k=0}^{m} \binom{m}{k} p^k (1-p)^{m-k} = 1$$

$$P_3(0) = (1-p) + p(1-p)^m + p \sum_{k=1}^{m} \binom{m}{k} p^k (1-p)^{m-k+mk};$$

$$P_3(m) = p^2 \sum_{j=1}^{m} jm \binom{m}{j} p^j (1-p)^{(j+1)(m-1)}$$

4. At two months,

T_1	$i, 0 \leqslant i \leqslant n$
P_1	p_i

at three months,

T_2	$j, 0 \leqslant j \leqslant 2n$
P_2	$p_j^{(2)} = \sum\limits_{\mu=0}^{j} p_\mu\, p_{j-\mu}$

Exercises 10-1c page 268

3. $g(s) = (1 - p) + ps^m$

$g_2(s) = (1 - p) + p((1 - p) + ps^m)^m$

$$= (1 - p) + p \sum_{j=0}^{m} \binom{m}{j} (ps^m)^j (1 - p)^{m-j}$$

Exercises 10-1d page 273

1. $g(s) = (1 - p) + ps^3$

$E(G_1) = m = 3p$

$E(G_2) = 9p^2$

$E(G_3) = 27p^3$

$V(G_1) = \sigma^2 = 9p(1 - p)$

$V(G_2) = 27p^2(1 - p)(1 + 3p)$

$V(G_3) = 81p^3(1 - p)(9p^2 + 3p + 1)$

$g(s) = \dfrac{3}{8} + \dfrac{1}{2}s + \dfrac{1}{16}s^2 + \dfrac{1}{16}s^3$

$E(G_1) = m = \dfrac{13}{16} = .8125$

$E(G_2) = .6601$

$E(G_3) = .5363$

$V(G_1) = \dfrac{167}{256} = .6523$

$V(G_2) = .9607$

$V(G_3) = 1.0650$

2. $E(G_k) = (mp)^k$, $k = 1, 2, 3, 4$.

$$V(G_k) = m^{k+1} p^k (1 - p) \left(\sum_{j=0}^{k-1} (mp)^j \right)$$

3. $E(G_n) = 1$ for all n; $V(G_n) = \left(\dfrac{3}{2} n \right)$ for all n.

Exercises 10-2b page 282

1. $g(s) = p + (1 - p)s$

$$g_n(s) = \sum_{k=0}^{n-1} p(1 - p)^k + (1 - p)^n s$$

$E(G_1) = 1 - p$, $P(G_n \to 0) = 1$

2. $g(s) = p + (1 - p) s^2$

$g_2(s) = p + (1 - p) p^2 + 2p(1 - p)^2 s^2 + (1 - p)^3 s^4$

$g_3(s) = \{p + p^2(1 - p) + 2p^3(1 - p)^2 + p^4(1 - p)^3\} + 4p^2(1 - p)^3 (1 + p)s^2$
$\qquad + 2p(1 - p)^4 (1 + 3p - 3p^2) s^4 + 4p(1 - p)^6 s^6 + (1 - p)^7 s^8$

$$E(G_1) = 2(1 - p); \ P(G_n \to 0) = \begin{cases} 1 \text{ if } p = \dfrac{2}{3} \\ \dfrac{1}{2} \text{ if } p = \dfrac{1}{3} \end{cases}$$

3. For $p = \dfrac{1}{4}$ and $p = \dfrac{1}{3}$, $P(G_n \to 0) = 1$

for $p = \dfrac{1}{2}$, $P(G_n \to 0) = \dfrac{\sqrt{5} - 1}{2}$

for $p = \dfrac{3}{4}$, $P(G_n \to 0) = \dfrac{\sqrt{21} - 3}{6}$

4. If $p \leqslant \dfrac{1}{m}$, $P(G_n \to 0) = 1$; if $p > \dfrac{1}{m}$, $P(G_n \to 0) < 1$

CHAPTER 11

Exercises 11-2b page 297

1. b and e satisfies R_1

2. e satisfies R_2 as well

3. e also satisfies R_3, hence in remaining problems $I(p) = -k \log p$.

4. (a) $-k \log \dfrac{1}{26}$ (b) $I(A) = -k \log \dfrac{9}{98}$, $I(F) = -k \log \dfrac{2}{98}$

     $I(Q) = -k \log \dfrac{1}{98}$, $I(S) = -k \log \dfrac{4}{98}$

5. $I(P) - I(FP) = k \log 3.33$

6. $-k \log \dfrac{11}{6}$

7. (a) $1.0000k$ (b) $.9182k$ (c) $.4690k$

8. (a) $2.0000k$ (b) $1.5000k$ (c) $1.0000k$ (d) $.8113k$

9. (a) $-k \log \dfrac{1}{26} = 4.696k$ (b) $4.319k$.

Exercises 11-2c page 304

2. $a < d < c < b$

3. $H\left(\dfrac{1}{N}, \dfrac{1}{N}, \ldots, \dfrac{1}{N}\right)$

4. $-\displaystyle\sum_{k=0}^{4} p_k \log p_k$ where $p_k = \dbinom{4}{k}\dbinom{48}{13-k}\dbinom{52}{13}^{-1}$

5. $-\displaystyle\sum_{j=0}^{8} p_j \log p_j$ where $p_j = \dbinom{8}{j} 2^{-8}$, $\log_2 2^8 = 8$

Exercises 11-3a page 306

3. $H_{\{4,5,6\}}(B) = 1.58493$

4. $H(A) = 3.3219$, $H_A(B) = .4236$, $H(A,B) = 3.7455$

Exercises 11-3b page 312

1. Joint distribution table of b and f (entries are $p(b_i, f_j)$).

f \ b	B	1	2	3	4	q_f
B	.083	.083	.013	0	0	.179
I	.050	.167	.050	.006	0	.273
II	.034	.066	.125	.025	.025	.275
III	0	.017	.050	.062	.038	.167
IV	0	0	.013	.031	.062	.106
p_b	.167	.333	.251	.124	.125	1

2. Table of values of $p_{b|f}$ on $b \times f$.

f \ b	B	1	2	3	4	$\sum\limits_{b_i \in b} p_{f_j}(b_i)$
B	.4637	.4637	.0726	0	0	1
I	.1832	.6117	.1832	.0219	0	1
II	.1236	.2400	.4545	.0909	.0909	1
III	0	.1018	.2994	.3713	.2275	1
IV	0	0	.1226	.2925	.5849	1

3. Distribution table of $A \times B$.

$B \backslash A$	a	e	i	o	u	q_B
a	$\dfrac{9}{64}$	0	$\dfrac{3}{256}$	$\dfrac{3}{384}$	0	$\dfrac{123}{768}$
e	$\dfrac{3}{128}$	$\dfrac{35}{128}$	$\dfrac{9}{256}$	0	$\dfrac{2}{256}$	$\dfrac{261}{768}$
i	0	$\dfrac{5}{192}$	$\dfrac{9}{64}$	0	$\dfrac{2}{256}$	$\dfrac{134}{768}$
o	$\dfrac{3}{128}$	0	0	$\dfrac{15}{96}$	$\dfrac{2}{128}$	$\dfrac{150}{768}$
u	0	$\dfrac{5}{384}$	0	$\dfrac{3}{128}$	$\dfrac{6}{64}$	$\dfrac{100}{768}$
p_A	$\dfrac{3}{16}$	$\dfrac{5}{16}$	$\dfrac{3}{16}$	$\dfrac{3}{16}$	$\dfrac{2}{16}$	1

Table of values of $p_{A|B}$.

$B \backslash A$	a	e	i	o	u	$\displaystyle\sum_{a_i \in A} p_{b_j}(a_i)$
a	$\dfrac{108}{123}$	0	$\dfrac{9}{123}$	$\dfrac{6}{123}$	0	1
e	$\dfrac{18}{261}$	$\dfrac{210}{261}$	$\dfrac{27}{261}$	0	$\dfrac{6}{261}$	1
i	0	$\dfrac{20}{134}$	$\dfrac{108}{134}$	0	$\dfrac{6}{134}$	1
o	$\dfrac{18}{150}$	0	0	$\dfrac{120}{150}$	$\dfrac{12}{150}$	1
u	0	$\dfrac{10}{100}$	0	$\dfrac{18}{100}$	$\dfrac{72}{100}$	1

Exercises 11-3c page 316

3. .0944

4. $R(A_1,B) = .0361$, $R(A_2,B) = .0343$, $R(A_3,B) = .0252$

Index